Register Now for Online Access to Your Book!

Your print purchase of *Physical Change and Aging, Seventh Edition,* **includes online access to the contents of your book**—increasing accessibility, portability, and searchability!

Access today at:
http://connect.springerpub.com/content/book/978-0-8261-5056-1
or scan the QR code at the right with your smartphone. Log in or register, then click "Redeem a voucher" and use the code below.

CK0DNBSM

Scan here for quick access.

Having trouble redeeming a voucher code?
Go to https://connect.springerpub.com/redeeming-voucher-code

If you are experiencing problems accessing the digital component of this product, please contact our customer service department at cs@springerpub.com

The online access with your print purchase is available at the publisher's discretion and may be removed at any time without notice.

Publisher's Note: New and used products purchased from third-party sellers are not guaranteed for quality, authenticity, or access to any included digital components.

PHYSICAL CHANGE AND AGING

Sue V. Saxon, PhD, is professor emeritus at the School of Aging Studies, University of South Florida. She received her PhD in developmental psychology and counseling from Florida State University and pursued additional graduate work in physiological psychology at the University of Wisconsin. She was a research psychologist for the National Institutes of Health and for the Laboratory of Perinatal Physiology at the University of Puerto Rico Medical School, before joining the faculty in behavioral sciences at the University of South Florida. As a charter faculty member in the Aging Studies program, she has developed and taught numerous courses in aging as well as presented workshops, given in-service training, and authored a number of books and book chapters on aging. She has been designated a gerontological pioneer for outstanding achievement and exemplary contributions to the field of gerontology by the Southern Gerontological Society. She can be reached by email at suevsaxon@gmail.com.

Mary Jean Etten, EdD, APRN, CMP, FT, has been a tenured full professor in the College of Nursing at St. Petersburg College, where she taught nursing, gerontology, and thanatology and developed an innovative curriculum teaching hospice care to nurses. She is currently an adjunct lecturer in thanatology in the School of Aging Studies at the University of South Florida. She received her doctoral degree in education from Nova Southeastern University and master's degrees in gerontology, counseling, and nursing from the University of South Florida. She is also a gerontological nurse practitioner. She is a fellow in thanatology and is board certified as a music practitioner. She has authored several books and manuals, as well as journal articles. She was a founder of Suncoast Hospice in 1978 and has remained on its boards for more than 43 years. In 2012, a building at Suncoast Hospice was named in her honor. She serves on the Empath Health board and several other boards. Recently she coauthored the book *A Caring Sanctuary: Forty Years of Caring.* In 2021, Dr. Etten was honored as a Distinguished Alumni by receiving the Dominican Mission Award from Edgewood College, Madison, Wisconsin. She received the Lifetime Achievement Award from the *Tampa Bay Business Journal* in 2013. She can be reached by email at ettenmj@gmail.com.

Elizabeth A. Perkins, PhD, RNLD, FAAIDD, FGSA, is a research associate professor and associate director of the Florida Center for Inclusive Communities—a University Center for Excellence in Developmental Disabilities—at the University of South Florida (USF). She has a PhD in aging studies from the School of Aging Studies at USF. She is also an RNLD (a registered nurse in the field of developmental disabilities) who trained at the Hereford and Worcestershire College of Nursing and Midwifery in England. She has authored numerous journal articles and book chapters, on topics ranging from aging, aging with intellectual and developmental disabilities, and caregiver quality of life. She was a guest editor for a special issue on aging and end-of-life for the *American Journal on Intellectual and Developmental Disabilities.* She is a fellow and past president of the American Association on Intellectual and Developmental Disabilities (AAIDD) and previously chaired AAIDD's Gerontology Division. She is also a fellow of the Gerontological Society of America and a former co-convener of their Special Interest Group on Lifelong Disabilities. In 2020, she was a national honoree of the National Historic Recognition Project for her significant contributions to the field of intellectual and development disabilities in the United States between 2000 and 2020. She can be reached by email at eperkins@usf.edu.

PHYSICAL CHANGE AND AGING

A Guide for the Helping Professions

Seventh Edition

Sue V. Saxon, PhD

Mary Jean Etten, EdD, APRN, CMP, FT

Elizabeth A. Perkins, PhD, RNLD, FAAIDD, FGSA

First Springer Publishing edition 2002; subsequent editions 2009, 2014

Springer Publishing Company, LLC
11 West 42nd Street, New York, NY 10036
www.springerpub.com
connect.springerpub.com/

Acquisitions Editor: Kate Dimock
Compositor: S4Carlisle Publishing Services

ISBN: 978-0-8261-5055-4
ebook ISBN: 978-0-8261-5056-1
DOI: 10.1891/9780826150561

SUPPLEMENTS:
Instructor Materials:

Qualified instructors may request supplements by emailing textbook@springerpub.com

Instructor's Manual ISBN: 978-0-8261-5057-8
Test Bank ISBN: 978-0-8261-5058-5
PowerPoint Slides ISBN: 978-0-8261-5059-2

21 22 23 24 / 5 4 3 2 1

Library of Congress Cataloging-in-Publication Data

Names: Saxon, Sue V., author. | Etten, Mary Jean, author. | Perkins, Elizabeth A. (Elizabeth Ann), 1969- author.
Title: Physical change and aging : a guide for the helping professions / Sue V. Saxon, PhD, Mary Jean Etten, EdD, APRN, CMP, FT, Elizabeth A. Perkins, PhD, RNLD, FAAIDD, FGSA.
Identifiers: LCCN 2021027718 (print) | LCCN 2021027719 (ebook) | ISBN 9780826150554 (cloth) | ISBN 9780826150561 (ebook)
Subjects: LCSH: Aging—Physiological aspects. | Geriatrics.
Classification: LCC QP86 .S29 2022 (print) | LCC QP86 (ebook) | DDC 612.6/7—dc23
LC record available at https://lccn.loc.gov/2021027718
LC ebook record available at https://lccn.loc.gov/2021027719

Printed in the United States of America.

To our parents:
Eston and Mildred Saxon
Henry and Amanda Etten
Richard and Ann Perkins

and to
Lori Perkins and Vanessa Harte

and lastly,
to all the individuals who died and the families and friends
affected by the COVID-19 pandemic

Contents

Preface

As educators in gerontology, it continues to be our experience that many available texts on the physical changes associated with aging either present their material in highly technical terms beyond the comprehension of those without extensive basic science backgrounds or else skim over this area superficially. In the years since the sixth edition of this book was published, additional research on the physical changes that occur with aging has become available. However, it continues to be difficult to separate "normal" aging from pathology, even though it is widely recognized that such a distinction is both useful and necessary in fully understanding the impact of the aging process on the human body and its functions. Data from healthy older adults clearly show that the aging process is not necessarily as devastating as earlier research had indicated, and more effort is currently being directed to the prevention or moderation of age-related changes previously thought to be inevitable. This presents a much more positive and realistic view of aging and allows for greater personal control over our individual aging process by directing attention to significant lifestyle modifications and preventive healthcare strategies.

Although this book focuses primarily on physical changes and the common pathologies associated with aging, it also considers the psychological and social implications of such changes for human behavior. Because aging is a complex process, it is impossible to consider biological or physical aspects without a comparable concern for the psychological, emotional, and social factors involved.

In this edition, a new chapter on gerontechnology is included and we have rewritten, updated, and elaborated on material throughout the book. We have also updated information on diagnosis and treatment because those who work with older adults will almost always become involved to some extent in their medical care. We have also continually stressed behaviors and interventions to promote more personal control over one's individual aging process. Appendix A includes practical hints for improving safety for older adults. Appendix B is a list of websites of relevant organizations. A glossary of medical terms used in the book is also included.

An instructor's manual, test bank, and PowerPoint slides to supplement the book are also available at https://connect.springerpub.com/content/book/978-0-8261-5056-1. **To obtain an electronic copy of these materials, faculty should contact Springer Publishing Company at textbook@springerpub.com.**

This book was written primarily for those in the helping professions—gerontologists, nurses, social workers, psychologists, rehabilitation specialists, clergy, counselors, and others who seek a better understanding of the physical aspects of aging and their implications for human behavior. It can be used as a textbook for academic courses, for workshops and in-service education, or as a resource for those who would simply like to know more about the aging process. We hope readers will find this book helpful in understanding both the human aging process and ways to improve quality of life in the later years.

Sue V. Saxon
Mary Jean Etten
Elizabeth A. Perkins

Acknowledgments

We are especially grateful to the late Sheri W. Sussman, Executive Editor at Springer Publishing Company, whose unrelenting encouragement to write the seventh edition was the impetus for us to undertake, once again, this labor of love. In addition, we thank Mehak Massand, our patient and ever helpful Assistant Editor, for her guidance and support. We also thank Dr. Stacy Orloff for her expert guidance on hospice and palliative care. Lastly, we thank Robert Singer Jr. for his continuing technical support and great patience. We thank them and all others who in any way contributed to the completion of this book.

1

Perspectives on Aging

Demographic changes in the United States have long indicated that we are an aging society. The fastest growing segment of the older population are the centenarians, those over age 100, while the second fastest growing numbers of people are those over age 85 (Jacelon, 2018). Many gerontologists agree that it is necessary to differentiate between those who are "young-old" (65–74), those who are "middle-old" (75–84), and those who are "old-old" (85+).

Most service providers agree that because needs are generally different in young-old and old-old individuals, services and programs should be planned, oriented, and delivered in different ways for each group. In general, the young-old need more programs and services to reintegrate them into meaningful roles and activities after retirement, whereas the old-old tend to need supportive and protective programs and services. However, the aging process is so highly individualized that some young-old need supportive and protective services and other older adults prefer and need reintegrative programs and services. Individuals age biologically, psychologically, sociologically, and spiritually, as well as chronologically (Touhy, 2012). Chronological age is not an accurate predictor of physical condition or behavior, but it is used for convenience and for certain legal purposes (such as voting, Social Security eligibility, and so on). Using the distinctions "young-old," "middle-old," and "old-old" serves to focus attention on the enormous diversity of older adults and suggests differentiations must be made in this large segment of the population if we are to provide effectively for all its needs.

People become more unique as they grow older, not more alike. Because of this, and because aging is a distinct part of the life cycle not yet personally experienced by most of those who work with older adults, understanding older people is difficult for many in the helping professions. In our attempts to understand others, we often lean heavily on our own personal experiences and can therefore empathize reasonably well with a child, adolescent, or young adult. To understand the behavior and perspective of older adults, though, it is necessary to project ourselves into an age context with which we have no personal experience. This is not easy to do and is one of the challenges in working effectively with older adults.

The academic study of the aging process includes gerontology, the broad study of the aging process, and geriatrics, a specialty concentrating on medical problems associated with growing older. Gerontology incorporates multidisciplinary concepts and approaches in an attempt to understand all aspects of the complex aging process. Three academic areas have traditionally contributed substantially to gerontology: *biological aging*, concerned with longevity and how (and why) the body changes as aging occurs; *psychological aging*, concerned with adaptive capabilities, including memory, intelligence, and how individuals cope with their own aging; and *social aging*, concerned with social roles and expectations for older adults in a particular culture or society.

However, many other academic disciplines also contribute to the study of aging because understanding aging is truly a multidisciplinary endeavor.

BASIC CONCEPTS IN PHYSICAL AGING

Research continues to indicate a much more optimistic picture of the aging process than previously presented. There are increased efforts now to differentiate "normal" aging from disease or pathology. It is clear that aging is not synonymous with illness or disease. True, certain aspects of the aging process make individuals more vulnerable to illness and disease, but no pathology is inevitable with age. Numerous physical changes historically attributed to aging are now recognized as more likely to be caused by lifestyle variables. For example, aches and pains traditionally attributed to aging more likely are due to a sedentary lifestyle or disuse of abilities rather than to aging per se. "Use it or lose it" is a common adage in gerontology and applies to physical, psychological, and social aging. Those skills and abilities that we continue to use will be maintained well into older age (barring accident or disease), whereas those we do not use will be lost.

Obviously, some factors associated with aging are nonmodifiable, such as genetics, gender, and age, but others can be modified by lifestyle, such as exercise, adequate nutrition, no smoking, and stress management. Because a substantial part of the aging process depends on lifestyle, we as individuals can make significant choices to increase the probability of healthy, positive aging. Three lifestyle factors that have a major impact on the manner in which we age are regular exercise, proper nutrition, and stress management. These and others will be discussed in this book.

LESSENED RESERVE CAPACITY

The major age-related change in the body is a lessened *reserve capacity*. All organ systems of the body have a substantial reserve capacity available to deal with high-demand or high-stress situations. With aging, there is a lessened reserve capacity in all the organ systems. Behavioral implications of lessened reserve capacity include the following:

1. *Slowness.* Although the aging process varies between one person and another, we all become slower with age. Most older adults are somewhat slower than they once were when taking in, processing, and acting on information. A fast-paced younger person will probably be a fast-paced older person but will be slower than when younger. Being slower in a fast-paced society is difficult, but it is important to realize that slowness is not synonymous with incompetence. Older adults who are allowed to pace themselves according to their own preferred schedule generally perform exceedingly well, whereas those who are forced into a schedule that is faster than they prefer are likely to perform much less well.
2. *Stress.* The body calls on its reserves to deal with high-stress or prolonged stress situations. The effect of stress tends to be greater on older adults because of their lessened reserve capacity. Being able to pace stressful situations appropriately helps older adults offset the effect of lessened reserve capacity.
3. *Homeostatic equilibrium.* Homeostatic equilibrium becomes more precarious as reserve capacity decreases with age. *Homeostasis* refers to a dynamic equilibrium that must be maintained in the body's internal environment. All the body's cells depend on a constant internal environment in order to function properly. Although a range of variation is possible in the internal environment, if homeostatic processes such as blood pressure, blood gases, acid–base balance (acidity or alkalinity of blood), and blood sugar become too high or too low, the individual will not survive. Highly complex regulatory mechanisms in the body help maintain homeostatic equilibrium, but with age and a lessened reserve capacity, it is easier

for homeostatic balance to be disrupted, and once disrupted it is difficult to restore. For this reason, older adults are more vulnerable to illness, disease, and accidents. Biological aging is sometimes considered to be a decline in the ability to maintain homeostatic equilibrium, leading to impaired functioning and ultimately to death. It is therefore necessary for older adults to be particularly attentive to health maintenance behaviors and healthy lifestyles.

4. *Pacing.* Being able to pace oneself, or doing things in one's preferred way and time frame, becomes increasingly important in older age as one way to decrease the effect of lessened reserve capacity. Those who work with older adults need to allow for pacing if they wish to help them perform effectively and competently.

PATTERNS OF DISEASE

Illness and disease are not uncommon in older adults, although no specific disease is inevitable in older age. As researchers become more concerned with the dynamics of normal aging, and as health promotion and education for older adults become more available, many diseases currently associated with aging may be prevented, or at least delayed until extreme old age, by healthy lifestyle choices and greater attention to health maintenance.

Diagnosing and treating illness and disease in older adults become complex for the following reasons:

1. Many older persons have several health problems that need to be treated concurrently (comorbidity). A new problem may be masked by one already existing, and medications desirable for one health problem may exacerbate an existing health condition.
2. The symptoms older persons describe may not be the classic symptoms characteristic of younger individuals. For example, older adults may have a "silent" heart attack and not experience the classic or usual symptoms, or a ruptured appendix may be reported as an upset stomach or abdominal cramps.
3. Older adults tend to expect pain and discomfort as they age and may not report symptoms until a medical problem is far advanced. Those who work with older adults need to encourage them to report unusual symptoms that arise and not assume they are just signs of "old age."

Because greater numbers of people are living into older age than ever before, an accurate understanding of how body functions change with age and the implications of those changes is becoming increasingly important for all who work with older adults and for those who wish to know more about the best kind of preparation for their own old age.

What this book attempts to convey can be summarized in four major perspectives:

1. *Recognition of aging as a highly diverse process.* Chronological age is not a reliable predictor of specific organ system efficiency. There is enormous variation in the rate of the aging process both among individuals of the same chronological age and also in the body systems of a given individual. This is because some organ systems age more rapidly than others depending on heredity, diet, exercise, and stresses caused by past illnesses and the environment.
2. *Importance of reserve capacity.* In spite of individual variations, the body organ systems, unless stressed, generally continue to function quite adequately in older age, although there is some loss of reserve capacity. Stress results in reduced efficiency or inability to cope. Proper nutrition, exercise, pacing oneself, and regulating the environment to be maximally supportive are all positive ways to help offset the effect of physical aging on the body.
3. *Accident prevention.* Age-related physical changes increase the possibility of accidents and injury. Older persons and those working with them need to become extremely sensitive

to and aware of situations that may contribute to accidents. Recovery time from accidental injury is usually longer for older adults, and often accidents are the first step in the transition from independence to dependence. Consequently, accidents have profound physical, psychological, and social significance for older adults' lives, and every effort should be directed toward preventing them.

4. *Increased health promotion and health maintenance activities.* Older adults are more susceptible to disease than the young. Physical changes associated with aging leading to loss of body reserves increase the older person's vulnerability to illness. Greater emphasis on health promotion, disease prevention, and health maintenance education for older adults allows them to become more actively involved in their own health and to feel they have more control over aging.

We do not intend, by the recitation of the physical changes and diseases associated with aging, to overemphasize decline and deterioration as an inevitable part of growing old. Indeed, many individuals are not drastically handicapped by age-related changes in their body systems. We believe others would be less impaired if they knew more about health promotion and health maintenance along with ways to adapt more efficiently to their individual aging process. The detrimental effects of age are a threat to self-image, to feelings of self-worth, and to independence, all of which are crucial to a satisfying and enjoyable life. Although physical changes are a reality of growing old, there are numerous ways to mitigate their effects and cope with them so as to at least partially offset the disabilities or limitations they impose. We suggest that gerontology and other helping professions place more research and educational efforts in helping older adults cope and adapt most effectively.

Although there are many different formulas for "successful aging," most include the following:

- Admit and accept the reality that aging imposes some limitations. Conserve energy; keep involved with life; make appropriate choices about use of time; and pace life realistically in accordance with needs, desires, and abilities.
- Be willing to change or modify lifestyle as necessary, especially physical activities and social roles. Remain flexible both mentally and emotionally. Reduce stress whenever possible. Plan a lifestyle to minimize disabilities and maximize remaining abilities.
- Develop new standards for self-evaluation and new goals. Measure self-worth by inner values such as the quality of human relationships, spirituality, appreciation of life, and not just by how much one can produce and achieve. Be a graceful receiver as well as a graceful giver. Older age can be a time of creativity and self-actualization if we choose to make it so.

REFERENCES

Jacelon, C. (2018). The aging population. In K. Mauk (Ed.), *Gerontological nursing: Competencies for care* (4th ed., pp. 23–40). Jones & Bartlett.

Touhy, T. A. (2012). Gerontological nursing and an aging society. In T. A. Touhy & K. Jett (Eds.), *Toward healthy aging: Human needs and nursing response* (8th ed., pp. 1–18). Elsevier Mosby.

2

Theories of Aging

INTRODUCTION

There is no consensus as to how or why biological aging occurs, and although numerous theories have been proposed, no single theory is acceptable as an adequate explanation of the complex aging process. All of the current theories are in need of further research verification. Explanations of biological aging range from genetic influences, to changes at the cellular level, to a consideration of entire organ system changes.

Some of the better known theories of biological aging, as well as selected relevant theories of psychosocial aging, are included in this brief overview of the topic. Biological theories can be grouped into two types: stochastic and nonstochastic (Eliopoulos, 2018).

STOCHASTIC THEORIES

Stochastic theories view aging as caused by a series of adverse changes in the cells that lead to replicative errors. These changes occur randomly and accumulate over time. Four theories of this type are the wear-and-tear theory; error theory; cross-linking, or connective tissue, theory; and free radical theory.

Wear-and-Tear Theory

The wear-and-tear theory is one of the earlier attempts to explain biological aging changes. It is based on the assumption that continued use leads to worn-out or defective parts of the body. This process is presumably further affected by the accumulation, over time, of by-products detrimental to the normal functioning of cells and tissues. This theoretical perspective ignores the various repair mechanisms available in the body and the fact that, in some cases (the muscles, for example), use contributes to increased strength and improved functioning. Recently the effects of stress have been studied but the outcome is not clear. Changes at the molecular level, especially the activity of the reactive oxygen species (ROS), can cause damage that is random and unpredictable (Jett, 2016).

Error Theory

This theory is primarily concerned with cumulative mistakes that occur in DNA and RNA with age. If random errors occur in the "copying" functions of RNA, inaccurate genetic information is copied and transmitted, thus impairing cell functions. Aging and death are presumed to be the result of errors that occur and are transmitted at the cellular level. Research has not yet provided adequate support for this theory; however, it has stimulated a great deal of research.

Cross-Linking, or Connective Tissue, Theory

Elastin and collagen (connective tissue proteins that support and connect body organs and structures) figure prominently in cross-linking, or connective tissue, theory. Collagen is the most variable and widespread of all body tissues. Both elastin and collagen tissues change with age from molecules that are loosely associated with each other (making the tissues flexible) to molecules that become more closely associated, or cross-linked (making the tissues less flexible and more rigid). Cross-linking not only lessens the flexibility of these tissues but also affects the accessibility of white blood cells to fight infection, decreases access to nutrition, inhibits cell growth, and reduces ability to eliminate toxins that are by-products of metabolism. Age-related changes in skin tissue are a good example of cross-linking. Although cross-linking occurs in some other proteins besides collagen and elastin (in DNA, for example), most of the available research has focused on these two particular tissues. Research on cross-linking continues, but how to prevent cross-linking and its actual impact on aging have yet to be clarified.

Free Radical Theory

Free radicals are chemical by-products of normal cell metabolism involving oxygen. They may also be produced by environmental pollutants such as pesticides and radiation, are extremely unstable, and last only a brief time (a second or less) but are highly reactive chemically with many other substances. Free radicals usually are quickly destroyed by protective enzyme systems or natural antioxidants in the body. However, some may escape and accumulate, damaging cell membranes, altering normal cell activity, and ultimately causing the death of cells. With age, the body's ability to neutralize free radicals decreases. The intake of supplemental antioxidants is not advised although diets that include natural antioxidants have been found to be useful.

NONSTOCHASTIC THEORIES

Nonstochastic theories view aging as caused by replicated errors in cells that are intrinsic and predetermined or programmed. Theories in this category are programmed aging theory and immunological or immunity theory.

Programmed Aging Theory

Hayflick and Moorehead (1961) raised the possibility that a biological or genetic clock may determine the aging process. Noting that human fetal fibroblastic cells (connective tissue cells) maintained in tissue cultures outside the body (in vitro) were able to divide approximately 50 times before deteriorating, they deduced that this is a form of programmed aging at the cellular level. Life expectancy is thought to be preprogrammed in a species-specific range.

Immunological Theory or Immunity Theory

Immunological or immunity theory deals with the immune system of the human body, which is composed of a series of responses by the body to protect itself against invasion of foreign materials, viruses, and bacteria. The bone marrow, spleen, thymus gland, and lymph nodes are the major organs of the immune system. Because the thymus gland, which is significant in the development of the immune system, decreases in size with age, and because immune system functioning declines with age, considerable research activity has been directed to understanding the significance and implications of thymic developmental changes. The thymus gland is of maximum size in late childhood or early adolescence and begins to atrophy (shrink) in the teens; by middle age, only remnants of the thymus remain. In old age the thymus is probably still functional, but the amount of tissue remaining is quite small.

Materials that initiate an immune response are called antigens. The body responds to antigens by producing antibodies (complex proteins), which combine with antigens to inactivate and destroy the invading material. The immune system is designed to recognize and ignore its own tissues but to attack and destroy invading foreign substances. With age, the immune system becomes less effective in warding off these invading substances, a process called immunosenescence. In addition, it loses the ability to distinguish between its own tissues and the invading materials and begins to attack and destroy its own tissues (an autoimmune response). Research has been concerned mostly with T lymphocytes and B lymphocytes, both of which decline in function with age. Therefore, if cells of the body are somehow changed with aging, these changed cells may not be recognized as body tissue and an autoimmune response will be triggered to destroy them. Autoimmune antibodies tend to increase with age. Older adults often have a decreased immune response, as evidenced by decreased resistance to disease, decreased ability to initiate the immune response, and, very likely, more autoimmune disorders. There is currently considerable research interest in the relationship of the immune system to aging.

Other areas of particular interest are the role of DNA in the aging process and the specific role of genetics in an individual's aging. The mapping, or identification, of the human genome will certainly add to our understanding of biological aging. Another research development of particular interest is the role of telomeres in determining the process of aging. Telomeres are areas at the ends of chromosomes that may act as "biological clocks." Each cell division in normal human cells results in a loss of part of the telomere; they shorten with age. In "abnormal" cell production, such as in cancer, an enzyme, telomerase, is produced that adds telomere sequences to the ends of chromosomes at each cell division. Research is currently focused on attempts to prevent the production of telomerase to stop cancer cells from multiplying.

Two other theories of biological aging are the neuroendocrine control (pacemaker) theory and the caloric restriction theory. The neuroendocrine theory suggests that aging is due to a programmed decline in nervous, immune, and endocrine system functioning. Specific hormones under investigation include dehydroepiandrosterone (DHEA), produced by the adrenal glands, and melatonin, produced by the pineal gland. Secretions of both decline with age. Still other hormones under investigation are estrogen, testosterone, and growth hormone. The hypothalamus–pituitary–endocrine gland feedback mechanisms are also currently under scrutiny for their roles in the aging process. The caloric restriction (metabolic) theory is based on earlier research indicating that caloric restriction in diet increases life span, slows metabolism, and at least delays the onset of a number of age-related diseases. A final theoretical view of aging is the apoptosis theory, which studies cell death as a noninflammatory gene-driven process occurring normally in the body. When regulated properly it is beneficial to the body, but dysregulation may cause aging (Yeager, 2019).

PSYCHOSOCIAL THEORIES OF AGING

Theories of psychosocial development consider the ways in which the experiences of earlier years contribute to behavior in later years. Even though there are insufficient empirical data for these views, a few attempts have been made to devise a series of developmental tasks encompassing the entire life span. These attempts are based on the assumption that specific tasks are expected to be encountered and learned at certain points or stages in the life cycle. Failure to master developmental tasks at the appropriate time presumably interferes with personal–social adaptation and with adjustment to the next stage and its specific tasks. Although these age–stage approaches are not acceptable to all, and supporting research is needed, a developmental perspective does provide guidelines to society's expectations for individual behavior at different ages and may therefore be useful in working effectively with older adults.

Maslow's Hierarchy of Basic Human Needs

Abraham Maslow (1968) proposed a hierarchy of basic human needs that motivate human behavior. As the needs of one level are met, the individual strives to meet the needs of the next level. According to the hierarchy, those needs necessary for survival are the most basic. Maslow's hierarchy of needs includes the following:

1. Physiological or survival needs (food, water, and oxygen), which must be met in order to live. These take priority over all other needs.
2. Safety and security needs. Once physiological needs are met, the individual is motivated to seek safety and security.
3. If physiological and safety/security needs are met, needs for belonging or affiliation become important. Humans have a basic need to belong, to be loved, and to be accepted, according to Maslow.
4. Esteem needs are next in this hierarchy. Once the previous needs are met, individuals need to develop a sense of self-esteem, or self-worth.
5. The final and highest level of Maslow's hierarchy is the need for self-actualization. This means to develop one's potential to the fullest and to be all that one can be. Some characteristics of self-actualization are acceptance of self and others, effective problem-solving, self-direction, appreciation of new experiences, identification with and concern for others, creativity, and strong personal values.

Maslow's hierarchy could conceivably be useful in planning programs and services for older adults. For example, if an older person is having difficulty meeting their safety and security needs, that person will have little energy or motivation to invest in a program promoting self-esteem or self-actualization. In this theoretical perspective, it is necessary to address each individual according to what personal needs are the most pressing at the moment in order to facilitate that person's growth and development toward the satisfaction of higher needs.

Erikson's Stage Theory of Development

Erik Erikson (1963) was one of the first theorists to suggest a psychosocial stage approach to the entire life cycle. He proposed a series of developmental "crises" that the individual resolves in either a predominately positive direction or a predominantly negative direction. For instance, the developmental crisis in a child's first year is to develop a sense of basic trust rather than mistrust. Obviously, few if any children are going to develop a sense of total trust, but children who have a preponderance of good, positive experiences with others rather than a preponderance of negative experiences will undoubtedly develop more of a sense of trust than of mistrust. The adjustments or attributes a person chooses at any stage may be reversed or altered later, depending on the nature of their interpersonal relationships and the environment. It is important not to view Erikson's stages as either–or phenomena or as adjustment choices that irrevocably determine the future direction of development. We will only consider middle age and older age stages here.

Middle Age: Generativity Versus Ego Stagnation. Middle age involves altering one's perspective of time in which individuals become more aware of the finiteness of life. The desire to leave a legacy or to leave some tangible evidence that one's life was lived becomes an important developmental concern at this time. Interests and concerns broaden to include social issues and succeeding generations rather than a focus exclusively on self and contemporaries. Failure to resolve earlier psychosocial demands, however, may result in increasing preoccupation with self and rigid adherence to the familiar.

Late Adulthood: Ego Integrity Versus Despair. In older age the major developmental task is to review one's life, reconcile successes and failures, and put it all into perspective. If this process is accompanied by feelings of self-worth and satisfaction in knowing one did the best one could in various life circumstances, ego integrity will be achieved. If life is viewed as a series of failure experiences and the individual feels they were inadequate in meeting most of life's demands, despair may well follow. How one met earlier life challenges or psychosocial crises will have a bearing on the resolution of this final stage.

Peck's Tasks of Middle and Old Age

Although Erikson covered the entire life cycle in his system of eight stages, the last two stages included the final 40 to 50 years of adult life. In an effort to address significant issues of later adulthood, Peck (1968) subdivided Erikson's last two stages into seven specific tasks. One difference between Erikson and Peck's positions is that Peck proposed four specific tasks for middle age and three for older age. The tasks for each age may be dealt with simultaneously rather than in a specific order. Sequence is not necessary.

Peck's four tasks for middle age are as follows:

1. *Valuing Wisdom Versus Physical Powers.* As physical strength and endurance decrease in middle age and the later years, it becomes necessary to shift one's value system to gain satisfaction and a sense of ego competence from mental activities rather than relying strictly on physical competence. Mental or intellectual abilities hold up well with age (barring accident or disease), whereas physical abilities peak in young adulthood and begin to decline gradually thereafter.
2. *Socializing Versus Sexualizing.* In middle age, Peck suggests people redefine their relationships with both sexes to stress friendship and companionship rather than "playing the sexual game" and relating to others primarily on the basis of physical attractiveness or sexual desirability. Obviously Peck does not suggest that sexual relationships should be replaced by companionship roles, but he urges the broadening of criteria for meaningful relationships to include other personal qualities as well as those specifically related to sexuality.
3. *Cathetic Flexibility Versus Cathetic Impoverishment* (emotional flexibility versus emotional rigidity). Emotional flexibility involves the ability to reinvest emotional energies in new relationships and new roles as older, well-established emotional attachments undergo changes with age. Those who are unable or unwilling to continue investing emotionally in new friendships, new social roles, or change in general may find themselves isolated because change is a prime ingredient in life.
4. *Mental Flexibility Versus Mental Rigidity.* As in emotional flexibility, it is also necessary to remain mentally or intellectually flexible in order to cope and adapt effectively.

Peck's tasks for older age are as follows:

1. *Ego Differentiation Versus Work Role Preoccupation.* Older adults who cling to previous lifestyles or work roles as measures of their self-esteem find these criteria inadequate if they are removed from such lifestyle roles or are unable to perform them satisfactorily. Older adults who value themselves as worthwhile, however, can enhance their self-esteem through a variety of continuing positive interactions with others.

2. *Body Transcendence Versus Body Preoccupation.* Those older adults who are able to rise above preoccupations with their health or the physical changes associated with their own aging process are better able to maintain an interest in and derive personal satisfaction from life in the later years compared with those who become preoccupied by or obsessed with evidence of poor health or physical changes.
3. *Ego Transcendence Versus Ego Preoccupation.* Older adults who are able to see beyond themselves and maintain an active interest in society and people are more likely to see themselves and their lives in a positive perspective.

Other Developmental Views

Adolescence is still an extremely significant age period in our culture because the transition from child to adult generally occurs (or is expected to occur) during this stage. Unfortunately, our culture has no clearly defined and universally accepted guidelines for determining when the transition is complete; thus adults, especially parents, may not behave consistently toward the adolescent who is considered to be an adult in one situation but may be treated as a child in another. Neither parents, friends, society, nor the adolescent knows when the transition to adult status has been completed, and this uncertainty increases the possibility of conflict. In addition, many important life decisions are made at this time, such as career choice, education, marriage, parenthood, establishment of an independent lifestyle, and personal identity. Establishing a firm sense of personal identity and independence becomes paramount.

Early adult years are an experimental stage in which young people test their decisions against reality. For many, this is the first opportunity for decisions made in the adolescent stage to be tried out in real-life situations. The young adult begins to establish themselves as an independent person testing self against the realities of work, home, civic, religious, recreational involvements, and interpersonal relationships.

Middle age is a consolidation stage, which for many is a time for intensive reevaluation of self and life. Middle age involves a changing time perspective with the realization that half of one's life is over and that one needs to set priorities for the last half of life. Emphasis here is specifically on coping with the physical and psychological implications of impending old age. Menopause in women and climacteric changes in men, gray hair, wrinkles, and lessened energy and stamina are all physical signs of age. Some middle-agers experience depression because a number of psychologically significant events often cumulate at about this point in the life cycle: children leave home, career and financial abilities peak, parents and friends begin to die, and one's time perspective changes to time left to live. For those who fear old age and death, depression and psychological problems are more likely to occur. Some make drastic changes in lifestyle (divorce after 20 or 30 years of marriage for example); others have a "last fling" of infidelity to substantiate their sexual prowess and attractiveness; and some experience actual emotional breakdown.

On a more positive note, however, middle age can be a highly satisfying period of life. Many find new interests, intensify current interests, and set new priorities for the meaningful use of time. For many, middle age is a time of competence and mastery, the prime of life: a very comfortable time of life.

Older age tends to be a time for evaluating one's life. A major task is to work through a life review—a purposeful, constructive effort to review one's life and put it into perspective and to cope with cumulating losses that usually occur with advancing age. A sense of personal integrity and the comfort of believing that one's life was well lived and was generally satisfying are important achievements during this period. There are other theories of aging but these have been chosen to represent the topics in this book.

SUMMARY

Human behavior involves complex interrelationships among physical, psychological, and social factors that are different for each individual. Both the nature and significance of biopsychosocial interrelationships change as aging occurs. Each individual remains a unique and complex being throughout life and can only be properly understood from a holistic perspective.

REFERENCES

Eliopoulos, C. (2018). *Gerontological nursing* (9th ed.). Wolters Kluwer.

Erikson, E. (1963). *Childhood and society* (2nd ed.). Norton.

Hayflick, L., & Moorehead, M. (1961). The serial cultivation of human diploid cell strains. *Experimental Cell Research*, *25*, 585–621. https://doi.org/10.1016/0014-4827(61)90192-6

Jett, K. (2016). Theories of aging. In T. A. Touhy & K. Jett (Eds.), *Ebersole & Hess' toward healthy aging* (9th ed., pp. 31–39). Elsevier.

Maslow, A. (1968). *Toward a psychology of aging* (2nd ed.). Van Nostrand Reinhold.

Peck, R. C. (1968). Psychological developments in the second half of life. In B. Neugarten (Ed.), *Middle age and aging* (pp. 88–92). University of Chicago Press.

Yeager, J. J. (2019). Theories related to care of the older adult. In S. S. Meiner & J. J. Yeager (Eds.), *Gerontologic nursing* (6th ed., pp. 18–30). Elsevier.

3

The Skin, Hair, and Nails

THE SKIN

The skin is the largest, most visible, and most complicated of the body systems. In an average adult it covers more than 3,000 square inches and weighs about 6 lb. It is served by one-third of all the blood circulating in the body. Aging is most apparent when observing a person's skin, hair, and nails.

The rate at which each person and organ system ages is highly individual and determined by both intrinsic and extrinsic factors. Intrinsic factors, specifically individual genes, determine sex, physical characteristics, hormonal balance, rate of aging, and propensity for disease. Extrinsic factors include diet, medications, and exposure to radiation and the sun. All of these determine the manner in which an individual's skin, hair, and nails age and show the signs of aging such as baldness, gray hair, or wrinkled skin. Those with dark-colored skin, which contains more melanin, are less sensitive to UV rays and do not readily show the visible signs of aging as do fair-haired and blue-eyed individuals.

PSYCHOSOCIAL IMPLICATIONS

The skin, hair, and nails reflect not only age but also hygiene habits, mental state, type of work, and even one's educational level. The appearance of the hair, skin, and nails often influences self-concept. Skin changes associated with aging may elicit negative reactions from others or even be the basis for prejudicial treatment in a youth-oriented society. Increasingly, middle-aged and older adults are spending considerable amounts of money on skin products, surgery, and other techniques promising to erase the signs of aging. For those who view age-related skin changes as unattractive, these alternatives may improve self-concept and enhance quality of life (Etten, 1996; Tabloski, 2014).

Through the skin we experience one of the most powerful and meaningful human senses, the sense of touch. Touch allows us to give and receive both positive and negative messages and feelings, so important to healthy development. The need to be touched begins at birth when the child is held and touched by parents and others, and continues throughout life. Older persons, though, are often deprived of this most important life-enhancing experience.

In addition to touch, sensations of heat, cold, pain, pressure, and vibration are felt through sensory receptors in the skin. They provide important information regarding health status, safety, and the environment (Figure 3.1).

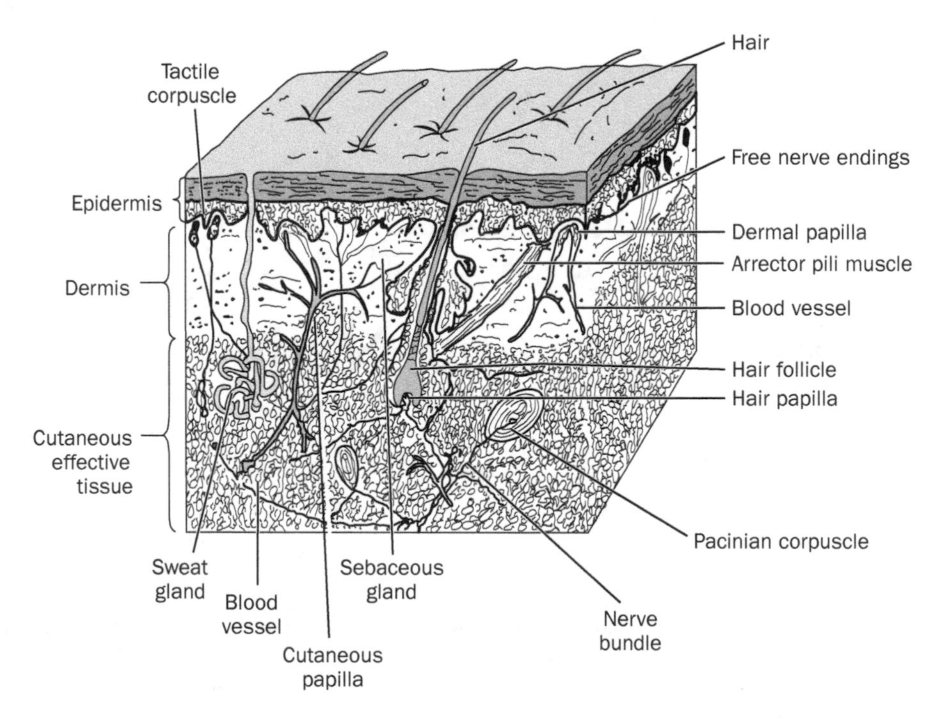

Figure 3.1 Structure of the skin and subcutaneous tissues.

COMPONENTS OF THE SKIN

The two layers of tissue comprising the skin are the epidermis and dermis. The hypodermis or subcutaneous fatty layer is not strictly speaking part of the skin, but it has some of the skin's protective functions (Marieb & Hoehn, 2019). Within these tissues are located sebaceous glands (oil glands), the sweat glands, blood vessels, nerves, hair follicles, and fatty tissue.

The Epidermis

The epidermis consists of four or five specific layers and four cell types, but it does not contain blood vessels or nerve fibers. Primarily made up of keratinocytes (cells that produce keratin, a tough protective substance) and an outer layer of stratum corneum (horny-like tissue), 20 to 30 cell layers protect the deeper cells and protect the body from chemical, physical, and biological trauma. The epidermis is continuously replaced by regeneration, cornification, and shedding. Each day millions of those cells rub off the skin, resulting in a new epidermis every 25 to 45 days. Thicker areas of this layer are found on the palms of the hands and soles of the feet, where the skin is continuously exposed to pressure. Melanocytes secrete melanin, which determines skin color; the more melanin, the darker the skin color. Langerhans cells in this layer help activate the immune system. Merkel cells, small in number, are located at the junction of the epidermis and dermis and are thought to act as sensory receptors for the sense of touch (Marieb & Hoehn, 2019).

The epidermis has an important function in the endogenous synthesis of vitamin D stimulated by exposure to UV light rays. This process is essential for calcium and phosphorous homeostasis (Lewis et al., 2017).

The Dermis

Located directly below the epidermis, the dermis is about 1 to 2 mm thick and contains strong but flexible connective tissue and elastic fibers that cover the body like a stocking. Blood vessels from the dermis bring nutrients to the epidermis. Pressure, touch, and pain receptors respond to touch, temperature, and pain stimuli. In addition, there are hair follicles, sebaceous (oil) glands, eccrine (water and salt) glands, and apocrine (sweat) glands. Apocrine glands are primarily located in the axillary, anus, pubic, and nipple areas and empty into the hair follicles. Their secretions are odorless until contacting the skin, at which time they decompose and provide the distinctive body odor of each individual. Ridges on the hands and feet allow for gripping and create our genetically determined fingerprints and footprints. The reticular layer makes up 80% of the dermis and contains dense connective tissues, thick bundles of collagen fibers that interlock, as well as elastin and reticular fibers. These fibers provide strength and elasticity to the skin. The basement membrane holds the epidermis and dermis together (Marieb & Hoehn, 2019).

The Hypodermis (Subcutaneous Layer)

Immediately below the dermis is the hypodermis or subcutaneous layer, which secures the dermis to the muscle tissue beneath. It contains loose connective tissue, blood vessels, nerves, and fat cells. Fat cells function as a storage place for calories, protect the body from injury, regulate heat loss, and help provide shape and form to the body.

FUNCTIONS OF THE SKIN

Functions of the skin include the following:

- Protection of underlying tissues and structures
- Temperature regulation through the mechanisms of sweating and shivering
- Excretion of water, salts, and organic wastes
- Prevention of tissue drying or excessive loss of water and electrolytes
- Physical, chemical, and biological barrier against harmful bacteria or foreign substances
- Metabolic functions in the formation of vitamin D in the presence of sunlight
- Protection of the body against UV radiation that can damage DNA in skin cells via the melanin in skin (Marieb & Hoehn, 2019)
- Provision of information from the environment through the sensations of touch, temperature, pain, vibration, and pressure

THE HAIR

Millions of hairs are found on the body except for the lips, soles of the feet, palms of the hands, nipples, and external genitalia. The hair consists of threadlike shafts of cornified cells that emerge through the epidermis from the papilla of the hair follicle located in the dermis. Hair color depends on the presence of pigment in the hair's cortex and on melanin production. It is continually shed and replaced, with eyebrows lasting 3 to 4 months and follicles of scalp hair active for an average of 4 years. As one ages hair is not replaced as fast as it is shed, thus causing a thinning of the hair and it becomes wispy. Hair serves as an insulator for the body, protecting it from sunlight, trauma, variations in heat or cold; plus it also protects underlying tissues (Marieb & Hoehn, 2019).

THE NAILS

The nails originate in the epidermis and are horny cell structures that are flat, curved plates forming a protective covering on the dorsal surface of the fingers and toes. A crescent-shaped white area near the base of the nail is called the lunula. Fingernails protect the fingers from trauma and assist with various tasks. Nail growth is continuous but slow. Toenails grow more slowly than fingernails. Overall, nail growth is affected by age, health status, climate, circulation of blood to and from the nails, and activity level of the individual (Miller, 2009; Tabloski, 2014).

AGE-RELATED CHANGES IN THE SKIN

The skin, hair, and nails all change with age. Two types of skin aging are intrinsic aging, based on genetic makeup and normal aging, and extrinsic factors such as smoking, exposure to UV light, and environmental pollutants. Most of the changes are not life threatening, but a few can result in serious conditions, such as pressure ulcers, which often require extensive medical treatment. Skin exposed to UV rays over time assumes a leathery wrinkled appearance, is fragile, and has impaired healing abilities. Smoking has been linked to increased wrinkling of the skin, grayish skin coloring, and a lessened ability of the skin to protect itself from UV rays (Tabloski, 2014).

Aging and the Epidermis

Age-related changes in the epidermis are more likely to occur in sun-exposed areas of the body. Epidermal thickness decreases somewhat, but the number of cell layers remains about the same. Cells of the epidermis have decreased moisture, contributing to a dry, itchy, and rough appearance. The elastic fibers clump together and fewer and stiffer numbers of collagen fibers are present. The skin's ability to act as a barrier against loss of body fluids or entry of substances through the skin is essentially stable. The rate of cell turnover declines between the third and the eighth decade from 30% to 50% and the keratinocytes are smaller and proliferate more slowly. These changes predispose older adults to greater sensitivity to chemical and mechanical trauma and also to slower healing (Friedman, 2011; Gilchrist, 1998).

Melanocytes decrease about 10% to 20% each decade, reducing skin protection from UV rays. Their regeneration rate varies, resulting in darkly pigmented areas called lentigo senilis (age spots). Langerhans cells in the skin decrease by nearly 50%, reducing the skin's immune response and increasing the incidence of skin tumors, allergic reactions, and infections (Mauk, 2018). Vitamin D3 production diminishes as we age as a result of lessened amounts of vitamin D3 precursor, plus elders tend to have less exposure to sunlight, which may result in osteoporosis and osteomalacia (Heineman et al., 2010). Decreasing numbers of melanocytes and dendritic cells potentiate a higher incidence of cancer among older adults.

Aging and the Dermis

The dermis loses about 20% of its thickness in older age, causing the skin to appear paper thin and transparent. Collagen decreases and assumes a disorderly arrangement, and elastin fibers thicken and fragment. Both these changes influence skin quality and elasticity with resultant skin wrinkling and sagging, especially in sun-exposed skin. There is greater fragility of the blood vessels, with hemorrhagic areas appearing on the skin; after only a mild bump, a tear injury may result (Linton, 2007). The loss of subcutaneous fat predisposes older individuals to wrinkles, as do our facial expressions, which over the years become entrenched through our characteristic expressions.

The nose and ears are composed mainly of cartilage tissue, which does not shrink or sag with age as much as skin tissue. An older person's nose and ears may appear larger because the loss of subcutaneous fat accentuates those features.

The size of sebaceous glands increases, whereas their sebum (oil production) tends to decrease. Cysts or blackheads in those glands may be caused by exposure to the sun. The sweat glands decrease in number and function, reducing perspiration and predisposing older individuals to hyperthermia. Older adults usually have a less distinct body odor because of diminished sweat gland secretions (Miller, 2009).

Small blood vessels that supply blood to the epidermis and play a role in regulating body temperature decrease in number with age. Lessened fat deposits, decreased sweat gland production, lessened muscle mass, and vascular changes in the dermis predispose elders to altered regulation in body temperature and a greater inability to adjust to extremes in environmental temperatures.

Both hypothermia (lowered body temperature) and hyperthermia (increased body temperature) are primarily disorders of older age. The thermoregulatory center located in the brain maintains the temperature of the body by controlling the constriction and dilation of blood vessels (vasoconstriction and vasodilation) as well as sweating, shivering, and chemical thermogenesis (heat production).

The shivering response decreases with age, as do vasodilation and vasoconstriction in response to heat and cold. The sweating response is reduced, and a lower metabolic rate caused by a decline in lean body mass lessens heat production. Insulation provided by body fat is less in those with smaller body mass (Jett, 2008).

Hypothermia may result in coma and hyperthermia may develop into a heat stroke unless reversed in time. Both can be fatal, with the risk increasing with age. Diagnosis can be difficult because temperature regulation is slower with age, as is the onset of temperature regulation problems, and symptoms are somewhat nonspecific. In addition, some medical situations decrease the body's ability to perspire, causing hyperthermia, such as cardiovascular, renal, and central nervous system diseases; dehydration; diabetes; and certain medications (e.g., lithium, phenothiazines, diuretics, beta-blockers, and anticholinergics). Hypothermia usually results from a contributing problem such as impaired thermoregulation; certain diseases; medications such as sedatives, hypnotics, antidepressants, and vasodilators; and alcohol use. Poor nutrition and immobility also increase the chances of hypothermia. When hypothermia or hyperthermia is present, immediate intervention is necessary to prevent death. At this time it is crucial to assess the client's body temperature, environmental temperature, existing health problems, medications taken, mobility, clothing, and fluid intake.

Older adults should avoid being exposed to heat and high humidity for long periods. Umbrellas, wide-brimmed hats, and loose light clothing are advised in the heat, as is drinking sufficient amounts of fluid unless contraindicated.

Nerve endings are only minimally affected by the aging process. However, tactile sensitivity is somewhat lessened. Pain and pressure perception decline slightly with age, causing a reduced ability to sense danger and react appropriately to protect oneself. Older adults are especially prone to burns, whether it is from a heating pad, cooking on a stove, hot water, or other heat source.

Aging and the Hypodermis

The layer of fat cells in the hypodermis becomes thinner with age, reducing both protection from trauma and the insulation that prevents loss of body heat. This is especially noticeable in the face, legs, and hands. On the other hand, there tends to be an accumulation of fat in the abdominal areas of men and the thighs of women despite no increase in caloric intake. Loss of fat padding on the feet seems to predispose elders to calluses, corns, foot pain, and ulcerations. Ambulation may be influenced by nerve changes in vibratory, light touch, and pressure sensations (Miller, 2009).

AGE-RELATED CHANGES IN THE HAIR

Some of the most visible signs of aging are graying of the hair, hair loss, and baldness. Graying and thinning of the hair begins around age 40 and is caused by sex-linked, genetic, and racial factors. Baldness, mostly a concern for men, usually begins with a receding hairline, but women too have thinner, finer hair with aging. Men often have increased hair growth in the ears, nostrils, and eyebrows, whereas women tend to have more hair growth on the chin and around the lips. In both men and women there is some loss of body hair with age (Miller, 2009; Plahuta & Hamrick-King, 2006).

AGE-RELATED CHANGES IN THE NAILS

Nails grow more slowly, often becoming lackluster, hard, thick, and brittle, and develop a gray or yellowish appearance. Longitudinal ridges and striations may cause the nails to split. Toenails become thicker with age and more difficult to cut. Vision, musculoskeletal flexibility, and eye–hand coordination problems all make toenail care challenging or impossible. Misshapen, untrimmed, thick nails may cause skin irritation and breakdown if they invade the surrounding skin areas, increasing the likelihood of foot infections and causing pain that may inhibit ambulation. Nails should be inspected and trimmed regularly, and those with diabetes or circulatory impairments require foot care by a podiatrist (Linton, 2007; Miller, 2009; Plahuta & Hamrick-King, 2006).

AGE-RELATED DISORDERS OF THE SKIN

Skin disorders are classified as noninflammatory, inflammatory, infectious, benign, premalignant, and malignant. Certain skin conditions such as herpes zoster (shingles) or pruritus (itching) may even indicate a systemic disease such as a malignancy or an endocrine or blood disorder. Skin reactions as a result of medications are commonly seen among elders. Accurate diagnosis of skin disorders by a primary care practitioner is essential because some lesions are not as identifiable in older adults as in younger people. Furthermore, the length of time a skin disease persists varies because of a slower rate of healing. Acute skin diseases may become chronic more rapidly, although at times certain lesions decrease in size and even disappear (Stanley et al., 2005).

Xerosis

Xerosis is a common condition in elders in which the skin becomes dry and rough, with a scaly appearance. Primarily occurring on the hands, forearms, anal, and genital areas, as well as the lower extremities, it commonly produces an itching sensation. Although the exact cause of xerosis is not known, it is thought to be due to reduced functioning of the sebaceous and sweat glands or irregularly aligned layers of corneocytes (cells in the corneum), as well as changes in the epidermis that may allow more fluid to escape from the skin. Other research suggests that changes in the lipid content of the skin as well as changes in the keratinization process are responsible (Tabloski, 2014). Additional contributing factors include cold weather, dry climates, indoor heating and cooling systems, use of harsh soaps, frequent bathing, and bed rest (Luggen & Jett, 2008).

Treatment involves decreased frequency of bathing and the application of emollients and moisturizing creams containing petroleum, glycerin, silicone, or lanolin. Itching may be alleviated by antihistamines and oral or topical steroids. Drinking eight glasses of fluid daily and increasing room humidity also help to keep the skin more moist. Applying a sunscreen when going outside and wearing soft, nonconstrictive clothing prevents drying and trauma to the skin (Linton, 2007; Miller, 2009).

Rashes

Rashes are commonly observed skin conditions. They may result from poor hygienic practices, allergic or chemical reactions, infections, and even being psychologically stressed. Most often they appear where two skin surfaces rub together such as under the breasts or arms or in the groin area. Attention to rashes is important because infection, skin breakdown, and even cellulitis may result.

Senile Purpura

With the loss of subcutaneous fat and connective tissue supporting the capillaries, even slight physical trauma can cause blood vessels to rupture into the surrounding tissues. These dark areas may last for weeks and be embarrassing, causing the individual to cover the areas with clothing or bandages.

Pruritus (Severe Itching)

Pruritus is a commonly observed and most annoying skin disorder in older adults. Trying to relieve the itch by scratching generates a counterstimulus stronger than the initial itch stimulus. Itching may be generalized or localized and can be aggravated by temperature changes, sweating, or clothing that irritates the skin. When it is generalized and not accompanied by skin lesions, pruritus may be symptomatic of other diseases such as a malignancy, thyroid dysfunction, liver disease, uremia, HIV infection, or psychiatric disorders and should not be ignored. Dry skin is perhaps the most common cause of pruritus, but reactions to medications may also cause itching. At times rubbing or scratching causes excoriation (abrasions), a secondary infection, thickened leathery skin, or nodules. Vigorous itching, especially at night, is a symptom of scabies, which often occurs in congregate living situations (Luggen & Jett, 2008).

Treatment involves the use of emollients containing alpha-hydroxy acids and menthol and camphor ointments to reduce itching. Topical corticosteroids and careful prescribing of antihistamines may also be effective. Cool compresses of saline (salt water), Epsom salts, or oatmeal baths may likewise help relieve itching. The body should not be rubbed vigorously because this stimulates the need to scratch even more (Friedman, 2011; Linton, 2007).

Solar Elastosis (Skin Aging Caused by Sun Exposure)

Solar elastosis, or photoaging disorder, occurs when the skin is repeatedly exposed to the sun, resulting in rough, leathery, wrinkled skin with irregular pigmentation, plaques, broken blood vessels, and actinic keratosis (premalignant lesions). Continued exposure to the sun dries the skin and causes it to age prematurely, increasing the risk for developing skin cancer. Photoaging can be prevented by staying out of the sun or, when exposed to the sun, covering the skin with clothing and applying a sunscreen that blocks UVA and UVB rays with a sun protection factor (SPF) of 30.

A monthly self-examination of the skin should detect questionable skin changes, and a yearly all-skin assessment by a primary care practitioner is advisable. Removal of a questionable lesion is quite effective when the lesion is identified and treated early.

Keratosis (Horny Growth)

Actinic keratosis (*solar keratosis*) is primarily caused by exposure to UV rays of the sun and is seen more commonly on sun-exposed skin, especially among fair-skinned elders. These usually appear on the face, on the backs of hands and arms, and on bald heads and upper trunks. Lesions are usually multiple, red, and rough to touch. Gradually they develop a yellowish-brown crust or scale, and some may become squamous cell cancers. Cryotherapy, topical chemotherapy, chemical peels, curettage, and laser resurfacing are possible treatment options. Older adults with this disorder should be encouraged to stay out of the sun or use a sunscreen of at least an SPF of 30 when exposed to the sun (Luggen & Jett, 2008).

Seborrheic keratosis usually appears as a raised, circumscribed, tan or reddish-brown to black papule or wart-like lesion covered with a waxy scale. Borders may be irregular, notched, round or smooth, or possibly elevated. These lesions are sometimes similar in appearance to malignant melanoma and may develop into cancer. They have a "stuck-on" appearance and are commonly found on the face, chest, back, and extremities. Treatment involves the use of curettage (scraping), freezing with liquid nitrogen, cryosurgery, or laser. These lesions should be removed if they become irritated or too unsightly.

Seborrheic dermatitis is a chronic inflammatory disorder of unknown cause most often affecting areas of high sebaceous activity such as the eyebrows, either side of the nose, around the ears, chest, armpits, and scalp. It appears as a reddened area with variable-colored white to yellow loose greasy scales. Scalp involvement is often called "dandruff." Treatment consists of topical hydrocortisone cream, antifungal agents, or systemic antibiotics. For affected hair areas, shampoos of keratolytic or tar preparations are recommended.

Psoriasis

Psoriasis is a noncontagious inflammatory autoimmune disease thought to be inherited and has periods of exacerbation. It is characterized by various-sized reddish-pink plaques over which are silver-white scales. These result from rapid replication of the cells of the epidermis and dermis. Areas most affected are the elbows, scalp, knees, hands, feet, lower back, and between the buttocks (Yaeger, 2019).

Individuals with psoriasis often are very self-conscious of the lesions and may have difficulty with intimacy issues, socialization, and rest and sleep. Treatment is usually topical with a variety of creams, such as corticosteroids, tar preparations, vitamin D derivatives, dithranol, and anthralin. UV light, photochemotherapy, or systemic chemotherapy may also be helpful in treating the disease (Cuzzell & Workman, 2013; Linton, 2007).

Skin Tags

Skin tags are pedunculated (stalk-like) lesions 1 to 5 mm long. They are soft, pink to brown papules found on the neck and trunk of the body, axilla, groin, or around the eyes. Although unsightly on the face, they are usually not serious and are easily removed with electrocautery or liquid nitrogen. Care should be taken to avoid chronic irritation caused by constant rubbing against clothing; if this occurs, the skin tags should be removed.

Herpes Zoster (Shingles)

Shingles is an acute viral infection that develops from the reactivation of the virus causing chickenpox and may stay dormant in dorsal nerve endings. When a child is originally exposed to the virus, chickenpox develops. Later in adult life the virus may be reactivated and shingles results. A higher occurrence of the disease is found among older adults thought to be due to decreased immune functioning. Although it is not contagious, if you have not had chickenpox or are immunosuppressed and are exposed to the fluid in the shingles blister, you may contract chickenpox. Therefore, a colloidal dressing covering the lesions is recommended for those with active shingles (Yaeger, 2019).

Shingles may be caused by a debilitating disease, malignancy, high stress, certain medications such as steroids or chemotherapy, or a compromised immune system. It appears as vesicles (fluid-filled lesions) along the nerve pathways of the skin. Initial complaints are tenderness, hypothermia, burning, itching, and a tingling pain in the affected area, after which the lesions appear. The area is painful when touched and the vesicle's crust falls off in 2 to 4 weeks. Older adults may develop postherpetic neuralgia causing chronic pain that could persist for years (Yaeger, 2019). Lesions do not cross the midline of the body, but they cause extreme pain that lasts 8 weeks or longer. When shingles invades the cranial nerves, blindness, severe pain in the face, palsy, dizziness, or deafness may result. Other commonly involved areas are the cervical (neck) and lumbar and sacral (lower back) regions of the body.

Antiviral treatment includes acyclovir or other antiviral medications and should be started within 72 hours of the rash onset to prevent postherpetic neuralgia, promote healing, and relieve pain. Wet compresses, antibiotic ointments, oral steroids, and pain medications may also be prescribed. A herpes zoster vaccine Shingrix is available in two doses that are administered 2 to 6 months apart to adults over age 60. It is very important to get both injections. This vaccine has proven to be up to 90% effective in preventing the incidence of herpes zoster.

Pressure Ulcers (Decubitus Ulcers)

Pressure ulcers (inappropriately referred to as bedsores) develop in body tissues that are situated directly over bony prominences. When prolonged pressure is exerted on the tissues, they are deprived of blood carrying necessary oxygen and nutrients to the tissues and also carrying waste products away. The result is tissue necrosis (dead tissue). The amount of necrosis depends on how long the pressure was unrelieved and on the initial health of the tissues. Tissue tolerance is influenced by extrinsic factors such as shearing, friction, and moisture from urine and feces. Intrinsic factors include circulatory impairment, poor nutritional status (especially lack of vitamins A and C), smoking, stress, dehydration, hypotension, immobility, and older age. Those older than age 70 are at greater risk for developing pressure ulcers because the epidermis becomes thinner with aging, subcutaneous fat decreases, blood vessels deteriorate, skin loses its elasticity, and there is lessened sensory perception. Furthermore, collagen fibers become more rigid and the immune response slows. Pressure ulcers can be categorized as category or stage 1, an intact unblanchable reddened area on the skin usually over bony prominences; category or stage 2, partial skin loss of the epidermis that looks like an open or ruptured blister or an abrasion; category or stage 3, full-thickness skin layer loss through the epidermis revealing the subcutaneous tissues as a deep skin crater, some tunneling in adjacent tissue involvement may be present; and category or stage 4, epidermis, dermis, and subcutaneous tissue destruction, with exposure of the muscle, bone, and supporting structures, that appears as a deep crater with neurotic tissue and tunneling may be present (Eliopoulos, 2018; Tabloski, 2014). On the surface a pressure ulcer may look small, yet it usually invades a much broader and deeper area than one would suspect (Linton, 2007). Certain chronic diseases predispose an individual to pressure ulcers. These include diabetes, spinal cord injuries, cardiovascular disease, fractures, arthritis, and an altered mental state.

Continued assessment of older adults who are at risk is most important. Redness of the skin is an indication that intervention must begin immediately to heal the lesion and to prevent further skin breakdown. Tools such as the Braden Scale for Predicting Pressure Sore Risk and the Norton Risk Assessment are invaluable in determining the risk for and monitoring the course of a pressure ulcer (Lewis et al., 2017). With adequate nursing care, most patients need not develop pressure ulcers. Prevention is the key. Continued assessments and prevention modalities are essential, such as use of special mattresses, skin-protective ointments, position changes, protective dressings, and a nutritious diet.

Foremost treatment includes eliminating every source of pressure, friction, or shearing. Various treatments used are dependent on the stage of the pressure ulcer. Among these are creating a warm moist environment, medications, topical antiseptics, polyurethane film, hydrocolloidal dressings, and absorptive gels. The use of whirlpool or medical or surgical debridement (removal of dead or damaged tissue) may be necessary. Vacuum-assisted closure devices and hyperbaric oxygen chambers are also employed, and an emphasis is placed on restoring nutritional and water balance to promote optimal healing (Cuzzell & Workman, 2013; Yaeger, 2019).

Management of a pressure ulcer demands a multidisciplinary approach. Not only does the person need physical care, but psychosocial issues involving both the individual and the family must be addressed. A pressure ulcer may take months to heal and often requires hospitalization, nursing home, or outpatient care. Those afflicted experience disfigurement, isolation, pain, and stupendous medical bills; thus the great importance of preventative care for those most vulnerable.

Venous and Arterial Stasis Ulcers

Venous Ulcers

Chronic insufficiency of the veins is thought to be one of the most prevalent health problems of older adults, affecting women more than men. Venous ulcers are caused by venous hypertension, malfunctioning valves in the leg veins, or by a blood clot in the deep veins, with the medial lower leg usually involved. The skin around the ulcer turns brown and bleeds easily; hair loss and scaling develop, and there may be atrophy of the leg (Yaeger, 2019). Stasis dermatitis occurs early in the disease, causing eczema-like eruptions and infection that may require weeks or months to heal. Treatment of venous ulcers includes oral or topical antibiotics, corticosteroids, consistent elevation of the leg, compression therapy, skin grafting, and nutrition therapy, with particular attention to zinc and vitamin C intake. Prevention involves patient education regarding excellent foot care by a podiatrist, adherence to the treatment regimen, keeping the legs elevated, and avoiding trauma to the legs.

Arterial Ulcers

Atherosclerotic arterial disease caused by arterial insufficiency or peripheral vascular disease often leads to ulceration of the skin on the leg or foot. Before ulcer formation, the individual may experience severe leg pain when walking. In more advanced stages, continuous severe leg pain occurs even when the leg is at rest, often interfering with sleep. The person may express concern over burning or cramping sensations. Gradually the legs may develop a bluish hue and feel cool to touch. The skin develops a shiny appearance, hair can be sparse, and the toe nails thicker (Yaeger, 2019). The individual often has a history of hypertension, peripheral vascular disease, coronary heart disease, or stroke. Treatment consists of bed rest, avoiding leg trauma, and vasodilator and adrenergic (adrenaline-releasing) drugs. Bypass surgery may reestablish the blood flow to the area, but at times amputation may be necessary.

Skin Cancer

Most cases of skin cancer are caused by exposure to the sun. In its early stages, skin cancer may appear as only a skin discoloration or toughening of the skin. Leukoplakia, a slightly raised, white translucent area, may also become cancerous. Leukoplakia is common in the mouth and caused by irritation from smoking, irregular teeth, or the continuous use of a pipe.

Basal Cell Cancer

Basal cell cancer is one of the most common skin tumors in Caucasians that occurs predominantly among older adults. Basal cell cancer often appears as a group of small, pearly, translucent nodules. Colors vary from white to red with purplish veins around the border. The center may be smooth, crusted, ulcerated, or even bleeding. Individuals with these lesions often state that the lesions are not healing. These ulcers are commonly located on sun-exposed areas of the neck, face, and ears and rarely spread to other areas of the body. Diagnosis must be confirmed by biopsy. Treatment varies from excision to radiotherapy, cryotherapy, and Mohs micrographic surgery. Regular assessment every 6 months by a primary care practitioner is recommended because there is about a 50% chance that a second lesion will develop within 3 to 5 years. Unnecessary exposure to the sun should be avoided.

Squamous Cell Cancer

Squamous cell cancer is the second most common malignant skin tumor in Caucasians and is the most common skin cancer in those who have dark skin (Tabloski, 2014). Usually observed as an isolated firm nodule, red or reddish-brown in color, with ulceration and scales, squamous cell cancer gradually begins to resemble a small cauliflower. It is painless, bleeds easily, and usually is found on sun-exposed areas or the mucous membranes of the mouth. This cancer may not be identified early on but can be aggressive and metastasize (spread) to the lymph nodes. Treatment includes excision, radiotherapy, deep cryotherapy, curettage, micrographic surgery, and topical 5-fluorouracil. More radical therapy may be necessary if the cancer has spread to other tissues. Individuals with this cancer should be assessed frequently by a primary care practitioner and be encouraged to stay out of the sun.

Malignant Melanoma

Most often seen in fair-skinned, blue-eyed individuals, malignant melanoma is increasing in incidence throughout the world and is the most serious type of skin cancer. Such an increase is directly attributable to repeated exposure to the UV rays of the sun. Other risk factors are family history of melanoma and light skin and hair (Tabloski, 2014). Superficial spreading melanomas are the most prominent type, with an incidence of 70% to 80%. Early in development, the lesions are pigmented papules flat or slightly raised and may be black, tan, brown, blue, or white, and asymmetrical. They tend to enlarge on the skin for some time before encroaching into the skin and are usually painless. If they show signs of growth, ulceration, or other changes, the prognosis is questionable. Those with a personal or family history of melanoma are especially at risk (Luggen & Jett, 2008).

Nodular melanomas account for 15% of all melanomas and are nodular, raised, brown lesions, often with ulceration. These may progress very rapidly, and the prognosis is grim. Elders with sun-damaged skin tend to develop lentigo malignant melanomas, which comprise up to 5% to 10% of melanomas. They are irregularly pigmented brown or tan lesions with varied pigmentation and notched asymmetric borders often on the face, upper trunk, and extremities. This type tends to occur more often among African Americans and Asians (Luggen & Jett, 2008).

Because melanomas grow rapidly and spread quickly to the lymph nodes and the liver, brain, or lungs, early detection and treatment are essential. The survival rate is related to the degree of tissue involvement. Treatments include deep excision of the lesion and regional lymph nodes, if necessary, chemotherapy, radiation therapy, and immunotherapy (Linton, 2007). Mohs micrographic surgery increasingly is being used for some types of melanoma (Tabloski, 2014).

PREVENTION OF SKIN CANCER

Most important in the prevention of skin cancer is the ability to assess lesions; therefore, a monthly self-assessment of the skin using a mirror is recommended. The ABCD method is a useful self-assessment tool.

A. *Asymmetry.* Is a mole asymmetrical—that is, does half of the lesion differ from the other?
B. *Borders.* Are the borders of the lesion irregular?
C. *Color.* Is the color irregular or uneven? Does the lesion contain shades of white, blue, red, gray, black, brown, or tan?
D. *Diameter.* Has the diameter of the lesion changed in size recently? Is it larger than the end of a pencil eraser?

If the answer is "yes" to any of these questions, immediate assessment by a primary care practitioner is advised. In any case, a skin examination should be part of the annual physical examination.

TIPS FOR MAINTAINING HEALTHY SKIN

The following tips can help to maintain healthy skin:

1. Drink at least 2 quarts of water daily.
2. Maintain a well-balanced diet with ample fruits, vegetables, and protein.
3. At least twice daily, use skin creams and lubricants that moisturize the skin.
4. Use mild or superfatted nonperfumed soaps.
5. Bathe in warm water and rinse the skin well before drying thoroughly.
6. Keep the air in the home humidified.
7. When outside use sunscreen that blocks both UVA and UVB rays with an SPF of 30 or higher and lip balm with sunscreen even on cloudy days.
8. When in the sun, wear cotton clothing that covers the body, such as long pants, long sleeves, a wide-brimmed hat, and sunglasses.
9. Limit time in the sun and exposure to the cold and wind.
10. Limit full-body bathing to two or three times a week.
11. If incontinent, make sure the skin is washed well and dried thoroughly and an emollient applied.
12. Avoid rough, irritating clothing, or laundering clothes with harsh soaps, bleach, or starch.

The American Cancer Society recommends examining the skin of the entire body every month and undergoing a yearly examination by a primary care practitioner. It is imperative that skin lesions be discovered and diagnosed early and treatment initiated to allow for the best chance for a cure.

REFERENCES

Cuzzell, J., & Workman, L. (2013). Care of patients with skin problems. In D. Ignatavicius & L. Workman (Eds.), *Medical-surgical nursing* (7th ed., pp. 470–510). Saunders Elsevier.

Eliopoulos, C. (2018). *Gerontological nursing* (8th ed.). Wolters Kluwer Health Williams & Wilkins.

Etten, M. J. (1996). Problems with the skin, hair, and nails. In A. Staab & L. Hodges (Eds.), *Essentials of gerontological nursing* (pp. 190–210). Lippincott.

Friedman, S. (2011). Integumentary function. In S. Meiner (Ed.)., *Gerontological nursing* (4th ed., pp. 596–627). Mosby Elsevier.

Gilchrist, B. C. (1998). Skin diseases and old age. In R. C. Tallis, H. M. Fillit, & J. C. Brocklehurst (Eds.), *Geriatric medicine and gerontology* (5th ed., pp. 1299–1308). Churchill Livingston.

Heineman, J. M., Hamrick-King, J., & Sewell, B. S. (2010). Review of the aging of physiological systems. In K. L. Mauk (ed.), *Gerontological nursing: Competencies for care* (2nd ed., pp. 128–231). Jones & Bartlett.

Jett, K. (2008). Physiologic changes with aging. In P. Ebersole, P. Hess, T. A. Touhy, K. Jett, & A. S. Luggen (Eds.), *Toward healthy aging* (7th ed., pp. 65–87). Mosby Elsevier.

Lewis, S. L., Bucher, L., Heitkemper, M. M., & Harding, M. M. (2017). *Medical-surgical nursing* (10th ed.). Elsevier.

Linton, A. D. (2007). Integumentary system. In A. D. Linton & H. W. Lach (Eds.), *Matteson & McConnell's gerontological nursing* (3rd ed., pp. 225–258). Saunders Elsevier.

Luggen, A. S., & Jett, K. (2008). Biological and maintenance needs. In P. Ebersole, P. Hess, T. A. Touhy, K. Jett, & A. S. Luggen (Eds.), *Toward healthy aging* (7th ed., pp. 157–193). Mosby Elsevier.

Marieb, E., & Hoehn, K. (2019). *Human anatomy and physiology* (11th ed.). Pearson.

Mauk, K. (2018). *Gerontological nursing: Competencies for care* (4th ed.). Jones and Bartlett Learning.

Miller, C. (2009). *Nursing for wellness in older adults* (5th ed.). Wolters Kluwer/Lippincott Williams & Wilkins.

Plahuta, J., & Hamrick-King, J. (2006). Reviewing the aging of physiological systems. In K. L. Mauk (Ed.), *Gerontological nursing* (pp. 143–264). Jones & Bartlett.

Stanley, M., Blair, K. A., & Beare, P. G. (2005). *Gerontological nursing: Promoting successful aging with older adults* (3rd ed.). F. A. Davis.

Tabloski, P. (2014). *Gerontological nursing* (3rd ed.). Pearson Education.

Yaeger, J. J. (2019). Integumentary function. In S. F. Meiner & J. J. Yaeger (Eds.), *Gerontologic nursing* (6th ed., pp. 279–310). Elsevier.

4

The Musculoskeletal System

INTRODUCTION

The musculoskeletal system allows us to actively respond to ever-changing demands in the environment. A complex system consisting of bones, cartilage, joints, muscles, tendons, ligaments, and bursae, its significance is often not appreciated until musculoskeletal limitation or impairment occurs. Mobility and independence depend, in large part, on the integrity of this system. Although age-related changes are not usually life threatening, musculoskeletal disorders and limitations cause substantial physical and psychological suffering and thus greatly affect quality of life in the later years. The skeletal and the muscular systems will be discussed separately.

THE SKELETAL SYSTEM

Components and Functions

The skeletal system is comprised primarily of bones but also composed of joints, cartilage, ligaments, and tendons. *Joints*, the junctures between two or more bones (articulations), make possible the wide range of movements and flexibility characteristic of the skeletal system. *Cartilage,* a nonvascular, tough, flexible connective tissue, assists in supporting the skeleton. *Ligaments,* bands of flexible connective tissue, bind bones together and reinforce joints. *Tendons* are fibrous connective tissue connecting muscle to bone or muscle to muscle.

Bones contain both organic and inorganic components and are classified according to structure or shape. Bone shape consists of four types: long bones, short bones, flat bones, and irregular bones. All these bones contain varying proportions of compact or cortical (hard, dense) bone or cancellous (spongy or trabecular) bone containing small cavities filled with marrow and usually enclosed by compact bone. Human bones range in size from the size of a pea (a small bone in the wrist) to the femur (thighbone), which is almost 2 feet long. Bone, an active and dynamic tissue, is constantly changing by "remodeling" or creating new bone (bone deposition) using osteoblasts and by bone resorption (removal by absorption) caused by osteoclasts. Bone remodeling is important in maintaining calcium balance in the body, but it is not uniform throughout the skeleton, although it goes on continuously.

The 206 named bones in the human skeleton are divided into axial and appendicular skeletons (see Figure 4.1). The skull, spinal (vertebral) column, and thorax (bony chest) comprise the axial skeleton. It forms the upright axis of the trunk and protects the brain, spinal cord, heart, and lungs.

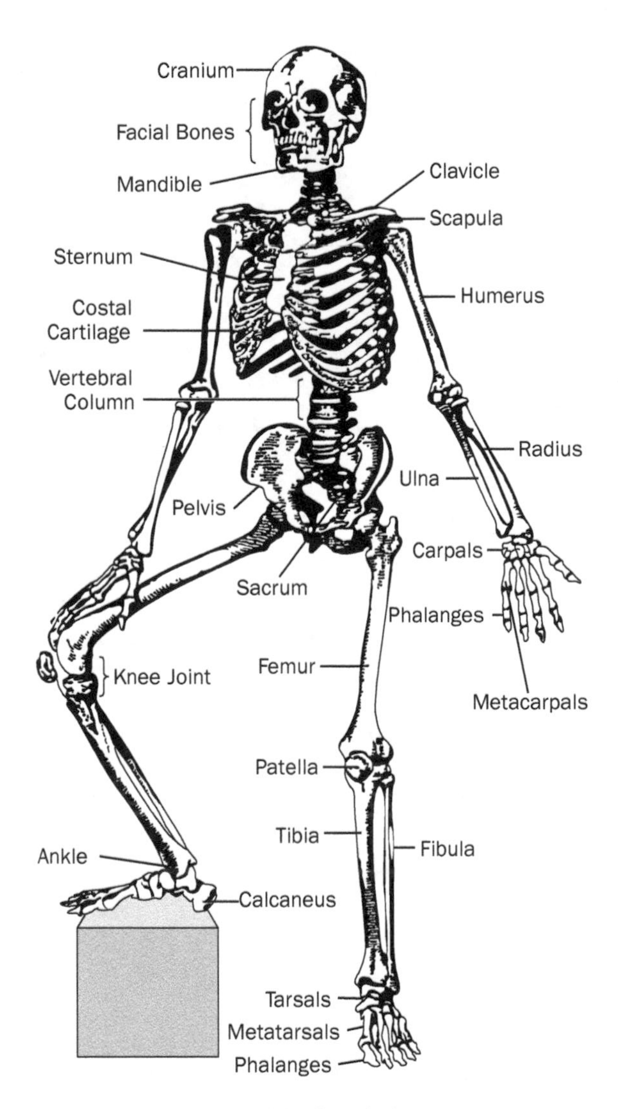

Figure 4.1 The skeleton.

The appendicular skeleton consists of the bones in the arms and legs and the shoulder and hip girdles attaching the limbs to the skeleton.

The spinal column deserves further discussion because it is a significant structure that is affected by the aging process. It contains 26 vertebrae (33 in infants, but several fuse by adulthood) and extends from the skull to the pelvis, where it ends in the coccyx (the "tail bone"). The spinal cord runs through a central cavity and is protected by the vertebrae, which are separated from each other by intervertebral discs, partially fluid-filled, cushion-like pads. Vertebral discs act as shock absorbers and provide flexibility for the spine. The divisions of the spinal column are as follows:

- Cervical, including seven vertebrae that are somewhat thin and light, allowing for flexibility
- Thoracic, including 12 vertebrae that are attachments for the ribs and heavier than the cervical vertebrae

- Lumbar, including five vertebrae that are dense and heavy for weight bearing and support the lower back
- Sacrum, formed by five fused vertebrae that strengthen the pelvis
- Coccyx, formed by three or four fused vertebrae

The functions of the skeletal system include the following:

- Provides support for all soft body organs
- Offers protection of brain, heart, and lungs
- Allows movement, in conjunction with muscles, as a leverage system to push, pull, and lift
- Provides storage, for fat and minerals (calcium, phosphate, sodium, sulfur, magnesium, and copper); stored minerals are released into the bloodstream and used by the body as needed; mineral withdrawals and deposits occur almost constantly
- Blood cell formation, which occurs within bone marrow (Marieb & Hoehn, 2019)

AGE-RELATED CHANGES IN THE SKELETAL SYSTEM

The primary and probably most significant age-related change in the skeletal system is a gradual loss of calcium from bone. Bone mass peaks at about age 35, after which there is a gradual loss of bone mass and bone density (osteopenia). Some of the factors influencing bone loss include genetics, smoking, excess consumption of alcohol, and hormonal factors. This is a nearly universal age-related change, irrespective of body size, race, or gender, although women tend to lose more bone mass than men (Marieb & Hoehn, 2019; Mauk, 2006). If loss of bone mass becomes great enough to produce fractures, unremitting pain, or immobility, the process is considered pathological and is called osteoporosis. In aging, the balance between new bone formation and bone resorption is disturbed and bone resorption begins to exceed bone formation, resulting in a loss of both bone mass and bone density. Consequently, bone strength declines with age. However, it varies both between and within individuals (Zychowicz, 2017).

Aging affects the cartilage in the body's joints. Cartilage surfaces become rougher in joint areas receiving the greatest stress. This reduces flexibility and the cushioning effect of normal cartilage, allowing bones to rub against bones, resulting in pain and restriction of joint movement. There is also decreased hydration or water content in cartilage with age. Some of the change in cartilage may be attributed to wear and tear over the years, but because individuals who have led sedentary lives also experience these changes, there appears to be some internal process also contributing to joint changes with age.

Normally, vertebrae are separated by intervertebral discs that both separate the vertebrae and act as shock absorbers in the vertebral column. Cartilage of the intervertebral discs changes with age by losing fluid and becoming stiffer or less compressible, restricting body flexibility. Low back pain, common in older age, may reflect these age-related degenerative changes. Compression of spinal discs and loss of height in individual vertebrae result in some loss of overall height, so older adults are somewhat shorter than when younger.

The strength of both tendons and ligaments decreases with age and shows some decrease in water content. These changes also contribute to decreased strength and mobility in the skeletal system (Tabloski, 2014).

Proprioception (the awareness of self in space) may also change, influencing balance and causing unsteadiness (Zychowicz, 2017).

AGE-RELATED SKELETAL SYSTEM DISORDERS

Osteoporosis

Osteoporosis, the most common metabolic bone disease in older adults, is characterized by a gradual, progressive change causing a rate of bone resorption greater than bone formation, resulting in reduced bone mass. Although bone mass is reduced, mineralization of bone does not change significantly. In other words, bones become porous, more fragile, and brittle, but the chemical composition of the bone remains normal. Reduced bone mass sufficient to cause fractures is called osteoporosis. Osteoporosis constitutes a major public health problem, especially for older women, and the costs of disability, dependency, and financial hardship are enormous. Major risk factors are advancing age, female gender, White or Asian race, family history of osteoporosis, estrogen deficiency, and small body build. Other risk factors include sedentary lifestyle, low calcium intake, excessive use of caffeine and alcohol, long-term use of corticosteroids, use of anticonvulsants and thyroid hormone, and excessive use of magnesium and aluminum-based antacids. Some nutritional factors that may contribute to osteoporosis are malabsorption of nutrients in the gastrointestinal (GI) tract, drinking large amounts of carbonated beverages, protein deficiency, and excessive tobacco use. Women are more likely to get osteoporosis, but men age 50 or over are also at risk. Osteoporosis affects all ethnic and racial groups, but thin White women tend to develop osteoporosis at an earlier age (Eliopoulos, 2018; Price, 2018).

Osteoporosis may be differentiated into primary and secondary osteoporosis. *Primary osteoporosis* accounts for most cases of osteoporosis. The cause usually is not completely clear, and no other disease state is present that could account for the osteoporosis. There are two types of primary osteoporosis. Type I, or postmenopausal osteoporosis, occurs between the ages of 51 to 75 and largely is responsible for vertebral fractures and fractures of the wrist. It primarily involves spongy or cancellous bone and is apparently related to estrogen deficiency. Type II, or involutional osteoporosis, is a slow, age-related bone loss, mainly in those older than 70, and often results in hip fractures and vertebral fractures. Age-related change in vitamin D synthesis resulting in decreased calcium absorption is thought to be primarily responsible for type II osteoporosis. It primarily involves compact bone. Both types I and II can be present simultaneously.

Secondary osteoporosis develops secondary to a number of factors or diseases that promote accelerated bone loss. It accounts for approximately 15% of all cases of osteoporosis. Some possible causes of secondary osteoporosis are drugs (especially corticosteroids), vitamin D deficiencies, thyroid disorders, GI disorders, alcoholism, and immobilization (Meiner, 2011). Osteoporosis results in skeletal instability caused by increasingly porous bones, which may not be capable of adequately supporting the body.

Fractures, the most serious problem associated with osteoporosis, are common, and a leading cause of disability and serious restriction of mobility in the elderly. It is called "the silent killer" since often no clinical signs are evident prior to the fracture occurring (Kennedy-Malone, 2019). Early fractures typically involve weight-bearing vertebrae, especially those in the lower back. Compression fractures of the vertebrae often go undetected because initially pain is usually minimal. Even simple activities such as bending, coughing, or sneezing may cause vertebral fractures in osteoporotic bones. As small fractures cumulate in the spine, though, pain ranges from mild to severe. These small vertebral fractures undoubtedly contribute to poor posture, chronic back pain, and a shortened stature in older age. Serious vertebral compression fractures cause severe pain and require immediate treatment. The first line of treatment for vertebral fractures usually involves hormone replacement therapy, bisphosphonates, calcitonin, and some degree of activity limitation. Surgical interventions may include vertebroplasty (injection of bone cement into the vertebral body) or kyphoplasty (using an inflatable balloon to expand the fractured vertebral body, removing the balloon, and then filling the cavity with cement). In older adults it is especially important to be attentive to preventing constipation, urinary retention, falls, and confusion during fracture treatment and in the recovery period.

Fractures of the wrist (Colles fracture) usually occur when an individual puts a hand out to break a fall. Hip fractures are usually the result of a fall, but in a few cases the weakened joint may break spontaneously without any apparent trauma such as a fall. When bones become too weak to withstand the force of gravity, they break and then the person falls. Optimal recovery from a fractured hip requires appropriate and intense medical care and rehabilitation. Even though men tend to have more bone mass than women, they do not have comparable hormone depletion such as occurs at menopause and also tend to fall less often. Caucasian men do have an increased likelihood of fracture caused by osteoporosis in their lifetime, especially in the later years. One-year mortality after hip fracture is nearly twice as high for men as for women. Factors contributing to osteoporosis in men include decreased testosterone levels, decreased calcium intake and calcium absorption, and lack of regular weight-bearing exercise. Other risk factors identified for women also apply to men. The consequences of hip fractures can include hospitalization, surgery, nursing home placement, and the real possibility of permanently restricted mobility.

Another serious consequence of osteoporosis involves postural or alignment problems commonly associated with aging. A "humpback" or flexed posture resulting from osteoporotic changes is called kyphosis; S-shaped curvature of the spine is called scoliosis; and swayback posture is called lordosis. These are relatively common phenomena in older age. Kyphosis and scoliosis, especially, interfere with stability and balance and may impede walking. A wide stance or waddling gait is adopted as an attempt to change the center of gravity and offset misalignment created by bone and muscle changes. Kyphosis can also interfere with breathing and digestion.

Diagnosis of primary osteoporosis is difficult because it has no early symptoms and is not apparent until quite far advanced. Often fractures are the first clear indication of osteoporosis. Other suspicious indicators are reduced height and kyphosis. There is currently no urine or blood tests to accurately diagnose osteoporosis, and standard x-rays do not show loss of bone density until at least 30% of bone mass has been lost. Diagnosis involves a thorough medical history, physical examination, selected laboratory studies to rule out other pathological conditions, and bone density tests. Dual-energy x-ray absorptiometry (DXA), assessing bone density of the lower spine, and hip, is considered to be the most accurate procedure for indicating osteoporosis. Individuals who have hip or spine bone mineral density of 2.5 standard deviations or more below peak bone mass are considered to have osteoporosis, whereas those with bone density between 2.5 and 1.0 standard deviations below peak bone mass are considered to have osteopenia and should be reevaluated every 2 to 5 years. Those within 1.0 standard deviation of peak bone mass are considered to have normal bone density. Other procedures to evaluate bone density are single photon absorptiometry, computed tomography (CT) scan, quantitative computed tomography (QCT), and peripheral ultrasound. An expert committee of the National Osteoporosis Foundation (Cosman et al., 2014) recommends bone density testing for all women older than age 65 and men aged 70 and older, and for those younger than age 65 who have one or more risk factors for osteoporosis.

Prevention is the best course. Understanding and modifying the various risk factors associated with osteoporosis (as early in life as possible) are necessary measures in reducing the chance of developing osteoporosis. Some risk factors are fixed and not subject to lifestyle modification, but those that can be modified should be. Education is necessary to encourage healthy lifestyle modifications. Currently, the two prevention and treatment strategies considered especially valuable are regular exercise and drug therapy.

Regular Exercise

Weight-bearing exercises stimulate bone growth and are considered significant preventive measures for osteoporosis. Walking is one of the easiest and safest forms of exercise. Brisk walking or other moderate-intensity weight-bearing exercise for 30 to 45 minutes 5 days a week is recommended, although even a shorter period over most days of the week is helpful. In addition, training with weights (resistance training) to strengthen and tone muscles and promote flexibility of the upper body contributes to better posture and reduces stress on the spinal column, a source of chronic back pain and body misalignment.

Drug Therapy

Controversy exists over the amount of calcium necessary and the form best used by the body, but adequate calcium intake is acknowledged as necessary to help prevent osteoporosis. Recommendations are for 1,200 to 1,500 mg/day. Calcium carbonate is the most widely used form of calcium, although in some individuals it may cause constipation, hyperacidity, and other GI complaints. Calcium lactate, calcium gluconate, and calcium citrate are possible alternatives to calcium carbonate. Calcium is absorbed better if taken with food. Vitamin D intake of at least 600 IU/day is essential for calcium supplementation to be effective, and many professionals are now recommending at least 800 to 1,000 IU daily for those older than 71 years (Ignatavicius, 2013a).

Hormone therapy, especially estrogen–progesterone combinations, has long been considered useful in preventing or treating excessive bone loss in postmenopausal women. It remains controversial, however, because of possible increases in the risk of developing breast cancer, gallbladder disease, blood clots, stroke, and heart disease. Risks and benefits need to be carefully assessed by each individual in consultation with a primary healthcare practitioner. For those who choose not to take hormones, alternative choices include bisphosphonates, such as risedronate (Actonel), alendronate (Fosamax), ibandronate (Boniva), and raloxifene (Evista), which have been shown to slow bone loss and increase bone density and are approved by the Food and Drug Administration (FDA) for both prevention and treatment of osteoporosis. Weekly or monthly oral doses can be taken, along with quarterly (Aredia) and yearly intravenous doses (Reclast), which may make compliance easier. However, a rare adverse effect of these medicines is osteonecrosis of the jaw (generalized death of bone tissue), which must be considered when prescribing bisphosphonates (Ignatavicius, 2013a). A newer drug romosozumab (Evenity) may be prescribed for those with a higher fracture risk who do not respond to other available treatment. Calcitonin, another drug used to treat osteoporosis, may be taken as a nasal spray. Other drugs are constantly being evaluated for treatment of osteoporosis.

Arthritis

Arthritis is a broad term referring to inflammation or degenerative changes in body joints, usually associated with the aging process. There are over 100 different types of arthritis and one in five persons are affected by it (Marieb & Hoehn, 2019). Joints, the junction between bones, involve various types of articulating surfaces that protect bones and maintain smooth joint movements. The ends of bones at most joints are covered with cartilage and enclosed in a capsule. Synovial fluid lines the capsule as a lubricant necessary for smooth movement. Tendons and ligaments also help support and protect joints. Any or all of these structures can be involved in arthritic changes. The three most common types of arthritis are osteoarthritis, rheumatoid arthritis, and gout.

Osteoarthritis

Osteoarthritis, also referred to as OA, or degenerative joint disease (DJD), is the most common form of arthritis and is one of the leading causes of disability in those older than 65. It is estimated nearly everyone by age 80 will develop OA (Marieb & Hoehn, 2019). Age alone does not cause OA, but age-related changes in cartilage predispose older adults to it. Factors in addition to age presumed significant in its development are obesity, trauma (wear and tear), lifestyle, smoking, and genetic predisposition (Kennedy-Malone, 2017). It primarily involves progressive loss of articular cartilage exposing the ends of the bones at the joint and allowing bones to rub together, resulting in pain, stiffness, and joint instability. Bony growths or bone spurs may appear at joint surfaces and cause enlargement of the joint. Eventually the joint capsule thickens, contributing to restricted movement and joint instability. Symptoms are generally mild early in the disease, with intermittent joint pain, stiffness on arising, and crepitation (creaking joints). As pain becomes more constant, limitation of movement and joint deformity occur, but inflammation is not usually present. Pain is relieved by rest and aggravated by movement or weight bearing, but eventually pain occurs at rest as well. Restriction of mobility because of joint pain increases stiffness, reduces muscle tone, and adds to weakness and joint instability.

Diagnosis includes a history and physical examination to detect limitation of motion, deformity of joints, pain, tenderness, and arthritic changes apparent on x-ray imaging or magnetic resonance imaging (MRI). However, the degree of arthritic change observed on x-ray imaging does not necessarily correlate with symptoms described by the person. Some individuals with severe degenerative changes apparent on x-ray imaging report few or mild symptoms. In addition, radiographs do not assess damaging changes in cartilage, whereas MRI techniques have greater sensitivity to changes in cartilage tissue (Ignatavicius, 2013b).

Specific treatment options include physical therapy, exercise, rest and reduced stress on joints, dietary modifications, drug therapy, and various surgical procedures. Symptom relief depends on rest balanced with an appropriate exercise program, including physical and occupational therapy. Pain control is essential. Exercise and physical therapy aimed at improving range of motion, strength training, increasing the individual's ability to carry out activities of daily living, and remaining independent are extremely important components of treatment (Kennedy-Malone, 2019). Dietary modification often revolves around weight loss because obesity is a definite risk factor for OA, especially in the knees. Weight reduction both reduces the risk of OA of the knees and also generally reduces symptoms in those who have it. Drugs most commonly used for OA are acetaminophen as a first choice and then nonsteroidal anti-inflammatory drugs (NSAIDs). Topical drug applications such as lidocaine patches may also relieve pain. Side effects of these drugs in long-term usage must be carefully monitored. For those with severe pain not responding to medications, corticosteroid or hyaluronic acid injections directly into the joint may be helpful, but these should not be used too frequently. Newer drugs are constantly made available for the treatment of arthritis, but these should be carefully evaluated by both individuals and healthcare providers. Side effects of these medications are common and caution is necessary, especially for prolonged use. GI bleeding may occur, so these medications should never be taken on an empty stomach and constant monitoring is essential. Nonpharmacological management techniques that may be useful are rest balanced with exercise, heat or cold applications, weight control, and various complementary or alternative therapies such as acupuncture, topical capsaicin, and dietary supplements like glucosamine and chondroitin. Educating older adults concerning the advantages, disadvantages, and side effects of various medications in managing OA is necessary, as is active involvement of the individual in the treatment program.

Management is crucial because there is no cure for OA. If nonsurgical treatment of OA is ineffective, surgical procedures may be indicated. Total joint replacement (arthroplasty) is an effective option if nonsurgical choices do not relieve severe pain and restriction of mobility and has a high success rate. Partial joint replacement (especially in the knee) is another possibility allowing for faster recovery and rehabilitation. Significant advances have occurred in joint replacement surgery, and artificial joints are now expected to last 15 to 20 years or longer and allow for a higher level of activity than previously experienced. In addition, surgeons now have more options for less invasive surgery, providing for fewer postsurgery complications, faster rehabilitation, and increased mobility.

Rheumatoid Arthritis

RA is an autoimmune disorder in which the body's antibodies attack body tissues. In contrast to OA, RA can involve the connective tissues of the entire body, but it is usually manifested in the joints. It is a systemic, chronic disease, with inflammation generally present, and it is the most prevalent type of inflammatory arthritis. Peak incidence is between 30 and 50 years of age, and it is a major cause of disability in later life. The incidence decreases after age 65, but it can still occur in older age. Women are more affected than men by a 3:1 ratio, and their symptoms tend to be more severe. One percent of all the population have RA (Marieb & Hoehn, 2019). The etiology is still unknown, but genetic, immunologic, and environmental factors and exposure to infectious agents such as viruses and bacteria are thought to be significant in RA.

In joints, the synovial membrane becomes inflamed and thickens; as the disease progresses, joint capsules and ligaments are stretched and destroyed. Tendons may shorten and move out of their usual position, producing deformity of the joints. Joints are usually involved symmetrically (i.e., the same joints on both sides of the body are involved). Hands and feet are often affected, but the knee, hip, ankle, shoulder, and elbow can also be affected. Other symptoms include malaise, fatigue, low-grade fever, weight loss, and morning stiffness lasting more than 1 hour. When RA occurs for the first time in older age, it often has a sudden onset but also tends to respond to treatment. Periods of remission are common.

Diagnosis is difficult because several diseases can masquerade as RA, but it generally involves a personal history and physical examination plus determination of an elevated erythrocyte (red blood cell) sedimentation rate (a laboratory test of speed at which erythrocytes settle) and the presence of a rheumatoid factor (an abnormal antibody) in the blood. However, both of these tests may also indicate other disorders in addition to RA. Other diagnostic tests include high-sensitivity C-reactive protein test, antinuclear antibody test, and other specific blood tests. X-ray examinations show some of the changes characteristic of RA, but MRI, bone scans with isotopes, and DXA scans provide more detailed information useful in assessing the progression of RA and response to treatment therapies.

Treatment is directed toward meeting three realistic goals: symptom relief, preserving joint function by learning to protect joints, and maintaining a reasonably normal lifestyle. Educating the individual and family and actively involving them in the treatment regimen is essential because RA is chronic, noncurable, and progressively disabling. Psychological support and pain management training are important in managing a long-term chronic disease. Every person with RA should engage in an appropriate program that balances exercise and rest. Neither excessive rest nor excessive exercise are therapeutic, and individuals must learn to monitor each of these to prevent further deterioration from inactivity (disuse) and also to prevent further exacerbation by excessive wear and tear on the affected joints.

Controlling inflammation and relieving pain are obviously primary goals of treatment. Medicines used for RA include steroidal and nonsteroidal anti-inflammatory drugs, with special attention to side effects in older adults. If these medications and physical therapies do not alleviate symptoms quickly, more potent drugs may be used because irreversible joint changes can occur within the first year of RA. Disease-modifying antirheumatic drugs (DMARDs) may be more effective treatments but may also have serious side effects. Various types of immunosuppressants and biological response modifiers (BRMs) are other possible treatment choices. New medications are constantly being developed, but all are potent medicines, have a variety of potentially serious side effects, and must be monitored closely. To further complicate the issue, combinations of these drug therapies are also commonly used. Corticosteroids are sometimes used as a bridging drug until another can be initiated, but long-term use is to be avoided when possible. If pharmacological interventions fail, surgical procedures, especially joint replacement, may be appropriate.

Lifestyle modifications recommended for those with RA include heat and rest, weight reduction, regular but appropriate exercise, and special modifications in the home such as utensils, door knobs, drawer handles, and so on designed for those with physical limitations. Exercise should incorporate stretching for joint flexibility, range-of-motion training for joint support, and aerobic exercise for overall health, weight control, muscle strength, and energy.

Some individuals opt for various complementary or alternative therapies such as acupuncture, hypnosis, homeopathy, selected nutritional supplements, mind–body therapies, and other approaches to help deal with the stresses of RA. Primary healthcare providers need to be aware when these options are being used in conjunction with more traditional treatments.

Rheumatoid arthritis is a long-term, debilitating disease that is physically, psychologically, and socially difficult to manage. It requires extensive education about the disease, realistic expectations, and ongoing management options for patients and families.

Gout (Gouty Arthritis)

Gout is a systemic disease of faulty metabolism in which there is an increased amount of uric acid in the blood and deposits of uric acid crystals in the joints. The increase in uric acid stems from an inherited defect in purine metabolism. Proteins in the body break down into purines (the end products of nucleoprotein digestion) and purine metabolism produces uric acid, which is usually excreted by the kidneys. Either increased production of uric acid or faulty elimination of it causes excess amounts to accumulate in the body. Excessive uric acid can form crystals in the joints called *tophi*, which produce inflammation of the joints and result in an attack of gout. Attacks are sudden and pain is excruciating, usually lasting from 5 to 8 days, during which time the individual is incapacitated. Although the attacks last only for a limited time, repeated attacks usually lead to chronic gout and damage to joints. The joints may eventually become deformed and disabled. Although any joint may be affected, the big toe seems to be a prime site. Gout often occurs for the first time in middle age and is more common in men, although women can get it, but rarely before menopause.

Diagnosis depends on the clinical presenting symptoms, a study of serum uric acid, urinary uric acid levels, renal function test, and a study of joint synovial fluid as well as the material in the tophi. A definitive diagnostic procedure is synovial fluid aspiration that shows crystals in the affected joint. Treatment involves medications such as NSAIDs, colchicine, allopurinol, probenecid, possible steroid injections into the joint, a diet free of purines, and drugs to lower the amount of uric acid in the blood. Aspirin is not recommended because it may increase uric acid levels. Individuals prone to gout attacks may need to take medications throughout their lives to reduce uric acid buildup in the blood. Secondary gout is not uncommon and is associated with certain medical problems (leukemia or cirrhosis, for example) and with other medications the individual may be taking. Diuretics in particular may cause attacks of gout. Other possible precipitating factors include being overweight; surgery; minor trauma; emotional upset; and ingestion of certain foods, especially foods high in purines such as shellfish and organ meats, alcoholic beverages, and drugs. If overweight, weight reduction is advised. Foods high in purines should be avoided such as shell fish and organ meats as well as alcoholic beverages (Kennedy-Malone, 2019).

Osteomalacia

Osteomalacia, a metabolic bone disease, is characterized by demineralized bone leading to bone softening, deformity, fractures, and bone pain. This disease may be easily confused with osteoporosis, in which mineralization of bone is essentially normal but bone mass is decreased. Symptomatology of both diseases is similar, but osteomalacia is not as common in older adults as osteoporosis. Because osteomalacia can be treated fairly easily, it is important to distinguish between these two bone pathologies. The primary cause of osteomalacia is vitamin D deficiency as a result of an inadequate diet (low in dairy products, fish, and fortified flour), lack of sunlight, liver disease, chronic kidney disease, phosphate deficiency (phosphate is essential to bone mineralization processes), and the use of some drugs (especially anticonvulsants). The classic symptom of osteomalacia is pain in skeletal areas (Linton, 2007). Other symptoms include muscle weakness, fractures, fatigue, and depression. Pain increases with movement.

The diagnosis depends on serum and urine laboratory studies. X-ray examinations are of limited value in distinguishing between osteoporosis and osteomalacia until the disease is far advanced. Bone scans may distinguish between them, but not necessarily in the early stages, and a bone biopsy may be necessary to firmly establish the diagnosis. Treatment is based on the cause of osteomalacia. If it is caused by a vitamin D deficiency, vitamin D replacement and calcium are effective. If it is primarily the result of phosphate deficiency, it will respond to phosphate salts. If caused by liver or kidney disease, these diseases must be treated as well as the osteomalacia. Those on anticonvulsant medications over long periods may also need vitamin D and calcium replacement. Older adults in nursing homes, or those with extremely limited mobility, should be monitored for appropriate vitamin D levels because they are likely to be at high risk for the development of osteomalacia.

Lyme Disease

Lyme disease is an inflammatory disease frequently causing arthritis and joint pain especially in the knees. It is caused by spirochete bacteria transferred by a tick that lives on deer and mice. Symptoms may also include an irregular heart beat, flu-like symptoms, and neurological disorders. Treatment involves antibiotics, which should be administered as soon as possible following the diagnosis (Marieb & Hoehn, 2019).

Paget's Disease

Paget's disease, a chronic metabolic bone disease, is characterized by excessive bone resorption and also excessive formation of abnormal bone that is extraordinarily vascular. The entire skeleton usually is not affected, but multiple localized sites may occur. Most commonly affected are the pelvis, spine, femur, tibia, or skull. Paget's disease rarely occurs in those younger than 40 and is most often diagnosed in those in their 60s or 70s. It affects between 1 million and 3 million Americans (Tabloski, 2014). Etiology is unknown, but it is likely there is a genetic influence that may be triggered by a slow virus or some environmental factor. Symptoms are usually minimal or nonexistent in early stages of the disease. Often it is detected by x-ray examination or an elevated serum alkaline phosphatase level. Bone pain occurs later in the course of the disease and varies from mild to moderate. Pain is primarily associated with deformities of the skull and weight-bearing bones. Severe bone pain generally indicates coexisting arthritis, acute fracture, neurological impairment, or bone lesions. Enlarged skull structures can result in headache, vertigo, tinnitus, and hearing loss. Bony enlargements at the base of the skull sometimes cause slurred speech, incontinence, visual difficulties, and problems in swallowing if the enlarged bones press on areas controlling these activities. If lumbar and thoracic vertebrae are enlarged, spinal nerves may be pinched or pressured. One of the most serious complications of Paget's disease is malignant bone tumors, which are difficult to detect in the early stages (Kennedy-Malone, 2019).

Diagnosis depends on finding elevated levels of serum alkaline phosphatase in the blood and on the results of urine studies, physical examination, MRI, radiographs, and isotope bone scans. Treatment is indicated when pain, bone deformities, neurological complications, or medical complications are present. Pain management is usually necessary. The most effective drug therapies include selected bisphosphonates. However, calcitonin may be given if a person cannot tolerate bisphosphonates (Ignatavicius, 2013a). NSAIDs are used to relieve pain and to reduce inflammation. Surgical interventions may be necessary if there are nerve compressions or fractures. Surgery may help fractures heal, replace damaged joints, realign deformed bones, and reduce pressure on nerves.

THE MUSCLES

Components and Functions of the Muscles

The three specific types of muscles in the human body are differentiated on the basis of histological (microscopic study of tissues) structure: skeletal, smooth, and cardiac (Marieb & Hoehn, 2019).

Skeletal Muscle

Skeletal muscle, also referred to as striated muscle because of its striped appearance, is attached to and covers the bony skeleton. Some skeletal muscles are attached directly to bones; in others, a band of dense, fibrous tissue (tendon) connects skeletal muscle to bone. In addition, striated muscle is also found in the tongue, soft palate, scalp, pharynx, and upper part of the esophagus and in extrinsic eye muscles. Skeletal muscles are the true voluntary muscles because they are the only type of muscle normally under conscious control. Mobility of the body depends on skeletal muscles. These muscles are able to contract rapidly and vigorously, but they fatigue easily and need to rest after even short periods of intense effort. Also, muscles must be exercised to maintain their strength and function. Exercising muscles increases the size of individual muscle fibers and promotes strength and endurance. Inactivity causes disuse atrophy when the muscles are immobilized. When this occurs the strength of the muscles may decrease at a rate of 5% each day. However, the rate of decline in strength in the later years may be slowed quite significantly by remaining physically fit (Marieb & Hoehn, 2019). Muscle atrophy tends to occur more rapidly in older age, so it is imperative for older adults to maintain strength and mobility with appropriate and regular exercise.

Skeletal muscles perform four important functions:

1. *Movement.* Essentially all movements of the body involve muscles. Muscles contract and relax to produce movement. A muscle that causes a specific motion is called the prime mover (agonist), and those assisting the agonist are called the synergists. Muscles causing movement opposite that of the agonist are called antagonists. The antagonist has to relax to allow the agonist to contract and produce movement. For example, bending the arm at the elbow to touch the face involves contracting the biceps muscle and relaxing the triceps muscles opposite the biceps. Muscles act on bones to create an efficient leverage system for pushing, pulling, and lifting. All body movement involves interrelationships among muscles, the bony skeleton, and the nervous system.
2. *Posture.* Skeletal muscles are crucial for maintaining posture against the force of gravity. Although we give little thought to this, skeletal muscles are constantly making necessary adjustments for us to maintain an erect or seated posture. Muscle tone (tonus) is also maintained constantly by some degree of muscle contraction in certain muscle fibers to keep muscles in a state of readiness to respond to contraction stimuli. Muscles with tonus less than normal (hypotonic) are flaccid, whereas muscles with greater than normal tonus are spastic (hypertonic).

3. *Stabilizing joints.* As muscles pull on bones for movement, they stabilize and strengthen joints.
4. *Heat production.* The fourth function of muscles is heat production. Body heat is a by-product of muscle metabolism and contraction and is essential in maintaining normal body temperature.

Skeletal muscle tissue has four special characteristics:

1. *Excitability,* the ability to respond to stimulation. The usual stimulus for muscle action is chemical, as when a neurotransmitter is released from a nerve cell, and the response is the transformation of chemical energy into mechanical energy.
2. *Contractility,* the ability to contract and become shorter when an appropriate stimulus is received.
3. *Extensibility*, the ability to lengthen (stretch). Muscle fibers shorten when contracting and lengthen when relaxing.
4. *Elasticity*, the ability to return to its original shape after having been stretched or contracted.

Smooth Muscle

Smooth muscle, so named because of its appearance, is found primarily in the walls of the digestive tract, trachea (windpipe), bronchi leading to the lungs, urinary bladder, gallbladder, ducts of the urinary and genital organs, walls of the blood vessels, spleen, iris of the eye, and hair follicles of the skin. The action of smooth muscle is typically slowed, sustained, and often rhythmic. It is mostly under the control of the autonomic nervous system and usually acts without conscious thought directed to its activity. Thus it is not necessary to will or command the smooth muscle of the digestive tract to begin digesting food. Digestion occurs without conscious attention, but thoughts and emotions do influence the process. For example, some body processes involving smooth muscles previously thought to be involuntary activities, such as digestion, blood pressure, and heart rate, can be brought under at least partial voluntary control by learning conditioning and biofeedback techniques. These interventions have been helpful in managing health problems such as chronic hypertension and muscle spasms. Consequently, the older clear distinction between voluntary and involuntary muscles has to be qualified somewhat in light of ongoing behavioral research and clinical applications.

Cardiac Muscle

Cardiac muscle is a special kind of muscle tissue found only in the heart. It has its own pacemaker system (a group of cells generating impulses to other areas of the heart), but additional stimulation is provided by the autonomic nervous system. Action is primarily (but not exclusively) involuntary, automatic, and rhythmic.

In summary, muscles are complex in both structure and function and are among the most remarkable of all body tissues. Although the distinction between voluntary and involuntary muscle action is not always clear, muscle activities primarily under *voluntary* control include (a) the maintenance of posture and (b) the majority of visible movements, such as facial expression, locomotion, chewing, and manipulation of objects. Muscle activities under *involuntary* control include (a) propulsion of material through the body (food and blood, for example), (b) expulsion of stored substances (bile from the gallbladder, urine from the bladder, and feces from the anus, although the latter two processes can also be under voluntary control), (c) regulation of the size of some body openings (such as the anus and urethral openings), and (d) regulation of the diameter of some tubes (as, for example, size of blood vessels and bronchioles).

SPECIFIC AGE-RELATED CHANGES IN MUSCLES

Age-related changes in muscles include the following:

1. Muscle strength tends to decline with age, partially as a result of loss of motor units and muscle fibers. However, a large body of evidence indicates regular appropriate exercise can slow loss of muscle strength and also increase strength, even in very old age.
2. There is some muscle atrophy with age, although how much is caused by the aging process and how much by disuse is not clear.
3. The decrease in muscle mass and in contractile force or weakness often noted in older adults is called sarcopenia. Sarcopenia increases fatigue, frailty, and disabilities; is a major risk factor for falling; and makes daily activities more difficult, therefore compromising independence in many older adults (Marieb & Hoehn, 2019).

SPECIFIC AGE-RELATED DISORDERS IN MUSCLES

Muscle Cramps

Muscle cramping, or sustained contraction of an entire muscle lasting anywhere from a few seconds to hours, increases with age. The muscle feels tight and painful. In older adults, muscle cramps often occur at night, especially after activity. Cramps may affect the thigh, calf, foot, hip, or hand. They result from peripheral vascular insufficiency and can be related to low blood sugar levels, dehydration, irritability of spinal cord neurons, and electrolyte imbalances (especially sodium and calcium). Stretching the muscles before sleeping and soaking in a hot tub may be helpful in relieving severe cramping. Quinine sulfate is sometimes prescribed but its effects are questionable (Kennedy-Malone, 2019).

Myasthenia Gravis

Myasthenia gravis, a progressive, chronic, acquired autoimmune disease, involves a defect in impulses transmitted from nerves to muscle cells. Antibodies attack and destroy acetylcholine receptors necessary for muscle contraction. Clinically it is characterized by an unusual susceptibility of muscles to fatigue and weakness. The incidence seems to be equal between women and men, but it seems to be more common in women younger than age 40 and in men older than age 60. Symptoms often first occur in the eyes, the bulbar muscles (those involved in chewing, swallowing, and talking), or the limbs. Ptosis (sagging) of the eyelids is a common early sign of myasthenia gravis. As the disease progresses, muscles of the face weaken and speech may be difficult to understand. Fluctuations in the progress of this disease occur unpredictably.

Diagnosis involves observation of signs of muscle weakness, testing for muscle strength, and a thorough neurological examination. A CT scan or MRI provides information on possible tumors (particularly of the thymus gland), and special blood tests assess other possible pathological states or other autoimmune disorders. Treatment is complex. Cholinesterase inhibitors are a major form of first-line treatment to improve muscle strength and contraction, but problems include the possibility of overdosing and the variability of medication effects. When this form of treatment is not effective, corticosteroids or immunosuppressive treatment is usually the next choice. Caution is advised because of the high incidence of undesirable side effects in older adults. Many drugs are contraindicated or must be used cautiously. Other treatment choices include plasmapheresis, which removes antibodies that block transmission of signals from nerve endings to muscle receptor cells. Yet another treatment possibility is intravenous immunoglobulin, which provides the body with normal antibodies. A surgical option is removal of the thymus gland even if there is no tumor in the thymus. This often reduces symptoms (Palmieri, 2013).

Polymyalgia Rheumatica

Polymyalgia rheumatica (PMR) is a rheumatic syndrome occurring most commonly in women older than age 50 and especially in women older than age 65. It is characterized by aching and stiffness in muscles of the neck, upper arms, shoulder girdle, and pelvic girdle. Often, PMR is accompanied by temporal arteritis (also known as giant cell arteritis), which is an inflammation of arteries, especially the arteries serving the temporal area of the brain. Major symptoms of this form of arteritis are headaches, changes in vision, and pain in the jaw. Giant cell arteritis may result in spontaneous blindness if not diagnosed and treated as early as possible. Etiology of both PMR and temporal arteritis is unknown, although genetic and environmental factors (such as a virus) are suspected. Both these conditions are generally self-limiting but still may last for months to several years. Treatment is necessary, though, to control pain and to prevent blindness.

Diagnosis depends on physical examination and laboratory tests, especially erythrocyte sedimentation rate and possibly C-reactive protein to assess inflammation. MRI and ultrasound also assist in assessing inflammation of tissues. A temporal artery biopsy may be necessary to confirm temporal arteritis. Treatment with corticosteroids produces a dramatic and immediate response, but for treatment to be most effective, early diagnosis is essential. The individual may need to be on this drug until lab values return to normal (Kennedy-Malone, 2019). NSAIDs may be used for pain management, and calcium and vitamin D are important to prevent the osteoporosis that is possible with high doses of steroids (Ignatavicius, 2013b).

Bursitis

Bursitis is a soft tissue disorder. In the joints where tendons or muscles pass over bones there are bursae, or sacs containing a small amount of fluid. Infection, calcium deposits, overuse, or trauma cause bursae to become inflamed and the fluid in them to increase, causing pain on movement of the joint. The most commonly affected sites are the shoulder and the elbow. Repetitive movements are a definite risk factor for bursitis. Common signs of bursitis are pain, warmth in the affected area, swelling, and range-of-motion limitations (Zychowicz, 2017).

Diagnosis involves physical examination; analysis of daily activities; x-ray examinations, MRI, or ultrasound; and laboratory tests to assess tissue injuries. Treatment depends on the cause and may include antibodies, rest, and application of ice, physical therapy, and pain medications as necessary. Gentle stretching and range-of-motion exercises prevent stiffness. In severe situations, extra fluid may be aspirated from the bursa. To prevent reoccurrences, daily range-of-motion exercises, modified activities that do not repeatedly strain muscles or joints, and protection of joints from excessive pressure are recommended (Tabloski, 2014).

SUMMARY

Aging in muscles and bones has a significant effect on the efficiency of a number of other body organs or organ systems.

1. Sharpness of vision decreases with age, partially because of weakening of the small muscles attached to the lens.
2. Skeletal and muscular changes associated with age affect the respiratory system when skeletal kyphosis (humpback) reduces overall volume of the lungs, whereas loss of muscle strength affects efficiency of breathing. Age-related changes in both bone and muscle contribute to reduced reserve capacity in the respiratory system.

3. Alterations in the musculature of the GI tract and the urinary system produce changes in the ability to digest food and to regulate defecation and urination. The embarrassment of partial or complete incontinence often has a severely deleterious effect on self-confidence and self-esteem in older age.
4. Muscles are one site of glycogen storage. Reduction in muscle mass results in reduced capacity to store glycogen, which is derived from carbohydrates and released when necessary to furnish quick energy in emergency situations. Thus, older adults may be expected to react more slowly to emergencies or fast-paced situations.

Pacing

As we age, physical activities definitely need to be paced more carefully to compensate for slower movements and decreased strength and stamina. The concept of pacing suggests that each individual should perform in their own way and in their own time frame. Attention to individual pacing schedules becomes much more important from middle age onward as an effective way to cope with age-related changes. Older adults, family members, caregivers, and healthcare professionals must all be more attentive to the need for pacing most behaviors and activities. Pacing can make the difference between competent performance and disorganized, inept efforts that may cause frustration for all involved.

Environmental Modifications

Whenever one is planning programs and activities for older age groups, it is especially important to allow for periodic "stretch" breaks if participants have been sitting or rest breaks if participants have been active. Sitting for long periods can result in painful joint stiffness, which lessens concentration on the activity or program.

In the home, furniture should accommodate the older person's less flexible muscular and skeletal systems. For example, low, overstuffed chairs without arms make it difficult to rise and at the same time maintain balance. Protruding furniture legs increase the probability of accidents, as do scatter rugs and waxed floors. Lighting must be adequate. In general, the home should be arranged so that accident hazards are reduced and safety devices increased. Such modifications can increase the competence of older persons and prolong independence (see Appendix A). Maintaining physical fitness throughout older age is essential to compensate for age-related changes in the musculoskeletal system. Aerobic exercise, strength building, and flexibility exercises are all absolutely necessary to preserve mobility and independence.

REFERENCES

Cosman, F., de Beur, S. J., LeBoff, M. S., Lewiecki, E. M., Tanner, B., Randall, S., & Lindsay, R. (2014). Erratum to: Clinician's guide to prevention and treatment of osteoporosis. *Osteoporosis International, 26*, 2045–2047. https://doi.org/10.1007/s00198-015-3037-x

Eliopoulos, C. (2018). *Gerontological nursing* (9th ed.). Wolters Kluwer.

Ignatavicius, D. (2013a). Care of patients with musculoskeletal problems. In D. Ignatavicius & L. Workman (Eds.), *Medical-surgical nursing* (7th ed., pp. 119–142). Saunders Elsevier.

Ignatavicius, D. (2013b). Care of patients with arthritis and other connective tissue diseases. In D. Ignatavicius & L. Workman (Eds.), *Medical-surgical nursing* (7th ed., pp. 318–356). Saunders Elsevier.

Kennedy-Malone, L. K. (2019). Musculoskeletal function. In S. E. Meiner & J. J. Yeager (Eds.), *Gerontologic nursing* (6th ed., pp. 449–478). Elsevier.

Linton, A. D. (2007). Musculoskeletal system. In A. D. Linton & H. W. Lach (Eds.), *Matteson & McConnell's gerontological nursing* (3rd ed., pp. 259–312). Saunders Elsevier.

Marieb, E. N., & Hoehn, K. (2019). *Human anatomy and physiology* (11th ed.). Pearson Education.
Mauk, K. (2006). *Gerontological nursing: Competencies for care.* Jones & Bartlett.
Meiner, S. (2011). *Gerontologic nursing* (4th ed.). Mosby Elsevier.
Palmieri, R. (2013). Care of patients with problems of the peripheral nervous system. In D. Ignatavicius & L. Workman (Eds.), *Medical-surgical nursing* (7th ed., pp. 986–1003). Saunders Elsevier.
Price, M. C. (2018). Musculoskeletal problems. In S. L. Lewis, L. Bucher, M. Heitkemper & M. M. Harding (Eds.), *Medical-surgical nursing* (10th ed., pp. 1496–1516). Elsevier.
Tabloski, P. (2014). *Gerontological nursing* (3rd ed.). Pearson Education.
Zychowicz, M. E. (2017). Assessment of musculoskeletal. In S. L. Lewis, L. Bucher, M. M. Heitkemper & M. M. Harding (Eds.), *Medical-surgical nursing* (10th ed., pp. 1446–1461). Elsevier.

5

The Nervous System

INTRODUCTION

The nervous system, certainly one of the most complex systems in the body, controls, coordinates, and integrates all body activities. Along with the endocrine system, it regulates and maintains homeostasis. The nervous system is fast acting and communicates with cells by electrical impulses, whereas the endocrine system is slower acting and is dependent on hormones released into the blood.

The nervous system allows us to adapt our behavior according to the stimuli we receive. Adaptive behavior requires

- complex and highly specialized sensory receptors receiving information from both the external and internal (body) environments
- processing and interpreting such information appropriately in the nervous system, a process called integration
- effectors (nerve endings in muscles and glands) that enable us to act on new information in an adaptive and life-sustaining manner

Sensory Receptors

Sensory receptors are nerve endings that respond to stimuli impinging on them. In lower animals a single receptor may be sensitive to all stimulation, but in higher animals and humans receptors have become highly specialized and react adequately only to very specific stimulation. For example, the specialized receptors for vision, hearing, and taste respond appropriately only to visual, auditory, and taste stimuli, respectively. Specialized sensory receptors in humans can be grouped into three types according to location:

1. *Exteroceptors.* These specialized receptors are located on or near the surface of the body and receive information from the external world. Examples include receptors for the "special senses" (i.e., vision, hearing, taste, touch, and smell).
2. *Interoceptors* (*visceroceptors*). These specialized receptors are located inside the body and receive information about the internal environment of the body. Receptors in the viscera (internal organs) supply information about sensations of pain, hunger, nausea, and other internal events. Such internal information is vital for general health and well-being. For instance, when the appendix becomes inflamed, pain localized in the lower right side provides significant warning that all is not well.

3. *Proprioceptors.* These specialized receptors are located in skeletal muscles, tendons, and joints and give continuous information about the body's position in space. To test this, close your eyes and extend your right arm horizontally. You are aware of the position of the arm solely because of "muscle information" received from proprioceptors.

The degree of specialization found in human receptors allows for a wide range of sensitivity to many different kinds of stimuli and for continuous awareness of conditions in both internal and external environments. Such a monitoring system provides fast, up-to-date information about one's life status.

Integration

The integrating parts of the nervous system are composed of the central nervous system (CNS), including the brain and spinal cord, and the peripheral nervous system (PNS), which includes both the somatic nervous system (nerves in peripheral parts of the body) and the autonomic nervous system (ANS). The ANS is composed of two divisions, the parasympathetic and the sympathetic.

The basic functional unit of the nervous system is the neuron, or nerve cell. Neurons are highly specialized cells that conduct nerve impulses throughout the body. Each neuron consists of a cell body with extensions called processes. The processes of a neuron consist of one extension called an axon and one or more (sometimes many) extensions called dendrites. Dendrites receive electrical nerve impulses and conduct them toward the cell body and the axon. Axons conduct nerve impulses away from the cell body. Neurons are closely associated with supporting cells, or neuroglia. Neuroglia comprise almost half the brain and spinal cord tissue and are much more numerous than neurons. They support, nourish, and protect neurons.

In general, CNS neurons are not able to replace themselves if injured or destroyed, although there may be a few exceptions. They also have an unusually high metabolic rate and require a continuous and adequate supply of oxygen; thus the nervous system is extremely sensitive to any lack of oxygen. Bundles of neuron processes in the CNS are called tracts; in the PNS they are called nerves. Neurons are characterized as

- motor (efferent) neurons, transmitting nerve impulses from the brain or spinal cord (CNS) to a muscle or gland, or from a higher center in the CNS to a lower center
- sensory (afferent) neurons, transmitting nerve impulses from sensory receptors to the brain or spinal cord, or from a lower center in the CNS to a higher center
- association neurons in the brain and spinal cord, transmitting nerve impulses from one neuron to another. About 99% of the neurons in the body are association neurons (Marieb & Hoehn, 2019).

Neurons are separated from each other by a space or junction called a synapse. Nerve impulses have to cross the synapse to pass from one neuron to another, or from a neuron to an effector cell. There are two types of synapses: electrical and chemical. In electrical synapses electric current flows between neurons, allowing the nerve impulse to pass from one neuron to another. In chemical synapses chemicals called neurotransmitters are secreted at the synaptic junction, either allowing the nerve impulse, or action potential, to cross the synapse (excitatory) or, in some cases, to block the transmission from one neuron to another (inhibitory). Some examples of neurotransmitters are acetylcholine, epinephrine, norepinephrine, serotonin, dopamine, gamma aminobutyric acid (GABA), glutamate, endorphins, and enkephalins. The electrochemical activity of the nervous system is highly complex and continues to be an exciting area of research.

Effectors

Effectors are nerve endings in muscles, glands, and organs that act to produce change. If you want to cross the street and you see a car coming, this information is transmitted to your brain, where the decision is made regarding whether you should cross or wait until the car passes. If you decide not to wait, should you run or walk when you cross? Such decisions depend on your knowledge of cars and speed factors, your fear of being hurt, how fast you can move, and other related information. Once a decision is made in the brain, muscles and glands are sent appropriate messages and action takes place. This brief explanation is much too simplistic because many systems of the body are involved in such behaviors, but it allows for some appreciation of the complexity and speed of decision-making made possible by means of the receptor–nervous system–effector circuit.

THE CENTRAL NERVOUS SYSTEM

The CNS includes the brain and spinal cord. Both are enclosed by bony structures (the skull and the spinal column, respectively) and cushioned by cerebrospinal fluid that completely surrounds them (Figure 5.1).

The Brain

The human brain represents the highest known form of development on the evolutionary scale. Some functions of the brain are as follows:

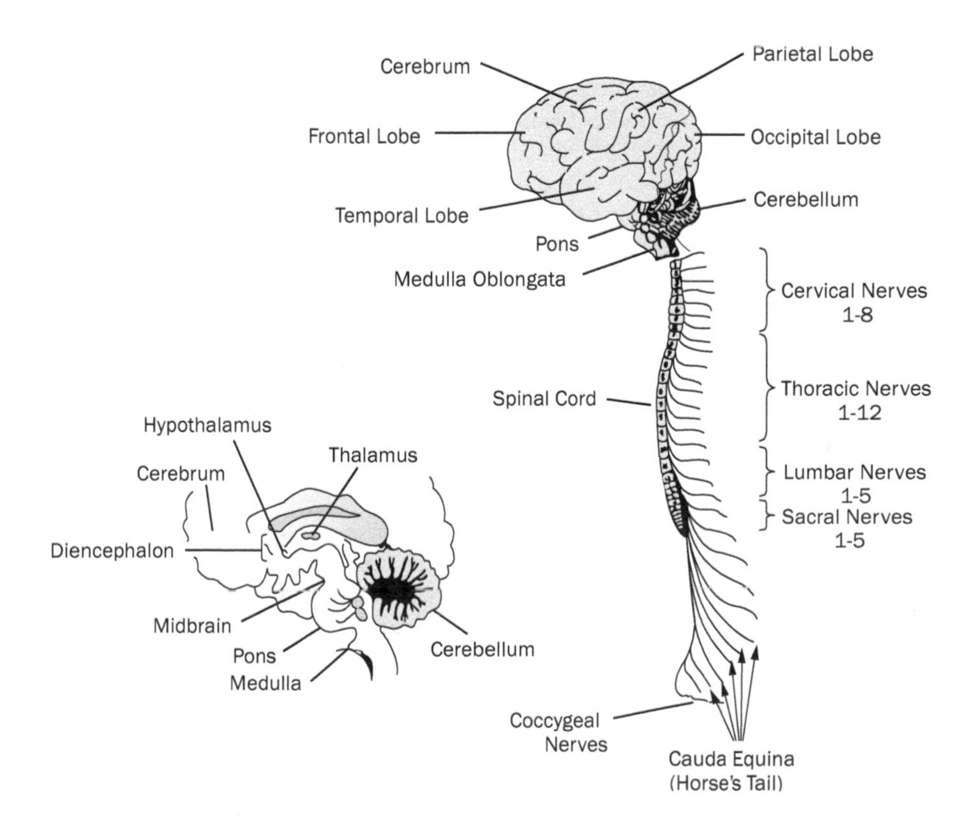

Figure 5.1 Central nervous system and cross-section of the human brain.

- to integrate and regulate body activities
- to initiate all voluntary acts of behavior
- to serve as the locus for learning, memory, thought, reasoning, and other complex mental activity
- to serve as the center for sensation and consciousness (awareness)
- to act as the locus of emotions and drives

The brain can be subdivided into four sections: cerebrum (composed of two cerebral hemispheres), diencephalon (thalamus, hypothalamus, and epithalamus), brainstem (midbrain, pons, and medulla oblongata), and cerebellum.

The Cerebrum

Together, the left and right hemispheres of the brain are referred to as the cerebrum. The two hemispheres are connected by a band of fibers called the corpus callosum. The cerebral hemispheres constitute more than 60% of total brain weight and are the most visible and obvious parts of the brain. The surface of the human cerebrum is called the cerebral cortex and is marked by many convolutions that consist of elevated ridges called gyri and shallow grooves called sulci; deeper grooves are referred to as fissures. Such a highly convoluted surface effectively triples the amount of brain tissue (Marieb & Hoehn, 2019). Thus, although the cerebral cortex is less than one-eighth of an inch thick, it accounts for about 40% of total brain mass (Marieb & Hoehn, 2019).

There are four major lobes in the cerebrum, which allow for some localization of function. They are the frontal, temporal, parietal, and occipital lobes. The frontal lobe mediates higher cognitive functions such as memory retention, thinking, and speech; the temporal lobe contains the auditory area, or center for hearing, and contributes to speech; the parietal lobe controls and interprets spatial information such as touch, pressure, and temperature; the occipital lobe contains the centers for vision. In addition, there is a motor area (skeletal muscle control) between the frontal and parietal lobes and a sensory area posterior to the motor area. In actuality, a number of CNS areas contribute to most activities rather than just one specific part of the brain being involved.

For the most part, the left half of the brain controls the right side of the body and the right half of the brain controls the left side of the body. Although we use both hemispheres for most activities, in approximately 90% of individuals the left hemisphere controls language, mathematical abilities, and logical reasoning, whereas the right hemisphere is concerned with motor skills, visual–spatial behaviors, emotions, intuition, art, and music appreciation.

The Diencephalon

The diencephalon is situated below the cerebrum. The three structures in the diencephalon are the thalamus, hypothalamus, and epithalamus.

The Thalamus

The principal function of the thalamus is to serve as a major relay center for sensory information between lower nervous system components and the cerebral hemispheres. Virtually all information ascending to the cerebrum passes through the thalamus, so it plays an important role in controlling information input to the highest centers of the brain.

The Hypothalamus

The hypothalamus serves as the major visceral control center and is crucial for the maintenance of homeostasis. Homeostatic mechanisms regulated by the hypothalamus include:

- ANS activities such as blood pressure, heart rate, force of heart contractions, gastrointestinal motility, respiration rate, and pupil size

- emotional responses such as perception of pain, pleasure, rage, fear, and sexual responsiveness
- body temperature regulation
- regulation of food intake, including perceptions of hunger and satiation
- regulation of water balance and thirst
- control of sleep–wake cycles
- regulation of many endocrine system activities

The Epithalamus

The epithalamus is the uppermost part of the diencephalon. It contains the pineal gland. The pineal acts with the hypothalamus in regulating sleep–wake cycles by secreting melatonin, a hormone significant in sleep–wake patterns that is also an antioxidant.

The Brainstem

The brainstem is the major pathway between the spinal cord and the cerebrum. It controls various reflex activities necessary to sustain life. Running through the brainstem are neurons and nerve fibers called the reticular formation, which receives and integrates information from many areas of the brain and from sensory (afferent) pathways mediating basic alertness. The brainstem includes the midbrain, the pons, and the medulla oblongata.

The Midbrain

Located between the diencephalon and the pons, the midbrain contains several large nuclei that relay information from lower centers to higher brain centers and assist in regulating balance, equilibrium, vision, and hearing.

The Pons

The pons is a part of the brainstem situated between the midbrain and the medulla. It also serves as a major relay center for ascending and descending nerve impulses and is the site of a number of larger nuclei for several of the cranial nerves. Other nuclei in the pons are involved in the maintenance of breathing.

The Medulla Oblongata

The medulla oblongata is the lowermost part of the brain continuous with the spinal cord. It is a crucial structure for reflexes necessary to sustain life. Three important centers in the medulla are the cardiac center, which mediates heart contraction rate and force; the vasomotor center, which regulates blood pressure; and the respiratory center, which maintains the appropriate rate and depth of breathing. Medullary nuclei also regulate reflex activities such as swallowing, coughing, sneezing, and vomiting.

Other structures of significance not relegated to one specific area of the brain include the following:

1. The basal ganglia, which connect to the cortex above and to the midbrain below, are also involved in the learning and programming of some behaviors.
2. The limbic system, part of the diencephalon and the midbrain, is involved in memory, emotional, and homeostatic activities.
3. The hippocampus, located in the inside fold of the temporal lobe and also part of the limbic system, stores information and memories, as well as being involved in emotional activities.

The Cerebellum

The cerebellum is located posterior to the medulla and pons. Primary functions of the cerebellum include the maintenance of muscle tone, posture, equilibrium, and the coordination of voluntary muscle movements. It is a highly significant integrative center for voluntary body activity and for coordination of certain reflexive behaviors related to body position and movement. It also appears to play a role in some cognitive activities, including thinking, language, and emotion (Marieb & Hoehn, 2019).

The Spinal Cord

The spinal cord, located within the vertebral canal of the spinal column, at the top is continuous with the medulla; at the bottom it tapers off in the region of the "tailbone," or coccyx, at the base of the spine. Thirty-one pairs of nerves emerge from the spinal cord, one pair at each spinal segment. Nerves leaving the cord at the base become the cauda equina (translated as the "horse's tail" because this is what they resemble). The basic functions of the spinal cord are

- to serve as a conducting pathway to and from the brain and the rest of the nervous system
- to act as a reflex center since some simple reflexes can be controlled at the spinal cord level. The knee-jerk response is an example of a simple reflex that is controlled primarily at the spinal cord level. Most human behavior, though, is infinitely more complicated; it involves nerves and muscles on both sides of the body as well as different spinal cord levels and the brain.

THE PERIPHERAL NERVOUS SYSTEM

The PNS includes the somatic nervous system and the ANS.

The Somatic Nervous System

The somatic nervous system includes 12 pairs of cranial nerves, which connect sensory organs; 31 pairs of spinal nerves, which bring information into the spinal cord and carry messages from the cord to the effectors; and various interconnecting nerves.

The Autonomic Nervous System

The autonomic nervous system activates involuntary activities in smooth muscles, glands, and cardiac muscle through its two principal divisions, the parasympathetic division and sympathetic division. These structures control activities essential to life. For example, the heart, blood vessels, respiratory organs, kidneys, bladder, reproductive organs, and endocrine glands are all activated by the ANS.

The Parasympathetic Division

The parasympathetic division of the ANS functions when the body is essentially at rest, and it tends to have specific effects on individual body organs rather than acting on the body as a whole. The nerve cell bodies of the parasympathetic nervous system are located in or near the structures they innervate. The PNS controls, in part, activities such as digestion, constriction of the pupils of the eyes, slowing of the heart, and increased storage of glycogen by the liver.

The Sympathetic Division

The sympathetic division generally opposes the parasympathetic division in function, thus maintaining balanced activity in the body. It is considered to be the body's arousal system; it is the emergency system for fast mobilization and quick energy. Activation of the sympathetic division affects organ systems rather than just individual body organs. Nerve cell bodies are located close to the spinal column. Sympathetic control increases heart rate, decreases gastric motility (allowing blood to be shunted to the muscles for immediate use), dilates pupils, temporarily stops metabolic body storage activities, and energizes the entire organism (fight-or-flight reaction). The autonomic nervous system is entirely motor; that is, it has some control over all effector organs of the body except voluntary muscles. The ANS is, in turn, controlled to some extent by higher brain centers, particularly the medulla, hypothalamus, and cerebral cortex.

Normally, homeostatic equilibrium is maintained effectively by the ANS in association with other body organ systems, and the reciprocal relationship between parasympathetic and sympathetic divisions generally prevents extremes in body regulatory activities. In a fear situation, for instance, one of the obvious body changes is increased heart rate (sympathetic activation). It would not be physiologically appropriate for heart rate to increase indefinitely, and because of parasympathetic intervention, heart rate usually returns to normal within a short time after excitation. The ANS thus keeps the body reacting appropriately to ongoing changes in the environment without constantly involving the highest brain centers (which focus on reasoning, intuitive thought, and decision making) over every bit of stimulation.

AGE-RELATED CHANGES IN THE NERVOUS SYSTEM

Certain problems arise in attempting to generalize about the significance of the aging process as it affects the nervous system. Because of difficulties involved in obtaining accurate data on nervous system functioning, much basic research has emphasized pathological rather than normal processes of aging. Also, a substantial amount of available information is based primarily on estimates of age-related change, with many conclusions derived from animal research. The process of human aging is highly individualized, and physiological, psychological, and social life experiences are enormously varied among people of the same age. Nevertheless, certain basic age-related changes do appear consistently even though their implications for human behavior are not always clear.

Researchers are especially interested in how the nervous system ages, because once a nerve cell dies, another apparently does not automatically regenerate to take its place. Red blood cells, for instance, live only about 120 days, but new cells are continuously produced so the total supply is constant. Thus, the age of blood cells varies. Nerve cells, though, may denote a truer age of the organism because they are present at birth and may not be replaced when they die.

Specific age-related changes in the nervous system include the following:

1. There is some loss of neurons with age, although the amount and location of the loss varies, and neuronal loss varies substantially among different regions of the brain (Burke & Barnes, 2006). Many neurons have increased dendrite growth into very old age, and this may partially compensate for neuron loss in some areas of the brain (Poirier & Finch, 1994). Some neurons may shrink with age rather than being completely lost (Burke & Barnes, 2006). Accompanying the loss of neurons is a decrease in brain weight and size. The nervous system

has large numbers of neurons, more than we probably ever use, so age-related neuron loss may not affect behavior appreciably until very old age.

2. Nonnervous or supporting tissue in the nervous system (neuroglia, for example) increases with age.
3. Earlier studies indicated general atrophy of the brain with age, but later imaging studies (primarily computed tomography [CT] and magnetic resonance imaging [MRI] studies) show highly selective atrophy with age rather than a general pattern.
4. Lipofuscin (fats or fatlike substances cross-linked with proteins) accumulates in the nerve cells, cardiac muscle, skeletal muscle, smooth muscle, thymus gland, pancreas, adrenals, liver, spleen, and parts of the sperm ducts. Some believe lipofuscin is related to cell activity, and the more active or normal the cell, the less accumulation of lipofuscin; however, research has not yet demonstrated this relationship satisfactorily.
5. Another change at the cellular level that occurs in older age is an increase in neuritic plaques (parts of degenerated neurons with an amyloid core), and neurofibrillary tangles (twisted nerve filaments in the cell bodies of neurons). These particular changes are also characteristic of the brains of those with Alzheimer's disease, leading to the suggestion that Alzheimer's may be an accelerated form of normal aging. Nevertheless, there is still no clear relationship among incidence of neuritic plaques, neurofibrillary tangles, and specific behavior changes.
6. Aging also initiates changes in transmission efficiency. If the central nervous system is viewed as a transmitting and receiving center, where messages are transmitted over very complex circuits from sensory organs to the brain, from the brain back to sensory effectors and organs, and from one part of the brain to another, the necessity for efficiency in the sending, transmitting, and receiving parts of the system is apparent. Several factors possibly affect transmission efficiency in older persons:
   a. Decreased numbers of functional nerve cells may reduce the strength of the message being transmitted.
   b. Fewer nerve cells result in more space to cross, and the coherence of the message may be disrupted, or random background noise (neural noise) could interfere with the clarity of the message.
   c. The motor part of the older cerebral cortex may continue to respond for a time after stimulation ceases, and such aftereffects could blur or interfere with subsequent incoming messages. These changes could then account for the increased time older people usually need to perform simple tasks as well as for their poorer retention and increased susceptibility to distraction in learning and memory tasks.
   d. Neurotransmitters are chemical substances facilitating or inhibiting the passage of nerve impulses across synapses. Their role in aging is not clear, but it is likely that less neurotransmitters are produced, along with a decline in receptors that detect them (Tabloski, 2014).
7. Older adults exhibit changes in brain wave patterns (as recorded on an electroencephalogram, or EEG), which reflect the electrical activity of the brain (Polich, 1997). Older persons' EEG patterns are likely to be slower and may resemble the EEG pattern of a child in the early developmental years, but individual differences in EEG patterns also increase with age. These particular age-related changes in healthy older people seem to have no apparent functional correlation with learning, memory, perception, or sensorimotor behaviors.
8. Sleep patterns change with age. Older adults generally report that although they spend more time in bed trying to sleep, total sleep time is reduced compared with the younger years and the time it takes to fall asleep increases (Touhy, 2016). Many awaken soon after going to sleep, or awaken often during the night, and daytime naps tend to increase with age, all of which can reduce the quality and efficiency of sleep (Ancoli-Israel & Martin, 2006).

Normal sleep includes two phases: nonrapid eye movement (NREM), or slow wave sleep, which accounts for about 75% to 80% of sleep, and rapid eye movement sleep (REM)

(Chokroverty, 2010). For many years NREM was thought to have 4 stages, but in recent years the consensus by researchers is 3 distinct stages (Chokroverty, 2010). Stage 1 is the lightest level of sleep. It is a transitional period of falling asleep during which an individual is easily awakened. In stage 2, sleep deepens, relaxation is greater, and an individual is not as easily awakened. Stage 3 is a period of deepest sleep during which vital signs, body temperature, and metabolism gradually decrease, and arousal is from this stage is the most difficult. This is considered to be the stage of the most restorative sleep. REM is similar to wakefulness and is associated with dreaming and increases in autonomic activity. Increased eye movements occur and vital signs (respiratory rate, heart rate, and blood pressure) are irregular and often elevated (Millsap, 2007). The usual pattern in sleeping is wakefulness, light sleep, deep sleep, and wakefulness (REM) again. During a night, individuals cycle through these stages several times, but NREM sleep stages are longer and REM sleep is shorter. This sequence, which ends with a REM period, is called a sleep cycle. A sleep cycle typically averages about 90 to 110 minutes in length and is repeated four to six times a night (Chokroverty, 2010).

In older adults, stage 1 NREM sleep is increased, stage 2 sleep is variable but often increased, and stage 3 is decreased and less deep. REM sleep is also reduced in older adults and interrupted more frequently (Eliopoulos, 2018). Changes in sleep patterns sometimes result in significant sleep deprivation and may have negative consequences for health. Some sleep disorders require specific evaluation and treatment because it is important to determine reasons for changes in sleep patterns that may be attributed to age. Possible reasons for altered sleep patterns include daytime napping, physical inactivity, drinking stimulating fluids before going to bed, drinking too much fluid requiring several visits to the bathroom during the night, eating a large meal near bedtime, anxiety, stress, and medications interfering with deep sleep. Obviously, lifestyle factors can greatly influence sleep patterns.

9. Age brings changes in the autonomic nervous system. Age-related changes in the ANS seem to be basically related to slower functioning and prolonged recovery time required after activation. There is substantial controversy over whether people become more or less easily aroused and activated by the environment as they grow older, and some evidence exists to support each point of view (Kaszniak & Menchola, 2012).

The nervous system may well be affected to a greater extent by decremental aging in other systems of the body than by intrinsic changes in nervous system tissue. Known nervous system changes associated with age can account, at least partially, for increased slowness that is so characteristic of older age, but beyond that there is too much individual variation to warrant substantial generalizations.

AGE-RELATED NERVOUS SYSTEM DISORDERS

Tremor

Tremors are identified by the activity that increases the tremor. They involve primarily the head, neck, face, and limbs. There are three types of tremors:

- *At-rest tremor,* most obvious when the affected limb is inactive or not moving
- *Postural tremor (action tremor),* most obvious when some posture is being maintained by the antigravity muscles, as in holding a glass or cup
- *Intention tremor,* the rhythmic movement of a limb when the individual attempts a specific, goal-directed movement such as moving the arm and hand to drink from a glass or cup

The causes of tremor in older adults are usually related to a number of factors rather than one specific causal agent. Most tremors of older adults, though, stem from Parkinson's disease, benign essential tremor, or metabolic/toxic tremor. Parkinson-related tremors are often postural tremors but may be combined with at-rest tremors. Benign essential tremor usually affects the head, voice, upper extremities, and, rarely, the legs. It is mainly a postural tremor but may also be present in intentional movements and occasionally at rest as well. This type of tremor is accentuated by fatigue and emotional stress, and although there are medications that reduce the tremor, they do not eradicate it completely. Older adults need to be reassured that benign essential tremor does not lead to Parkinson's disease. A number of metabolic or toxic factors can cause tremors, usually postural tremors. Examples of causative agents are lithium, tricyclic antidepressants, caffeine, steroids, hyperthyroidism, uremia, liver failure, and alcohol withdrawal.

Parkinson's Disease (a Form of Parkinsonism)

Parkinsonism refers to a group of symptoms involving abnormal movements, including slowness of voluntary movement (bradykinesia), resting tremor, increased muscle tone or rigidity, postural instability, and flexed posture (Millsap, 2007). Parkinson's disease, a chronic and slowly progressing disease, is the major cause of parkinsonism. Parkinson's is also the most common neurological movement disease in older adults, currently affecting more than 1 million Americans and projected to affect 1.6 million by 2037 (Yang et al., 2020).

The actual cause of Parkinson's is not known, although aging changes, environmental influences, and genetic factors are all being researched. It has been referred to as "shaking palsy" because the most obvious symptoms are pronounced tremor (both postural and at rest), muscle rigidity, and bradykinesia, which results in an inability to carry out normal activities of daily living and drastically curtails independence. As the disease progresses, standing, walking, and balance are impaired, with falls increasingly likely. Other symptoms besides muscle rigidity and tremors include a masklike facial expression, a fine "pill-rolling" tremor beginning in the hands and fingers, drooling, and difficulty swallowing. Psychological concomitants include depression, anxiety, social withdrawal, sleep disturbances, and emotional lability (unpredictable emotional responses). Dementia may be associated with Parkinson's in up to 31% of cases, especially in the later stages (Aarsland et al., 2005). As the disease progresses, other complications arise and need to be recognized and treated. The progression of symptoms can take 20 years or more.

The hallmark of Parkinson's disease is degeneration of dopamine-producing receptors in the substantia nigra part of the midbrain, resulting in dopamine deficiency. Why these neurons are lost or destroyed is not yet clear. Normally, the neurotransmitters acetylcholine and dopamine act together to balance nervous excitation and inhibition necessary for smooth and coordinated motor function. When an imbalance occurs, motor activity is significantly affected. A secondary form of Parkinson's may be caused by encephalitis, poisoning, or toxicity, or may be drug induced.

Diagnosis of Parkinson's disease is difficult because some other disorders closely resemble it. If Parkinson's exists, it is necessary to differentiate types that have known causes from the more common, primary type in which the specific cause is unknown. Accurate diagnosis depends on thorough physical examination, personal history, and clinical features present. A more definite diagnosis is made if at least two of the three classic symptoms are present: tremor, rigidity, and bradykinesia. Diagnosis is further confirmed if the individual responds to antiparkinsonian medications.

There is currently no cure for Parkinson's disease, but symptoms may be controlled reasonably well with an individualized medication program based on severity and type of symptoms, amount of impairment, other diseases present, and expected benefits and risks from available medications (Imperio & Pusey-Reid, 2006). All medications used to treat Parkinson's disease have significant side effects, and because this is a long-term disease, medications must be adjusted periodically as the individual's needs change. Medications are used to deal with symptoms, not provide a cure. The advent of levodopa (L-dopa) revolutionized treatment of Parkinson's. Dopamine does not cross the blood–brain barrier and is therefore an ineffective medication in treating

Parkinson's, but L-dopa does cross the blood–brain barrier and is converted to dopamine in the brain, exerting a therapeutic action in the brain itself. L-dopa does not prevent further deterioration, but it assists those with Parkinson's to function more effectively for a longer time. Most of those with idiopathic (primary) Parkinson's will experience at least some symptom relief from levodopa or levodopa plus carbidopa (Sinemet), but dosage must be carefully monitored in older adults to prevent undesirable or dangerous side effects such as confusion and hallucinations. Catechol O-methyltransferase inhibitors (COMTs) are commonly used in conjunction with levodopa because COMTs increase the effectiveness of levodopa.

Other medications used to help control symptoms of Parkinson's, especially in the earlier stages, are anticholinergic medications (to reduce the amount of acetylcholine in the brain), monoamine oxidase (MAO) inhibitors, dopamine agonists that activate dopamine receptors, anti-histamines, and antidepressants with anticholinergic actions that restore a more appropriate balance between the neurotransmitters dopamine and acetylcholine. Surgical procedures available if medications do not alleviate symptoms include ablation (destruction) of brain tissue and deep brain stimulation. Deep brain brain stimulation involves having electrodes implanted into the brain connected to a small electrical device called a pulse generator, that can help block the brain signals that cause many of the motor symptoms (National Institute for Neurological Disorders and Stroke [NINDS], 2020). Stem cell transplantation is currently an experimental procedure though major clinical trials are underway. Stem cell therapies have the potential to restore normal brain function, unlike current medications that treat symptoms but are unable to prevent the progression or cure the disease. Cells derived from embryonic neural tissue or induced pluripotent stem cells (cells from adults that are genetically reengineered to an embryonic cell-like state) are used to provide new dopamine-producing cells in the brain (NINDS, 2019).

Rehabilitation should be a continuing part of the total treatment regimen for Parkinson's disease so that functional abilities may be maximized for as long as possible. Multidisciplinary approaches are necessary to intervene and treat the disease as it progresses. Physical therapy, occupational therapy, and speech therapy are all very useful adjuncts to the management of Parkinson's disease. Persons with this illness and their families need continued emotional support and understanding from those around them.

Tardive Dyskinesia

Tardive dyskinesia is an increasingly common form of movement disorder in older adults who are on long-term antipsychotic medication regimens. Prolonged therapy involving these neuroleptic medications has been implicated as a cause of this disorder. Symptoms include "fly catching" movements of the tongue, lip smacking, grimacing, athetoid movements (recurring changing of position) of the head and extremities, and rocking movements of the trunk. The movements involving the mouth interfere with chewing, communicating, and the ability to use dentures.

Prevention, early detection, and medication manipulation are essential. Treatment involves stopping or reducing the dosage of the drug causing the tardive dyskinesia, although in many instances even this does not help to relieve symptoms. As more and more drugs are produced and used with older adults in long-term drug therapy regimens, tardive dyskinesia may be expected to increase in the older population.

Brain Tumors

The incidence of brain tumors is highest in older adults , especially in those older than age 75 (Nayak & Iwamoto, 2010). Primary tumors are those arising from tissues in the brain, but the most common form of brain tumor is actually a tumor arising from a malignancy elsewhere in the body. Approximately 25% of cancers in the body metastasize to the brain or surrounding structures. The most common metastases to the brain come from lung cancer, breast cancer, melanoma, colon and rectal cancers, and kidney cancer. The most common primary brain tumor is malignant glioma (Millsap, 2007).

Symptoms depend on location and size but generally include headache, seizures, gait changes, short-term memory loss, and some cognitive decline, especially in memory. Diagnosis in older adults may be delayed because the presenting symptoms are somewhat vague in the early stages. Diagnostic procedures use MRI, CT, and positron emission tomography (PET) scans as well as a thorough physical examination and personal history. Other diagnostic techniques are used to investigate the possibility of a primary tumor elsewhere in the body. Tissue biopsies are used to provide more definitive information.

Treatment depends on tumor location and size as well as neurological status and age. Surgery is the preferred treatment choice, but complete removal of the tumor is not always possible if it is inaccessible or involves vital parts of the brain. Radiation therapy is often used as a follow-up after surgery, and chemotherapy is another option, although it is often difficult because of the blood–brain barrier. New techniques are currently being researched, but the outlook for those with primary brain tumors remains rather poor (Laskowski-Jones, 2007).

Stroke

Cerebrovascular accidents (CVAs or strokes) are discussed in the chapter on the cardiovascular system. Additional discussion is not included here except to remind the reader that a CVA involves varying degrees of neurological damage to the brain itself.

Dementia

Dementias are discussed in Chapter 6, Dementia and Delirium.

Sleep Disorders

Recognition of sleep disorders has been an important advancement in medicine. These disorders affect quality of life and in some instances present life-threatening situations. Those who work with older persons need to be aware of the ramifications of sleep disorders and assist them in obtaining appropriate intervention. Sleep disorders affecting older adults may be classified as disorders in initiating and maintaining sleep, disorders of excessive sleepiness, and disrupted sleep–wake cycles (Haponik, 1994). There are several disorders that make initiating and maintaining sleep difficult. Approximately half the population of older adults complain of sleep disorders (Eliopoulos, 2018).

Insomnia

The most common sleep disorder is insomnia. As many as 40% of older adults have insomnia, and increasing age is the greatest risk factor (Foley et al., 2004). Insomnia is difficulty falling or staying asleep or premature waking (Eliopoulos, 2018). Insomnia can be short term (transient) and occurs as a result of illness, stress, or changed environments (Eliopoulos, 2018). Chronic insomnia (lasting more than 4 weeks) can be caused by physical or mental illness, can be a side effect of medications, can be caused by substance abuse, and/or can result from environmental factors (Eliopoulos, 2018). Sedatives can assist in reestablishing a normal sleep pattern, though it is always preferable to address and treat underlying causes wherever possible (Eliopoulos, 2018).

Sleep Apnea

Sleep apnea is a breathing disorder typically defined as the cessation of airflow through the nose and mouth for at least 10 seconds and with more than five episodes occurring per hour of sleep. Older adults are at greater risk for developing sleep apnea (Townsend-Roccichelli et al., 2010). Two types of sleep apnea are central and obstructive. In central sleep apnea, respiration efforts cease because of either decreased responses to central chemoreceptors (as in chronic obstructive pulmonary disease [COPD], for example) or increased responsiveness (as in high altitudes). Obstructive sleep apnea is more common in older adults and involves airway obstruction that prevents airflow. If sleep apnea is so severe that it impairs adequate oxygenation of body cells and tissues, it can cause impaired cardiac function and possibly death. Obstructive sleep apnea is associated with increased incidence of heart failure, stroke, coronary heart disease and atrial fibrillation (Drager et al., 2017).

Those with sleep apnea complain of excessive daytime sleepiness and of not feeling rested after a night's sleep. Men are more prone to sleep apnea than women, and it is also associated with snoring and obesity (Townsend-Roccichelli et al., 2010). Individuals with degenerative diseases such as Alzheimer's or vascular dementia and those with clinical depression often experience sleep apnea.

Restless Legs Syndrome and Nocturnal Myoclonus

Restless legs syndrome is characterized by a strong urge to move the legs repeatedly, which interferes with falling asleep and interrupts sleep. Nocturnal myoclonus involves periodic leg twitching or flexion of the leg muscles during sleep, which wakes the individual. These movements occur more frequently in older adults and may be caused by metabolic, vascular, or neurological factors.

Psychophysiological Problems

The presence of untreated emotional problems can also cause sleep disorders in older adults. Psychiatric disorders, depression, and dementias are all possible causes. Other issues to be evaluated are symptoms associated with physical disease, such as pain, renal disease, congestive heart failure, neurological disorders, and nocturia (excessive urination at night). Personal or social changes in older age, such as bereavement, retirement, financial concerns, and institutionalization, are also major contributors to altered sleep patterns (Haponik, 1994).

Disorders of Excessive Sleepiness

Disorders of excessive sleepiness, another category of sleep disorders, are characterized by excessive sleepiness that interferes with daytime activities. Possible causes include obstructive sleep apnea, nocturnal myoclonus, and narcolepsy (a chronic ailment of recurring attacks of drowsiness and sleep occurring at any time). Drugs (especially sedatives) are often major offenders in patterns of disrupted sleep.

Disrupted Sleep–Wake Cycle

The third category of sleep disorders is a disrupted sleep–wake cycle. If the sleep–wake cycle is disturbed by institutional routine, travel, work schedules, or other changes in the usual pattern, a sleep disorder may result.

Assessment of all sleep disorders depends on comprehensive evaluation, including a detailed personal history to determine previous sleep patterns, use of drugs and alcohol, exercise, diet, medications, emotional status, and other lifestyle variables. A sleep diary may be a source of helpful information, as are family members, especially a spouse. Physical examination and laboratory tests assist in screening potential causes of sleep problems. It is important to ascertain the actual severity of the complaint because many older adults are extremely concerned about losing sleep and the potential implications for their health. If a disorder exists and its causes cannot be identified by the primary care practitioner, referral to a sleep disorder clinic is appropriate.

Treatment involves providing an environment conducive to restful sleep, medication adjustment by discontinuing offending drugs if possible (or reducing dosages), counseling the individual about the desirability of establishing a regular routine for going to bed and arising, and specific treatment of any physical difficulties contributing to interrupted sleep patterns. In addition, certain activities that may negatively affect sleep should be avoided. These include eating a heavy meal before bed, exercising at night, daytime napping, discussing emotional issues before bedtime, drinking alcohol, and ingesting stimulants (such as caffeine). Other options for treating sleep disorders are behavior modification techniques, relaxation training, and continuous positive airway pressure (CPAP), in which a mask is worn during sleep and air is delivered under positive pressure to keep the airway open. Oral appliances are also available to improve airway flow during sleep by moving the tongue or mandible forward. Drug therapy may be prescribed, and surgery is an option for those who do not respond to other treatment choices (Linton, 2007). For example, obstructive sleep apnea may be improved by surgery, such as tonsillectomy, and by removing excess tissue in the throat to make the airway wider.

Dangerous Consequences of Sleep Disorders

There is a growing recognition of the profound importance of good quality sleep for overall health and well-being. Persistent lack of adequate sleep may lead to substantial adverse social, psychological, and physical health consequences. Accidents and injuries (e.g., traffic accidents, falls) can often be caused by drowsiness, reduced alertness, and inattention. Routine activities and work performance may be affected by changes in cognitive performance making it harder to focus and complete everyday tasks. Relationships may become strained due to irritability and frustration. Furthermore, inadequate sleep can lead to specific health outcomes, including increased blood pressure, reduced immune system functioning, increased risk of type 2 diabetes, and be associated with depression and anxiety (Benca, 2017). These adverse outcomes can be avoided if sleep disorders are diagnosed and treated quickly, especially in the aging population.

LEARNING AND MEMORY

One area of special concern is whether the aging of the nervous system affects learning and memory. Common myths and stereotypes have long implied that older adults are not able to learn new material and that poor memory is a part of normal aging. Such beliefs are overly simplistic, do not take other important variables into consideration, and are often based on observations of older people who have neurological pathologies. Nevertheless, these beliefs greatly influence the negative way some older adults think about themselves and for many become self-fulfilling prophecies.

Learning is usually defined as the acquisition of new information or a new skill through practice or experience. Memory is being able to retrieve or recall information once it is learned. The learning-memory process consists of acquisition (learning), storage (memory), and retrieval (memory). Variables with a definite effect on learning in older adults include the following:

1. The pace at which the material to be learned is presented and the speed with which a response has to be given affect learning. A fast-paced learning situation may interfere with acquisition, storage, and retrieval of information.

2. Cautiousness tends to increase with age, and older adults in learning situations may not respond to a new stimulus until they are certain about its meaning. Numerous learning experiments have shown that older adults more often commit errors of omission (not responding) than errors of commission (making an incorrect response).
3. Previously learned material may be more likely to interfere with a new learning situation in older adults than in younger adults, primarily because older adults have much more stored information than younger adults.
4. Older adults need to continue to practice and use learning and memory skills all through life. There is substantial research to indicate that "use it or lose it" applies also to cognitive skills in older age.

All in all, the ability to learn does not change significantly with age, although the process tends to be slower. When there is motivation to learn, freedom from cognitive-impairing disease or trauma, appropriately presented learning opportunities, and the continued use of previously acquired skills and strategies, the majority of older persons learn very effectively. Additional information on learning and memory may be found in the chapter on teaching older adults.

SUMMARY

The nervous system is one of the major integrating systems of the human body. It controls not only thinking, reasoning, and other cognitive processes but also all body movements. With age, nervous system activities are slowed. Thus, older adults are usually somewhat slower in receiving information through sensory receptors; slower in transmitting, processing, and interpreting information; and somewhat slower in acting on information. Being slower, however, does not imply incompetence; it does imply the need for proper pacing as a coping strategy to help offset age-related slowness. Older adults also show wide individual variations in both the rate of slowing and the age at which slowness becomes a significant behavioral issue. Generally, fast-paced younger people become fast-paced older adults (barring accident or disease) but will be slower than when younger.

Although there are changes in nervous system tissues with aging, such as decreasing numbers or shrinkage of neurons, increases in lipofuscin, and increased neuritic plaques and neurofibrillary tangles, certain characteristics of the nervous system may well serve to mitigate much of the behavioral effects of these aging changes. One such characteristic is redundancy, or having many more neurons than we use; another is compensatory mechanisms in the nervous system that may appear after damage to nervous tissue to take over some of the functions formerly performed by the damaged area. Finally, as some neurons die, others lengthen and increase their dendrites, making possible new connections and implying a greater plasticity of the nervous system than previously believed. In addition, many other variables influencing behavior changes in older age, such as medications, other organ system dysfunctions, environment, lifestyle, motivation, and disuse, make it virtually impossible to separate such related factors from actual nervous system changes.

REFERENCES

Aarsland, D., Zaccai, J., & Brayne, C. (2005). A systematic review of prevalence studies of dementia in Parkinson's disease. *Movement Disorders, 10*, 1255–1263. https://doi.org/10.1002/mds.20527

Ancoli-Israel, S., & Martin, J. L. (2006). Insomnia and daytime napping in older adults. *Journal of Clinical Sleep Medicine, 15*, 333–342.

Benca, R. (2017). Five surprising health risks of poor sleep. https://www.ucihealth.org/blog/2017/06/sleeping-health-issues

Burke, S. N., & Barnes, C. A. (2006). Neural plasticity in the ageing brain. *Nature Reviews Neuroscience, 7*, 30–40. https://doi.org/10.1038/nrn1809

Chokroverty, S. (2010). Overview of sleep & sleep disorders. *The Indian Journal of Medical Research, 131*, 126–140.

Drager, L. F., McEvoy, R. D., Barbe, F., Lorenzi-Filho, G., Redline, S., & INCOSACT Initiative (International Collaboration of Sleep Apnea Cardiovascular Trialists). (2017). Sleep apnea and cardiovascular disease: Lessons from recent trials and need for team science. *Circulation, 136*(19), 1840–1850. https://doi.org/10.1161/CIRCULATIONAHA.117.029400

Eliopoulos, C. (2018). *Gerontological nursing* (9th ed.). Wolters Kluwer.

Foley, D., Ancoli-Israel, S., Britz, P., & Walsh, J. (2004). Sleep disturbances and chronic disease in older adults: Results of the 2003 National Sleep Foundation Sleep in America Survey. *Journal of Psychosomatic Research, 56*(5), 497–502. https://doi.org/10.1016/j.jpsychores.2004.02.010

Haponik, E. F. (1994). Sleep problems. In W. R. Hazzard, E. L. Bierman, J. P. Blass, W. H. Ettinger, & J. B. Halter (Eds.), *Principles of geriatric medicine and gerontology* (3rd ed., pp. 1213–1228). McGraw-Hill.

Imperio, K., & Pusey-Reid, E. (2006). Cognitive and neurological function. In S. E. Meiner & A. G. Lueckenotte (Eds.), *Gerontologic nursing* (3rd ed., pp. 653–692). Mosby Elsevier.

Kaszniak, A. W., & Menchola, M. (2012). Behavioral neuroscience of emotion in aging. In M. Pardon & M. Bondi (Eds.), *Behavioral neurobiology of aging: Current topics in behavioral neuroscience* (Vol. 10., pp. 51–66). Springer.

Laskowski-Jones, L. (2007). Acute intracranial problems. In S. L. Lewis, M. M. Heitkemper, S. R. Dirksen, P. G. O'Brien, & L. Bucher (Eds.), *Medical–surgical nursing* (7th ed., pp. 1467–1501). Mosby Elsevier.

Linton, A. D. (2007). Respiratory system. In A. D. Linton, & H. W. Lach (Eds.), *Matteson & McConnell's gerontological nursing* (3rd ed., pp. 353–405). Saunders Elsevier.

Marieb, E. N. & Hoehn, K. (2019). *Human anatomy and physiology* (11th ed.). Pearson Education Limited.

Millsap, P. (2007). Neurological system. In A. D. Linton & H. W. Lach (Eds.), *Matteson & McConnell's gerontological nursing* (3rd ed., pp. 406–441). Saunders Elsevier.

National Institute for Neurological Disorders and Stroke. (2019). *Focus on stem cell research: NINDS program description*. https://www.ninds.nih.gov/Current-Research/Focus-Tools-Topics/Stem-Cell

National Institute for Neurological Disorders and Stroke. (2020). *Parkinson's disease information page: What research is being done?* https://www.ninds.nih.gov/Disorders/All-Disorders/Parkinsons-Disease-Information-Page#disorders-r2

Nayak, L., & Iwamoto, F. M. (2010). Primary brain tumors in the elderly. *Current Neurology and Neuroscience Reports, 10*(4), 252–258. https://doi.org/10.1007/s11910-010-0110-x

Poirier, J., & Finch, C. E. (1994). Neurochemistry of the aging human brain. In W. R. Hazzard, E. L. Bierman, J. P. Blass, W. H. Ettinger, & J. B. Halter (Eds.), *Principles of geriatric medicine and gerontology* (3rd ed., pp. 1005–1012). McGraw-Hill.

Polich, J. (1997). EEG and ERP assessment of normal aging. *Electroencephalography and Clinical Neurophysiology, 104*, 244–256. https://doi.org/10.1016/S0168-5597(97)96139-6

Tabloski, P. (2014). *Gerontological nursing* (3rd ed.). Pearson Education.

Touhy, T. A. (2016). Sleep. In T. A. Touhy & K. Jett (Eds.), *Ebersole & Hess' toward healthy aging* (9th ed., pp. 221–232). Elsevier.

Townsend-Roccichelli, J., Sanford, J., & VanderWaa, E. (2010). Managing sleep disorders in the elderly. *The Nurse Practitioner, 35*(5), 30–37. https://doi.org/10.1097/01.NPR.0000371296.98371.7e

Yang, W., Hamilton, J. L., Kopil, C., Beck, J. C., Tanner, C. M., Albin, R. L., Ray Dorsey, E., Dahodwala, N., Cintina, I., Hogan, P., & Thompson, T. (2020). Current and projected future economic burden of Parkinson's disease in the U.S. *NPJ Parkinson's Disease, 6*, 15. https://doi.org/10.1038/s41531-020-0117-1

6

Dementia and Delirium

It is now well established that dementia is not an inevitable part of the aging process. Although statistics vary considerably, in the United States researchers estimate that approximately 8.2 million people have some form of dementia, with a substantial number of those older than age 85 (Alzheimer's Association [AA], 2020a). Approximately 5.8 million Americans age 65+ have Alzheimer's disease, and death most often occurs because of complications associated with the long-term debilitation it causes (AA, 2020a). More women develop Alzheimer's disease than men, perhaps because women live longer. Although Alzheimer's is the dementia most publicized, other forms of dementia also affect older adults, and a thorough differential diagnosis is necessary to correctly identify the various forms of dementia and differentiate them from delirium states. Differentiation between delirium and dementia is essential because many delirium states are reversible, whereas true dementias are essentially irreversible.

CHANGES IN CATEGORIZING DEMENTIAS AND DELIRIUM

The Diagnostic and Statistical Manual of Mental Disorders (5th ed.; *DSM-5*; American Psychiatric Association [APA], 2013), widely used by clinicians, replaced the category "Dementia, Delirium, Amnestic, and Other Cognitive Disorders" with "Neurocognitive Disorders." *Neurocognitive disorders* (NCDs) is a more inclusive term that can be applied to cognitive disorders across the life span (e.g., those arising from a traumatic brain injury or an infection), not only those from intellectual disability, whereas the term *dementia* has traditionally been associated with cognitive deficits in older adults (APA, 2013). Another change is that the term *mild cognitive impairment (MCI)* is no longer used, having been replaced by *mild NCD.* MCI was used to describe an intermediary phase to capture the fact that pathological decline had occurred from normal aging but was not severe enough to warrant a diagnosis of dementia; it was almost always used in the context of Alzheimer's disease.

The change in the general diagnostic process is advocated in an effort to make earlier diagnoses and to be more specific between whether it is a mild and major NCD (Blazer, 2013). First, a clinician needs to determine that cognitive functioning (a) is normal, (b) is a mild NCD, or (c) is a major NCD. Second, if known, the cause of the dementia will be assigned for both mild and major NCDs (Blazer, 2013). Thus, a person may now be diagnosed with a mild or major NCD caused by Alzheimer's disease, Parkinson's disease, and so on. However, even though dementias are newly named as major/mild NCDs, it is understood that the term *dementia* is still commonly used and will continue to be used by clinicians and the general public (APA, 2013).

DELIRIUM

The NCD of delirium disrupts brain function and usually has multiple causes. In clinical delirium there are major behavioral changes not explained by a preexisting dementia. These include an inability to shift or maintain attention to external stimuli, disorganized thinking, perceptual disturbances such as hallucinations, disturbances in the sleep–wake cycle, increased or decreased psychomotor activity, reduced level of consciousness, disorientation, and memory impairment. Symptoms vary from person to person both in severity and progression and develop more rapidly (hours to days) than those of dementia. There are fluctuating periods of rationality over 24 hours rather than the gradual, steady deterioration characteristic of dementia (APA, 2013). Delirium is sometimes categorized as hyperactive, hypoactive, or mixed psychomotor behaviors (APA, 2013). Individuals who have hyperactive delirium are agitated, irritable, and restless. Those with hypoactive delirium are lethargic and apathetic and are often misdiagnosed as having depression. In the mixed form of delirium individuals fluctuate between hypoactivity and hyperactivity (Hain et al., 2016). Delirium is considered acute if it lasts hours to days; it is considered persistent if it continues for weeks to months (APA, 2013).

Delirium is commonly encountered in older adults who become hospitalized but is often missed or misdiagnosed. It is estimated that as many as one-fourth of all older patients in a hospital for treatment of an acute physical illness develop delirium, and the incidence is even higher in intensive care units. In spite of this high incidence, too many healthcare professionals do not recognize and diagnose delirium correctly (Hain et al., 2016). The term *sundowning* refers to older adults who become confused and disoriented in evening and night hours when sensory stimulation and activity are reduced. It is common for people with delirium to exhibit this nighttime-related worsening of symptoms (APA, 2013). Reduced sensory stimulation, along with the effects of illness, medications, social isolation, and unfamiliar surroundings, often leads to delirium behaviors. To further complicate diagnosis, delirium may be superimposed on a dementia or on a major mental illness. Nevertheless, those with delirium do not usually develop dementia.

There are numerous possible causes of delirium in older adults, including infections, dementia and other cognitive impairments, immobilization, dehydration, malnutrition, electrolyte disorders, coexisting medical problems, surgery, sensory deprivation, and medications. Medications are probably the most common cause of delirium, and no drug is above suspicion. Some of the common medications clearly associated with delirium states in older adults are antihypertensives, anti-inflammatory agents, muscle relaxants, anticonvulsants, hypnotics, sedatives, psychotropics, antihistamines, decongestants, analgesics, and anesthetics. Psychosocial factors of special significance include sleep deprivation, sensory deprivation or overload (especially in vision and hearing), and severe stresses such as bereavement or cumulative losses.

Diagnosis and Treatment

Diagnosis depends on a detailed history (including information from relatives, friends, and caregivers), a thorough physical examination with special attention to all medications used, and laboratory tests to identify possible infections or other medical problems that exist and contribute to the delirium. Genetic assessment may be appropriate if other family members have had Alzheimer's disease. X-ray examinations and CT or MRI scans are ordered if head injury is suspected. Cognitive assessment is a crucial part of diagnosis. Various screening tests, such as the Mini-Mental State Examination (MMSE; Folstein et al., 1975), are often used to assess cognitive functioning. Other commonly used assessment tools specific to identifying delirium are the Confusion Assessment Method (Inouye et al., 1990) and the NEECHAM Confusion Scale (Neelon et al., 1996). It is extremely important to differentiate delirium from true dementia, functional psychosis, and psychogenic dissociative states. Early identification of delirium is imperative if treatment is to be effective in reversing delirium. Once the underlying causes are identified, treatment revolves around multidisciplinary attempts to reverse the delirium behaviors. Environmental and behavioral modifications are paramount in importance. Social interventions include communication directed toward orientation and cognitive stimulation, facilitating mobility, preventing sensory misperceptions, and providing adequate rest and sleep.

MILD NEUROCOGNITIVE DISORDER

Mild NCD refers to cognitive changes greater than that of normal aging but not severe enough to warrant a diagnosis of major NCD. According to the *DSM-5* (APA, 2013), the diagnostic criteria for mild NCD are as follows:

- Clear evidence of mild cognitive decline from previous performance in one or more of the following domains of cognitive function: attention, learning, memory, executive functioning (such as organizing, planning, and the correct ordering of activities), language, social cognition, or perceptual-motor activity. This is based on concern from the individual, someone else with in-depth knowledge of the individual, or the healthcare provider/clinician that there has been a slight decline combined with modest cognitive performance impairment that has ideally been documented using standardized neuropsychological testing or clinician assessment.
- The cognitive deficits are not caused by delirium.
- The cognitive deficits do not interfere with the individual's ability to act independently in everyday life.
- The cognitive deficits do not manifest from other mental health disorders or issues.

MAJOR NEUROCOGNITIVE DISORDER

Major NCD is characterized by multiple disturbances in neurological, psychological, and social functioning. According to the *DSM-5* (APA, 2013) diagnostic criteria for major NCD are as follows:

- Clear evidence of major cognitive decline from previous performance in one or more of the following domains of cognitive function: attention, learning, memory, executive functioning, social cognition, language, or perceptual-motor activity. This is based on concern from the individual, someone else with in-depth knowledge of the individual, or the healthcare provider/clinician that there has been a major decline combined with significant reduction in cognitive performance that has ideally been documented using standardized neuropsychological testing or clinician assessment.
- The cognitive deficits are not caused by delirium.

- The cognitive deficits significantly impact the ability for everyday independent living, necessitating substantial help with activities of daily living such as managing medications.
- The cognitive deficits do not manifest from a different mental health issue or disorder.

The diagnostic criteria for major NCD and mild NCD (i.e., mild or moderate dementia) are very similar, with the exception of the severity of deficits. The onset of dementia is usually more gradual than that of delirium and involves irreversible and progressive deterioration of functioning often over a long period. Diagnosing dementia is difficult, especially in the early stages. To differentially diagnose dementia it is essential to rule out (a) delirium; (b) schizophrenia; (c) mood disorders, especially depression; (d) any other disorder with psychological symptoms similar to dementia; and (e) intellectual disability. The most common forms of dementia are the primary dementias (those caused by pathological conditions of the brain). Primary dementias include Alzheimer's disease, frontotemporal lobar dementias, dementia associated with Huntington's disease (HD), dementia with Lewy bodies (DLB), Creutzfeldt–Jakob disease (CJD), and vascular dementia. Secondary dementias are those associated with normal pressure hydrocephalus, drugs and alcohol, major depressive disorder, metabolic disturbances, thyroid and nutritional deficiencies, brain tumors, head injuries, Parkinson's disease, and AIDS or other infections (Bello & Schultz, 2011). Many of the secondary dementias are reversible.

PRIMARY DEMENTIAS

Alzheimer's Disease

The most prevalent primary dementia is Alzheimer's disease, and it accounts for 60% to 80% of all cases of dementia (AA, 2020a). Alzheimer's disease is a form of dementia that can affect middle-aged, young-old, or old-old individuals, but older individuals constitute the largest group. Memory loss, particularly for recent events, is one of the primary symptoms of Alzheimer's and is especially obvious in the early stages of the disease. The person may forget appointments, be repetitious in conversations, experience episodes of confusion, and forget usual obligations. It is important, however, not to interpret normal age-related changes in memory and cognitive behavior as early indices of Alzheimer's disease. Other common signs, noted by the AA (2020a), include the following:

- Challenges in planning or solving problems
- Difficulty completing familiar tasks at home, at work, or at leisure
- Confusion with time or place
- Trouble understanding visual images and spatial relationships
- New problems with words in speaking and writing
- Misplacing things and losing the ability to retrace steps
- Decreased or poor judgment
- Withdrawal from work or social activities
- Changes in mood and personality

The person with dementia becomes less spontaneous, loses interest in things previously found interesting, and begins to withdraw from social interactions. Even as these changes occur, the individual may be alert enough to cover up behavioral difficulties to a great extent, and only the immediate family or very close friends may be aware of changes in the person's usual behavior.

In the intermediate stage of Alzheimer's disease behavioral changes are more consistently obvious. Individuals often get lost, even in their own homes, and are essentially unable to learn or use new information. Memory for remote events is affected as well as memory for recent events. Disorientation becomes more pronounced when familiar objects or people are no longer recognized. These individuals require assistance with activities of daily living (e.g., bathing, eating, dressing, toileting). Speech problems include the inability to name objects or to choose appropriate words in speech, and speech may be limited to stereotypic word usage and repetitiveness. Behavior changes difficult for caregivers include wandering, hostility, agitation, uncooperativeness, possible physical aggressiveness, and a greatly increased risk of falling or having accidents. Apathy is common, and those with Alzheimer's disease become more self-centered, lose interest in others, and have little consideration for the needs of others. Anxious, clinging behavior and greatly increased dependency are also likely to occur.

The last, or terminal, stage of Alzheimer's is characterized by an inability to walk and talk, incontinence, and a complete inability to take care of oneself. These behaviors are difficult for caregivers to understand and manage. As the affected person becomes totally dependent on caregivers, there is increased risk for developing pneumonia, malnutrition, and pressure sores. The end stage of Alzheimer's disease is coma and death, often from pneumonia. The course of this disease can range from 3 to more than 20 years from onset of symptoms to death, but the average duration is about 8 to 10 years after diagnosis. Although Alzheimer's often follows the progression described here, there is great variability among individuals, and it does not always progress in the orderly way these stages suggest.

As for pathology, studies of the brains of those who died of Alzheimer's have found cortical atrophy, reduced total brain weight, and granulovacuolar degeneration in which granular material accumulates and is surrounded by spaces (vacuoles) in the cytoplasm of brain cells. Neuron loss occurs, especially in the cerebral cortex, hippocampus, and subcortical structures of the brain. Amyloid plaques (nerve cells surrounding a core of an abnormal brain protein called amyloid) are found primarily in the amygdala, hippocampus, and cerebral cortex.

Neurofibrillary tangles (tangled filaments or abnormal collections of a protein called tau) have been identified in the cerebral cortex, hippocampus, amygdala, and brainstem nuclei. Amyloid plaques and neurofibrillary tangles are considered to be the hallmark anatomical characteristics of Alzheimer's disease. Some neuronal loss, amyloid plaques, and neurofibrillary tangles occur in the brain as part of the normal aging process, but they are far fewer in number than those in the brains of individuals who have Alzheimer's disease. These similarities have led some to speculate that Alzheimer's may be a form of accelerated aging. In Alzheimer's disease, there are also multiple neurotransmitter deficiencies and imbalances, as well as a significant decrease in an enzyme important in acetylcholine production, learning, and memory. The amount of aluminum in the brain increases, but the significance of this is not understood. It is not yet clear whether these changes are partly the causes of Alzheimer's or the result of it.

The diagnosis of Alzheimer's disease is difficult and complex. A complicating factor is that those with the disease may also have delirium, other forms of dementia, or a severe psychiatric disorder. Diagnosis essentially remains dependent on excluding everything else that might cause the behavioral changes in question. Although diagnostic accuracy has improved over the years, definitive confirmation of Alzheimer's is currently only possible at autopsy. In the early stages of the disease it is important not only to rule out major depression or delirium but also to carefully assess normal age-related cognitive changes. Alzheimer's disease has become so widely publicized and so feared that older adults tend to look at any forgetfulness or slowness in recall of learned material as evidence of impending Alzheimer's disease. A thorough differential diagnosis depends on a complete history, physical examination, neurological and mental status examination, blood chemistries, electrolyte evaluation, thyroid function tests, folate and vitamin B_{12} levels, and urinalysis. CT or MRI scans may be used to exclude other causes of brain dysfunction. Newer imaging techniques and functional brain imaging procedures such as PET scans and magnetic resonance spectroscopy (MRS) allow for early identification of brain changes and for monitoring response to treatment. Recent research has identified that routine nasal discharge screenings show potential for identifying cognitive decline, which may lead to a simple, cheap, fast, and far less invasive diagnostic technique (Yoo et al., 2020).

Also vital to diagnosis are ongoing interviews with those close to the individual, such as family, friends, and caregivers, who can provide invaluable information about the severity and progression of symptoms. A complete evaluation of Alzheimer's disease is costly, time consuming, and mainly available in medical centers or facilities specializing in dementia. Because of this, many older adults are labeled as having Alzheimer's disease without benefit of a thorough assessment. The risk of misdiagnosis is substantial, especially because a diagnosis of Alzheimer's involves excluding every other physical or psychological basis for the behavior changes under consideration.

Regarding etiology, no one cause of Alzheimer's disease has been identified, and most researchers believe it has multiple causes. Age is the most significant risk factor, although there are other possible causes.

Etiology

Genetic—Early Onset

Familial Alzheimer's disease, in which multiple members of a family are affected by Alzheimer's in consecutive generations, is a rare early-onset form of the disease affecting less than 10% of those with Alzheimer's and usually occurring between ages 30 and the mid-60s (National Institute on Aging [NIA], 2019a). Familial Alzheimer's is associated with gene mutations on chromosomes 1 (abnormal presenelin 2), 14 (abnormal presenelin 1), and 21 (abnormal amyloid precursor protein [APP]). The protein beta-amyloid (involved in amyloid plaques) is formed from APP. Overproduction of beta-amyloid is considered to be a significant risk factor for Alzheimer's. Down syndrome is caused by an additional copy of chromosome 21, and many individuals with Down syndrome do develop Alzheimer's disease at an earlier age compared with the general population. Unfortunately, clinical trials targeting beta-amyloid have not been successful in patients with mild or moderate Alzheimer's disease, and although trials continuing with people with earlier stages are ongoing, initial results have also been disappointing (Panza et al., 2019).

Genetic—Late Onset

Considerable research has been invested to identify genetic risk factors associated with later-life Alzheimer's disease, especially because it is the most common form. The most common gene associated with Alzheimer's is apolipoprotein E (ApoE) on chromosome 19. There are several variations of ApoE, whose main purpose is to carry cholesterol in the blood. Those who inherit the ApoE e4 type appear to be at greater risk for developing Alzheimer's but do not inevitably develop it (NIA, 2019a). Those with ApoE e2 appear to have a reduced risk of Alzheimer's (NIA, 2019a). There are several other genes implicated in cholesterol metabolism, including the clusterin gene (CLU), SORL1, and ABCA7 (Mohan et al., 2016). There are also several genes associated with immune response including CR1, which may contribute to chronic brain inflammation, and others including CD33, MS4A, and TREM 2 (Mohan et al., 2016). Other genes that regulate communication between neurons and response to neural damage, including PICALM, BIN1, EPHA1, and CD2AP, are also being investigated (Mohan et al., 2016). All these and several others are currently being studied. Such research is incredibly complex, but the potential of identifying the mechanisms of Alzheimer's disease may result in actual treatment that prevents progression. Indeed, the ultimate goal would be to prevent the disease from ever developing.

Proteins

Other research is concerned with the role of tau in Alzheimer's disease. Tau is a protein in the neurofibrillary tangles associated with Alzheimer's disease. Normally, tau stabilizes microtubules in neurons and supports cell nutrition, but in Alzheimer's disease tau becomes twisted into the neurofibrillary tangles.

Viral

Slow-acting viruses are thought to be responsible for two other forms of dementia: Creutzfeldt–Jakob and kuru. Some research suggests Alzheimer's may be, at least in part, caused by a slow-acting or reactivated virus or some other infectious agent, including herpes simplex virus (HSV-1), which causes cold sores (Cheng et al., 2011). More recently, some researchers have concluded that the herpesvirus family (including HSV-1) does not cause Alzheimer's disease but may accelerate the progression (McKeehan, 2019).

Toxins

Environmental contaminants are considered a possible cause of Alzheimer's but research with human subjects is still needed to determine and provide clinical evidence in many cases (Manivannan et al., 2015). It is speculated that 30% of AD risk includes environmental factors and lifestyle patterns (Manivannan et al., 2005). Some of the suspected environmental toxins include organic and inorganic hazards, exposure to toxic levels of metal (e.g., aluminum, copper), pesticides, industrial chemicals (e.g., flame retardants), and air pollutants (Manivannan et al., 2005). According to Manivannan and colleagues, it is speculated that long-term exposure of contaminants that accumulate over a lifetime can trigger neuropathology and neuroinflammation that paves the way for the development of Alzheimer's disease.

Neurotransmitters

Changes in various neurotransmitter substances have been identified in the brains of those with Alzheimer's disease. For example, those with Alzheimer's have lower levels of acetylcholine, a neurotransmitter important in memory and most commonly found in the hippocampus and cerebral cortex, which are two areas especially affected by Alzheimer's disease. Further research is necessary before these complex relationships can be clarified.

Other avenues of research regarding possible causes of Alzheimer's disease include the role of inflammation in the body, significance of nerve growth factors that regulate nerve cell development, vascular risk factors (such as smoking, diabetes, serum lipids, and fat intake), and the specific role of early-life risk factors.

Treatment

As for treatment, there is currently no cure for this disease, but it is extremely important to treat symptoms as they develop to promote a higher quality of life for as long as possible. Cholinesterase drugs that increase the amount of available acetylcholine, a neurotransmitter important in memory, slow the progression of Alzheimer's for a limited time. For mild to moderate Alzheimer's disease, donepezil (Aricept), rivastigmine (Exelon), and galantamine (Razadyne) are commonly used. For treatment of moderate to severe Alzheimer's, memantine (Namenda) is used to block abnormal glutamate activity that is related to neuronal cell death and cognitive dysfunction. This medication may be used in conjunction with the cholinesterase drugs. Other drugs designed to help slow the progression of Alzheimer's are constantly being developed. Medications such as narcoleptics, anticonvulsants, and antidepressants are also useful in treating the various behavioral problems associated with this disease. Use of atypical antipsychotic medications to treat behavioral problems in older adults who have dementia has now been linked to increased deaths in older adults and must be used with extreme caution, if at all (Livingston et al., 2017). All these medications have side effects that need careful monitoring (Livingston et al., 2017). Other medications and over-the-counter supplements for the prevention or treatment of Alzheimer's continue to be explored. These alternatives include nonsteroidal anti-inflammatory drugs (NSAIDs), cholesterol-lowering drugs, vitamin B, E, folic acid, and ginkgo biloba. However, none of these have yet been proven to be effective in the prevention or treatment of Alzheimer's disease (Livingston et al., 2017).

Environmental interventions and modifications are important aspects of treatment regimens for Alzheimer's disease. Treatment revolves around physical care, safety, management of symptoms, support, and psychosocial caregiving strategies to help maintain the individual's self-esteem. Patients should be involved in their treatment and decision-making as long as possible to promote autonomy and give crucial information that can be helpful for clinicians to monitor the progression of the disease. Family members, who are by and large the primary caregivers, can also provide valuable information to clinicians and other health care providers. The active involvement of such caregivers is often vital to maintaining the quality of life of the person with dementia. Psychosocial rehabilitative techniques are used with varying degrees of success in the management of this disease, especially in the early stages. Comprehensive care and treatment plans must be flexible to deal with the variability the disease exhibits from one person to another and from time to time in the same person.

General Guidelines for Caring for Those with Alzheimer's Disease

General guidelines for caring for those with Alzheimer's disease include the following:

1. Maintain an unchanging routine for everything involving the patient. Create a calm, orderly, predictable environment with no surprises, including not changing the arrangement of furniture.
2. When any sort of change is necessary, introduce it very gradually and in small steps or increments. Make no abrupt changes without advance preparation.
3. In the early stages, help maintain the patient's orientation by use of clocks, calendars, newspapers, magazines, reality boards, and other types of orientation cues.
4. Monitor the person's health continually. Pay special attention to nutrition, dental care, foot care, and exercise. Exercise, as a vital part of healthcare, should be regular and systematic. Walking and stretching exercises are especially beneficial. Research indicates that those who exercise not only find it enjoyable but are calmer and sleep better than those who do not. Soft music may be used in exercise routines, but it should be calming, soothing music. Those with Alzheimer's are often more willing to participate in exercise if the caregiver exercises with them.
5. Make a concerted effort to maintain communication and affectional ties. Use all sensory systems possible in communicating, but do not introduce too much because sensory overload occurs easily in those with dementia. When communicating, use direct eye contact, use the person's name frequently, and use touch (as appropriate) as a therapeutic adjunct to verbal communication. Keep sentences brief. Sit down with the person and give them your undivided attention for a certain amount of time each day. Listen carefully to words and phrases the person uses as these may be important cues about their fears and anxieties.
6. Use humor and laughter appropriately as a therapeutic tool.
7. Use a lot of positive reinforcement. Arrange situations so the person will experience success. Praise and encourage generously.
8. Agitation and panic are signs of helplessness and loss of control. Help the person feel safe and secure by maintaining a calm attitude and consistent behavior.

Prevention

It appears that those who continue to engage in mentally stimulating and challenging activities may lower their risk of developing Alzheimer's or other forms of dementia. In addition, some other possible preventive actions, although not yet definitively proven to be effective, are to lower homocysteine levels, lower cholesterol, lower blood pressure, increase regular exercise, increase education, and control inflammation.

Frontotemporal Lobar Degeneration

Frontotemporal lobar degeneration (FTLD) is a syndrome associated with a group of neurodegenerative disorders involving primarily frontal and temporal lobe pathological conditions. FTLD is the most common cause of dementia in people younger than 60, and an estimated 60% of people with FTLD are aged between 45 and 64 years (NIA, 2019b). The disease progresses much faster than Alzheimer's disease (NIA, 2019b). Clinically, there are distinct behavioral and language changes. Behaviorally, individuals demonstrate impulsive or inappropriate social behavior early in the disease; later, judgment is impaired, and lack of initiative and apathy is apparent. Language changes include difficulty with expressive language initially, followed by reading and writing difficulties. Other cognitive changes appear as FTLD progresses.

Diagnosis is difficult, primarily because the behavioral changes are similar to other disorders such as depression, schizophrenia, and antisocial personality. Pathologically, the brains of those with FTLD show atrophy of the frontal and temporal lobes and virtual absence of neuritic plaques and neurofibrillary tangles. Confirmation of diagnosis depends on autopsy, as is true for Alzheimer's. There is no specific treatment except for supervision and safety measures, and the cause is unknown.

Dementia Associated With Huntington's Disease

HD is an inherited disease caused by a faulty gene for a protein called huntingtin (Huntington's Disease Society of America [HDSA], 2020). According to the HDSA approximately 41,000 Americans have HD, with symptoms usually appearing between the ages of 30 and 50 and progressing over a 10- to 25-year period (HDSA, 2020). Initially, those with this disease exhibit uncontrollable writhing movements known as chorea, and HD is sometimes identified as Huntington's chorea. Gait changes and lack of coordination develop. There are gradual personality changes, and finally dementia occurs. HD patients develop slurred speech and have difficulty with swallowing. Occasionally, dementia appears before the movement disorder. Frontal lobe functions are impaired and affect the ability to reason, plan, and organize. The cause is a deficiency of the neurotransmitters acetylcholine (ACh) and gamma-aminobutyric acid (GABA), especially in the basal ganglia. Medications often help control the movement disorder, but there is no cure and no effective medication for the dementia. The progression of motor, cognitive, and psychiatric symptoms ultimately results in the death of the individual (HDSA, 2020). Every child of a parent with HD has a 50/50 chance of inheriting the gene that causes the disease. HD had previously been considered to be quite rare; however, 200,000 Americans are currently at risk of inheriting the disease, making it one of the more common hereditary diseases (HDSA, 2020).

Dementia With Lewy Bodies and Parkinson's Disease Dementia

About 5% of people with dementia have DLB, though most also have it along with Alzheimer's disease (AA, 2020a). DLB is characterized by the presence of Lewy bodies (abnormal clumps of the protein alpha-synuclein within neurons) that can accumulate in several areas of the brain, including the cortex, hippocampus, basal ganglia, and the brainstem (NIA, 2018). There are two diagnoses of LBD, namely DLB and Parkinson's disease dementia (NIA, 2018). One important distinction is that with DLB, the dementia always appears first, whereas with Parkinson's disease dementia, the individual would have many years of living with Parkinson's disease before the onset of the Parkinson's disease dementia, and not all people with Parkinson's disease will develop Parkinson's disease dementia (Davis Phinney Foundation, 2018).

As symptoms of DLB can resemble those of Alzheimer's and Parkinson's diseases, and psychiatric disorders like schizophrenia, a differential diagnosis is imperative. There are typically fluctuations in attention, communication difficulties, and severe psychiatric symptoms (especially hallucinations). Motor difficulties usually occur as the disease progresses. No specific diagnostic procedures are available for DLB, so personal history, hallucinations, attention fluctuations, and motor behaviors must be assessed as possible indicators. Management of symptoms is the primary treatment and may involve some of the medications used for Alzheimer's disease. However, side effects are common and may be serious. Environmental support is necessary for safety and fall prevention. Caregiver education is a necessity because this form of dementia is not well known.

Creutzfeldt-Jakob Disease

CJD is a rare, fatal neurodegenerative disorder that progresses very rapidly; death usually occurs within a year after symptoms appear. In the United States, there are about 350 cases per year (National Institute of Neurological Disorders and Stroke [NINDS], 2020a). It is caused when normal prion proteins are converted into abnormal prion proteins in the brain, although it is not known why this occurs. A variation of CJD known as vCJD was identified in 1996 when humans consumed beef infected with bovine spongiform encephalopathy, or "mad cow disease." Nevertheless, the risk of contracting vCJD is very small. Symptoms include memory and behavior changes early in the disease, followed rapidly by involuntary movements, weakness, mental deterioration, blindness, and finally coma. There is no diagnostic tool specific for CJD, and no specific treatment is available. Only examination of the brain after death can confirm the diagnosis.

Vascular Dementia

Vascular dementia (mild/major vascular NCD [APA, 2013]) is caused by accumulated damage to brain tissue from successive small or moderate strokes. The symptoms are similar to those of Alzheimer's disease, but in vascular dementia, there is a history of heart disease, previous strokes, specific neurological signs, and hypertension. Neurological changes do depend on the areas of the brain affected. However, in vascular dementia there is a sudden onset of symptoms and fluctuating periods of rationality, or a stepwise progression of the dementia rather than the slow, steady course of deterioration characteristic of Alzheimer's. Unlike in Alzheimer's, impaired judgment or the inability to make plans tends to be the initial symptom rather than memory loss (AA, 2020a). Problems with balance and coordinated movement seem to occur earlier than they do in Alzheimer's, and psychomotor slowness is more characteristic of vascular dementia than other types of dementia. This type of dementia is more common in men than in women. Diagnosis is difficult because symptoms are similar to Alzheimer's disease, and vascular dementia can coexist with Alzheimer's. Useful diagnostic techniques are a thorough history, physical examination, cognitive testing, and CT or MRI scanning. Testing for ischemia using the Hachinski ischemic scoring system is also helpful in identifying vascular dementia (Kim & Kwon, 2014). Treatment techniques include medications to prevent further strokes, reduce hypertension, and modify other risk factors for cerebrovascular disease.

Mixed Dementias

As diagnostic and research techniques improve, and larger scale autopsy studies are conducted, studies have recently found that the likelihood of having more than one type of dementia is much more common than previously thought (AA, 2020a). Researchers investigating neuropathologies (brain abnormalities such as beta-amyloid plaques, neurofibrillary tangles, Lewy bodies, vascular changes etc.) of over 1,000 older people (who had undertaken cognitive assessments and consented to brain autopsy upon death) found 94% of participants had one specific type of neuropathology; however, 78% had two or more, 58% had three or more, and 35% had four or more (Boyle et al., 2018). They found Alzheimer's disease was the most frequent diagnosis (65%) but only 9% had Alzheimer's disease alone. Quite remarkable was the finding that there were over 230 different combinations of neuropathologies observed, leading to their conclusion that there is far greater variety in the comorbidity of age-related neuropathologies than currently known, and novel approaches need to be developed that consider the complexity of dementia causing diseases (Boyle et al., 2018). The most common form of mixed dementia is when Alzheimer's disease is present along with vascular dementia (AA, 2020b).

SECONDARY DEMENTIAS

Unlike primary dementias where the disease itself directly causes progressive and irreversible dementia, secondary dementias can develop from an existing illness. In many cases, secondary dementias can actually be reversed if appropriate and timely treatment is given.

Dementia Associated With Parkinson's Disease and Other Movement Disorders

Diseases that damage neurons affecting motor areas of the brain can ultimately create further brain damage that leads to dementia. Dementia may be associated with Parkinson's in up to 31% of cases, especially in the later stages (Aarsland et al., 2005). Most often when dementia symptoms do occur, they appear several years after the movement disorder symptoms. However, in more severe cases mental and motor symptoms may occur simultaneously. The behavior patterns of affected persons are similar to those with Alzheimer's disease. Individuals with multiple sclerosis and amyotrophic lateral sclerosis are also at risk of developing cognitive impairment and dementia with the progression of these diseases (NINDS, 2020b, 2020c).

HIV-Associated Dementia

HIV is the virus that causes AIDS. HIV-associated dementia (HAD) leads to damage of the brain's white matter, resulting in a dementia that is characterized by poor concentration, social withdrawal, and memory problems. The development of more effective combination of antiretroviral therapies for the underlying HIV infection has resulted in a substantial drop of HAD in recent years (Saylor et al., 2016). AIDS in older adults sometimes is acquired through blood transfusions from years ago, although AIDS as a result of unprotected sex is more prevalent in older adults than most realize. Those who work with older adults need to be aware of an increase of AIDS in the older population and possible dementia in the later stages of AIDS.

Reversible Dementias

Several conditions that can cause or mimic dementia are able to be reversed if timely and appropriate treatment is given (Bello & Schultz, 2011).

Normal Pressure Hydrocephalus

Normal pressure hydrocephalus occurs when the ventricles of the brain become enlarged and compress the surrounding areas. This obstructs the flow of cerebrospinal fluid and the fluid builds up in the brain. The classic triad of gait disturbances, incontinence, and dementia helps to differentiate this form of dementia from others. Additional symptoms are similar to the cognitive impairments of Alzheimer's disease. Etiology is often unclear, but in some instances it results from subarachnoid hemorrhage, meningitis, tumor, and other head trauma. Surgical shunting of fluid away from the brain usually decreases symptoms significantly and in some cases can completely reverse symptoms.

Substance/Medication-Induced Dementia

Older persons often use many medications, both prescription and over the counter. The sheer number of drugs used plus changes in how the aging body handles drugs make drug-induced dementia, a leading cause of secondary dementia, much more likely in older adults than in younger people. Although any medication can potentially cause cognitive impairment, the most likely causes of drug-induced dementia are long-acting tranquilizers, analgesics, digitalis, antihypertensives, antiarrhythmics, antidepressants, antihistamines, nonprescriptive sedative hypnotics, anti-Parkinson's drugs, and psychotropic drugs and their cumulative side effects. Those who work with older adults should be aware of the possibility of medication-induced dementia and opt for shorter-acting medications or reduced dosages whenever possible.

Chronic substance abuse of alcohol and recreational drugs often produces dementia symptoms in older adults. People who drink excessive amounts of alcohol are often deficient in vitamin B_1 (thiamine). Korsakoff's syndrome is caused by lack of thiamine (vitamin B_1) from chronic alcoholism. A cardinal sign of Korsakoff's syndrome is the loss of short-term memory. Another form of alcohol-induced dementia involves a gradual progression of dementia behaviors, including cerebral dysfunction in many areas. These dementias will usually partially remit if abstinence can be maintained for several months. However, in some individuals damage will be permanent—a condition known as substance-induced persisting dementia (APA, 2013).

Dementia Caused by Major Depressive Disorder

A major depression in older adults may interfere with cerebral functioning and produce a dementia syndrome. The prefix "pseudo" has been applied to this dementia, but that is very misleading because the dementia is as real as any other form of secondary dementia. A complete history and clinical examination will usually help to separate clinical depression from dementia caused by a major depressive disorder. Antidepressant treatment often relieves many of the symptoms associated with this form of dementia.

Dementia Caused by Metabolic Disturbances

Metabolic disorders can produce dementia if they are not identified and treated early. Common causes of metabolic-induced dementias are liver failure; kidney failure; thyroid problems; hyperglycemia (high blood sugar); disorders of fluid and electrolytes, especially hypercalcemia (excessive calcium in the blood); and alterations in sodium. Many hormone alterations and vitamin B_{12} and folate deficiencies can also produce dementia.

Dementia Caused by Space-Occupying Lesions

Major causes of dementia as a result of space-occupying lesions are tumors and hematomas produced from stroke or trauma. Clinical history often distinguishes these types of dementia from other forms. Most of these dementias are treatable if diagnosed early enough.

Cardiovascular-Anoxic Dementia

Numerous myocardial diseases, such as myocardial infarction, heart block, and arrhythmia, can produce dementia, as can advanced pulmonary disease. Because of reduced amounts of oxygen being delivered to the brain, early diagnosis is especially important for effective treatment of these conditions.

SUMMARY

Dementia and delirium are not a part of the normal aging process, and most older adults do not have these disorders. When symptoms appear, intensive efforts should be launched to identify causal factors. If these factors cannot be identified, efforts must be directed to treating symptoms as they appear. Just because a disease cannot be cured, there is no reason to resist initiating treatment options. Dementias are difficult for everyone—the patient, family, friends, caregivers, and healthcare professionals. However, symptoms can often be relieved as they occur with judicious use of medications, various supportive therapies, and education of patient, caregivers, and family.

REFERENCES

Aarsland, D., Zaccai, J., & Brayne, C. (2005). A systematic review of prevalence studies of dementia in Parkinson's disease. *Movement Disorders, 20*(10), 1255–1263. https://doi.org/10.1002/mds.20527

Alzheimer's Association. (2020a). Alzheimer's Association report: 2020 Alzheimer's disease facts and figures. *Alzheimer's & Dementia, 16*(3), 391–460. https://doi.org/10.1002/alz.12068

Alzheimer's Association. (2020b). Mixed dementia. https://www.alz.org/media/Documents/alzheimers-dementia-mixed-dementia-ts.pdfAlzheimer

American Psychiatric Association. (2013). *Diagnostic and statistical manual of mental disorders* (5th ed.). Author. https://doi.org/10.1176/appi.books.9780890425596

Bello, V., & Schultz, R. R. (2011). Prevalence of treatable and reversible dementias: A study in a dementia outpatient clinic. *Dementia & Neuropsychologia, 5*(1), 44–47. https://doi.org/10.1590/S1980-57642011DN05010008

Blazer, D. (2013). Neurocognitive disorders in DSM-5. *American Journal of Psychiatry, 170*(6), 585–587. https://doi.org/10.1176/appi.ajp.2013.13020179

Boyle, P. A., Yu, L., Wilson, R. S., Leurgans, S. E., Schneider, J. A., & Bennett, D. A. (2018). Person-specific contribution of neuropathologies to cognitive loss in old age. *Annals of Neurology, 83*(1), 74–83. https://doi.org/10.1002/ana.25123

Cheng, S.-B., Ferland, P., Webster, P., & Bearer., E. L. (2011). Herpes simplex virus dances with amyloid precursor protein while exiting the cell. *PLoS ONE, 6*(3), e17966 https://doi.org/10.1371/journal.pone.0017966

Davis Phinney Foundation. (2018). *The difference between Lewy body dementia, Parkinson's disease, and Alzheimer's disease.* https://davisphinneyfoundation.org/difference-lewy-body-dementia-parkinsons-disease-alzheimers-disease/

Folstein, M. F., Folstein, S. E., & McHugh, P. R. (1975). "Mini-mental state". A practical method for grading the cognitive state of patients for the clinician. *Journal of Psychiatric Research, 12*(3), 189–198. https://doi.org/10.1016/0022-3956(75)90026-6

Hain, D., Ordonez, M., & Touhy, T. A. (2016). Care of individuals with neurocognitive disorders. In T. A. Touhy & K. Jett (Eds.), *Ebersole & Hess' towards healthy aging: Human needs and nursing response* (9th ed., pp. 152–169). Elsevier.

Huntington's Disease Society of America. (2020). *Overview of Huntington's disease.* https://hdsa.org/what-is-hd/overview-of-huntingtons-disease/

Inouye, S. K., van Dyck, C. H., Alessi, C. A., Balkin, S., Siegal, A. P., & Horwitz, R. I. (1990). Clarifying confusion: The confusion assessment method. A new method for detection of delirium. *Annals of Internal Medicine, 113*(12), 941–948. https://doi.org/10.7326/0003-4819-113-12-941

Kim, Y. H., & Kwon, O. D. (2014). Clinical correlates of Hachinski ischemic score and vascular factors in cognitive function of elderly. *BioMed Research International, 2014*, 852784. https://doi.org/10.1155/2014/852784

Livingston, G., Sommerlad, A., Orgeta, V., Costafreda, S. G., Huntley, J., Ames, D., Ballard, C., Banerjee, S., Burns, A., Cohen-Mansfield, J., Cooper, C., Fox, N., Gitlin, L. N., Howard, R., Kales, H. C., Larson, E. B., Ritchie, K., Rockwood, K., Sampson, E. L., ... Mukadam, N. (2017). Dementia prevention, intervention, and care. *Lancet, 390*(10113), 2673–2734. https://doi.org/10.1016/S0140-6736(17)31363-6

Manivannan, Y., Manivannan, B., Beach, T. G., & Halden, R. U. (2015). Role of environmental contaminants in the etiology of Alzheimer's disease: A review. *Current Alzheimer Research, 12*(2), 116–146. https://doi.org/10.2174/1567205012666150204121719

McKeehan, N. (2019). *Can herpesvirus increase your risk for Alzheimer's disease?* Alzheimer's Drug Discovery Foundation. https://www.alzdiscovery.org/cognitive-vitality/blog/can-herpesvirus-increase-your-risk-for-alzheimers-disease

Mohan, G., Zhang, M., & Lu, Y. (2016). Genes associated with Alzheimer's disease: An overview and current status. *Clinical Interventions in Aging, 11*, 665–681. https://doi.org/10.2147%2FCIA.S105769

National Institute on Aging. (2018). *What is Lewy body dementia?* U.S. Department of Health & Human Services. https://www.nia.nih.gov/health/what-lewy-body-dementia

National Institute on Aging. (2019a). *Alzheimer's disease genetics fact sheet*. U.S. Department of Health & Human Services. https://www.nia.nih.gov/health/alzheimers-disease-genetics-fact-sheet

National Institute on Aging. (2019b). *What are frontotemporal disorders?* U.S. Department of Health & Human Services. https://www.nia.nih.gov/health/what-are-frontotemporal-disorders

National Institute of Neurological Disorders and Stroke. (2020a). *What is Creutzfeldt–Jakob disease?* https://www.ninds.nih.gov/Disorders/Patient-Caregiver-Education/Fact-Sheets/Creutzfeldt-Jakob-Disease-Fact-Sheet

National Institute of Neurological Disorders and Stroke. (2020b). *Multiple sclerosis information page: What research is being done?* https://www.ninds.nih.gov/Disorders/All-Disorders/Multiple-Sclerosis-Information-Page

National Institute of Neurological Disorders and Stroke. (2020c). *Amyotrophic lateral sclerosis (ALS) fact sheet* https://www.ninds.nih.gov/disorders/Patient-Caregiver-Education/Fact-Sheets/Amyotrophic-Lateral-Sclerosis-ALS-Fact-Sheet

Neelon, V. J., Champagne, M. T., Carlson, J. R., & Funk, S. G. (1996). The NEECHAM confusion scale: Construction, validation, and clinical testing. *Nursing Research, 45*(6), 324–330. https://doi.org/10.1097/00006199-199611000-00002

Panza, F., Lozupone, M., Logroscino, G., & Imbimbo, B. P. (2019). A critical appraisal of amyloid-β-targeting therapies for Alzheimer disease. *Nature Reviews Neurology, 15*, 73–88. https://doi.org/10.1038/s41582-018-0116-6

Saylor, D., Dickens, A. M., Sacktor, N., Haughey, N., Slusher, B., Pletnikov, M., Mankowski, J. L., Brown, A., Volsky, D. J., & McArthur, J. C. (2016). HIV-associated neurocognitive disorder-pathogenesis and prospects for treatment. *Nature Reviews Neurology, 12*(4), 234–248. https://doi.org/10.1038/nrneurol.2016.27

Yoo, S. J., Son, G., Bae, J., Kim, S. Y., Yoo, Y. K., Park, D., Baek, S. Y., Chang, K. A., Suh, Y. H., Lee, Y. B., Hwang, K. S., Kim, Y., & Moon, C. (2020). Longitudinal profiling of oligomeric Aβ in human nasal discharge reflecting cognitive decline in probable Alzheimer's disease. *Scientific Reports, 10*(1), 11234. https://doi.org/10.1038/s41598-020-68148-2

7

The Sensory Systems

INTRODUCTION

All knowledge of the world in which we live comes to us through our sensory systems. To survive, we must constantly be aware of the environment and changes taking place within it. We must also be able to interpret incoming information, integrate it with knowledge about our body state at the moment, and act on it adaptively. Adaptive behavior, in fact life itself, depends on the integrity of the receptor–nervous system–effector system. Inaccurate or partial information received in the nervous system results in distorted or inappropriate behavior. Such behavior is particularly significant in older persons attempting to maintain independence and control in the face of the various decline factors and cumulating losses associated with advancing age. In older age, both the amount and the quality of sensory input are vital factors in adaptive and adjustive behavior. Various research and clinical data suggest that humans need both an adequate amount and an adequate variety of stimulation in order to remain mentally intact and in contact with the real world. The behavioral implications of sensory deprivation resulting from the aging process are complex and intriguing.

Sensory systems of major concern in the study of aging are vision (sight), audition (hearing), gustation (taste), olfaction (smell), tactile (touch), vestibular (balance), and kinesthetic ("muscle sense"). Each contributes a specific type of information necessary for continuing adaptation and adjustment. Sensory changes usually begin in the 40s and 50s with a gradual reduction in acuity or sharpness of discrimination, but they do not appreciably limit behavior until about the 70s or 80s. For example, it is common to observe a 40-year-old person holding a newspaper at arm's length because of the increasing farsightedness of middle age, but age-related poor vision probably will not curtail their driving until many years later. Having to hold a paper at arm's length may be a nuisance, but it does not limit behavior or change total lifestyle as not driving does. No longer being able to drive in our mobile society has far-reaching psychological and social consequences for older persons. The best programs and services ever devised will be of little use if lack of transportation makes them inaccessible, as is often the case among older adults. Certainly, not being able to comfortably and safely negotiate one's day-to-day environment limits independence, self-sufficiency, and one's sense of personal competence.

Simply measuring the decrease in the functioning of a given sensory system cannot enable one to predict an older person's unique behavior capabilities or limitations associated with the particular sensory loss. First, there is significant variation among individuals in the rate of aging. Second, the amount of loss is highly variable from one organ system to another within a given individual. Third, humans have an amazing ability to adapt and compensate for gradual changes. For some, compensation and adaptation to a large sensory loss may be so effective that activities of daily living (ADL) are only minimally affected, but for others, a minimal sensory loss will produce major changes in lifestyle and possibly even result in the individual becoming housebound and requiring substantial support. Using more effective ways to assist people in adapting and compensating efficiently to gradual age-related changes would probably eliminate a number of common problems besetting many older adults as well as prolong their personal independence and self-maintenance. Fourth, some sensory systems are obviously more important in everyday functioning than others. We live primarily in a visual and auditory world and are very dependent on the integrity of our sight and hearing in dealing with day-to-day needs. Loss of smell, for instance, does not hinder a person nearly as much as loss of vision.

One generalization that can be safely made about sensory changes and age is that as we age, it takes stronger stimuli to activate sensory receptors; lights need to be brighter, sounds louder, and smells stronger for the aging person to obtain the quality and quantity of information from the environment needed for effective, adaptive action. This fact has enormous practical implications for creatively improving and modifying living and working environments to make them more supportive and appropriate for older individuals.

VISION

The main structures of the eye are as follows:

1. *Sclera.* The sclera is the outer layer of the eyeball, which is the "white" of the eye. It protects and shapes the eyeball.
2. *Cornea.* The cornea is the transparent avascular (without blood vessels) surface of the eyeball through which light rays enter the eye. The primary function of the cornea is to bend (refract) light rays, so they will come to a focal point directly on the retina for maximal stimulation of visual receptors. It is the most exposed part of the eye and therefore most vulnerable to damage.
3. *Anterior and posterior chambers.* The anterior chamber is the space between the cornea and the iris. The posterior chamber is the space between the iris and the lens. Both are filled with aqueous humor, a clear liquid transporting nutrients and waste products to and from the lens and cornea. Aqueous humor is continually produced and drained away. Usually, the production and drainage are equal and a constant intraocular pressure is maintained, but if the drainage of aqueous humor is blocked, intraocular pressure in the eye increases, causing compression of both the retina and the optic nerve, possibly resulting in glaucoma.
4. *Iris.* The iris is a thin, pigmented, circular, muscular sphincter suspended between the cornea and the lens. The opening at the center is the pupil. The amount of pigment in the iris gives color to the eyes, and the function of the iris is to regulate the amount of light entering the eye through dilation (opening) and constriction (closing) actions, which change pupil size. When illumination is low or dim, the pupil opening becomes large or dilated, allowing a maximum amount of light to stimulate visual receptors, but in bright light, pupils constrict and the opening becomes smaller, so receptors will be stimulated but not damaged by intense light rays.
5. *Lens.* The transparent, flexible, avascular crystalline lens helps to focus light rays, so they converge, or come to a focal point, precisely on the part of the retinal surface producing the sharpest vision at different viewing distances. The lens is enclosed in a capsule and arranged in concentric layers that continue to grow by adding layers throughout life. The relatively

flexible lens, suspended in place behind the iris by ligaments and ciliary muscles, can flatten or bulge to change its shape as necessary for sharp vision. In distance vision, the ciliary muscles relax and the lens is as flat as it can get; in near or close vision, ciliary muscles contract, making the lens bulge to focus light rays from a nearby object, so they fall on the retina correctly. This process is called visual accommodation. Changing lens shape to bring converging light rays to a focus directly on the retinal surface allows for very sharp and precise vision at both near and far distances.

In some people, the shape of the eyeball, the cornea, or the lens brings light rays to a focus at a point behind the retinal surface, resulting in hyperopia (farsightedness). Similarly, the shape of either the eyeball, cornea, or lens may produce myopia (nearsightedness) when light rays come to a focus at a point in front of the retinal surface rather than directly on it. Astigmatism, or irregularities in the curvature of the cornea or lens, is another common visual problem; it causes blurred or indistinct visual images. Hyperopia, myopia, and astigmatism can usually be corrected by prescription eyeglasses, contact lens, or various surgical procedures. Surgical options commonly used are laser surgeries or implants.

6. *Vitreous humor.* The vitreous humor is a clear, gel-like material contained in the area behind the lens and in front of the retina of the eye. It helps to maintain the shape of the eyeball, contributes to intraocular pressure, and transmits light.
7. *Retina.* The retina, consisting of several distinct layers of cells, is photosensitive tissue at the back of the eye. Visual receptors (rods and cones) and nerve pathways are contained in the retina. There are millions of photoreceptors in the retina that convert light energy (Marieb & Hoehn, 2019). Both rods and cones manufacture pigments, which are changed by light rays and result in the initiation of a nerve impulse. Rods mediate dim light or night vision and peripheral vision, whereas cones are responsible for day vision and color vision. Rods and cones are distributed differentially on the retinal surface—cones are densely clustered at the back of the retina, and rods are located predominantly along the sides of the retina. Thus, to see an object most distinctly at night or under very low illumination, look slightly off to the side of the object rather than directly at it, because more rods will be stimulated than cones. The human retina is estimated to contain about 125 million rods and 6 million cones. Approximately 50,000 cones are concentrated in the macula, the area of sharpest and most distinct vision. In the center of the macula is a small area called the fovea.
8. *Optic nerve.* The optic disc, a blind spot with no sensory receptors, is at the back of the eyeball where the optic nerve leaves the eye. The optic nerve contains more than a million nerve fibers (see Figure 7.1).

AGE-RELATED CHANGES IN VISION

Cornea

The cornea becomes thicker and less curved with age. This affects its refractive ability and causes older adults to be more prone to astigmatism. Often in those older than age 60, a gray ring, arcus senilus, forms around the outer edge of the cornea, but this does not affect vision appreciably.

Anterior Chamber

As the lens thickens with age, the anterior chamber decreases in size. Sometimes the growth of the lens puts pressure on the canal of Schlemm at the junction of the iris and the cornea, the point at which aqueous humor drains from the eye. An increase in intraocular pressure results, which can lead to glaucoma.

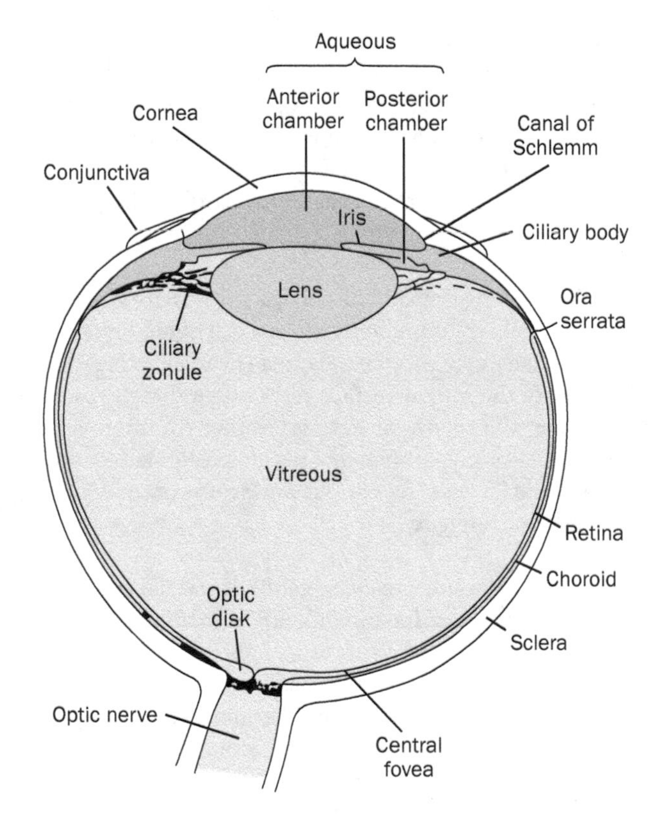

Figure 7.1 Structure of the eye.

Iris

Eye color is determined by the pigmentation of the iris. With age, eye color fades somewhat and older eyes may not appear as lustrous as those of younger individuals.

Pupil

The pupil is the aperture or opening in the iris. With age, pupil diameter decreases and the amount of light reaching the visual receptors by age 70 may be only a third of the amount reaching a younger person's eye. This reduction in pupil size is called miosis and results from age-related changes in the smooth muscle of the iris. There is also some decrease in ability to dilate, resulting in poorer adaptation to darkness (Workman, 2013).

Lens

The lens has two extremely important functions in vision—refraction and accommodation. Refraction requires a crystalline clear lens, whereas accommodation requires the lens to be elastic and able to change shape. Both these processes are affected by aging. The lens becomes thicker, denser, and less elastic. As the lens becomes more dense or cloudy (more opaque), it also becomes more yellow, impairing refractive ability.

Yellowing of the lens ("yellow filter effect") eventually produces changes in color vision, although many older adults remain unaware that color discrimination has changed. Older persons with distorted color perception are often able to discriminate between bright colors such as reds and yellows better than between darker colors such as blues and greens. Color coding is one effective way to use color perception information. For example, painting restroom doors one bright color and dining room doors another color can likely increase mobility and social interaction among nursing home residents, because distinct color coding improves space and place orientation and residents are more willing to move about. Color coding can be effective in many other situations, such as public buildings, transportation systems, or housing areas, to maximize legibility and effectiveness of visual messages. Color contrast also aids in-depth perception and ability to judge distances. Because color perception changes are often subtle, behavior changes related to deficient color perception may be misinterpreted. For instance, people may assume an older person who is not able to discriminate medications by color is just careless or not paying attention when medication errors occur. Similarly, mismatched or dirty clothing may be interpreted as lack of interest in self-care or even as early stages of dementia, when in reality the older person simply cannot discriminate colors accurately and is unaware of this.

If the lens becomes increasingly cloudy or opaque, a cataract results. Cataracts are the most common disability of the aging eye and will be discussed under age-related disorders of the eye. As the lens becomes less elastic and more dense, visual accommodation is affected. Farsightedness increases with age, and near-vision tasks become more difficult, an age-related change called presbyopia. Most adults in their 40s and 50s need corrective lenses for reading or other near-vision activities. Although decreased elasticity of the lens is the major reason for presbyopia, lessened tone in the suspensory ligaments and ciliary muscles holding the lens in place also contributes to a decline in visual accommodation.

Vitreous Humor

The vitreous humor becomes less gel-like and more liquid with age. It also becomes less transparent and causes light rays to scatter as they pass through the vitreous humor to the retina, leading to less distinct, more blurred vision. Older adults are often aware of brief flashes of light or of opacities in the eyes called floaters. Floaters are due to changes in the vitreous humor and are generally loose cells and tissues casting shadows on the retina. Floaters are at times annoying but are not usually dangerous.

Retina

Age-related changes in the retina include the following: (a) Blood vessels and capillaries narrow and may atrophy; (b) there is some loss of rods and cones; and (c) light and dark adaptation processes decrease as changes occur in the chemical sensitivity of the rods and cones.

Specific Implications of Age-Related Visual Changes

Age-related changes in the visual system have numerous significant implications for behavior.

1. *Decreased visual acuity.* The decreased sharpness in vision occurs because of (a) changes in refraction of light rays by the cornea and lens; (b) decreased accommodation ability; (c) less light admitted to the eyes as a result of smaller pupils; and (d) reduced numbers of visual receptors (rods and cones). Increasing the illumination, eliminating glare, and using larger-print materials help to offset the behavioral effects of age-related lessened visual acuity.
2. *Light and dark adaptation.* There is a decrease in light and dark adaptation processes with age. Dark adaptation is the process by which the eyes become maximally sensitive to the dark after having been in the light, and light adaptation is the converse, when eyes become

maximally sensitive to the light after having been in the dark. A good example of these processes is the experience of walking out of bright sunlight into a dark movie theater. Initially it is impossible to see anything, but after a few minutes the eyes become sensitive to the dark (they become dark adapted) so that empty seats can be identified and even individuals can be recognized. Dark adaptation is a chemical and neural process that takes time for completion. Reasonable sensitivity is usually attained in 2 to 4 minutes, although the chemical process is not complete for about 20 minutes. Conversely, when coming out of the dark theater into the light, the brightness is very uncomfortable for a few minutes before the eyes become light adapted. Dark and light adaptation are both mediated by photosensitive pigments contained in the rods and cones bleached out and restored according to prevailing levels of illumination.

3. *Visual threshold.* There is a higher visual threshold of sensitivity with age requiring more light to adequately stimulate visual receptors. *Visual threshold* refers to the minimum amount of light that will stimulate visual receptors and trigger a nerve impulse to the brain, thereby registering visual information in the highest cortical centers of the nervous system. A higher threshold means greater illumination is needed to obtain the maximum amount of visual information from the environment, a fact important in designing optimal living–working situations for middle-aged and older people.
4. *Increased sensitivity to glare.* Exposure to glare is more difficult for older adults because of age-related changes in the cornea, lens, and vitreous humor. Light is scattered throughout the eyeball rather than being focused at a precise point on the retinal surface and thus interferes with sharp vision. Common sources of glare include bright sunlight, exposed light bulbs, and light reflected off white, shiny surfaces. Even bright and shiny walls and floors can produce disturbing glare.
5. *Peripheral vision.* A significant loss of peripheral vision is common in aging and may influence both physical activity and social interactions. Older adults may be more likely to spill food or drinks placed in their visual periphery or not see objects out of their range of vision. Similarly, they may be unable to see people who are outside their range of vision. Loss of peripheral vision also has implications for driving (Linton, 2007).

General Implications of Age-Related Visual Changes

Behavioral implications of changes in vision are primarily associated with older adults' decreasing efficiency in responding to the visual world and with accident prevention. For instance, driving at night and coping with the glaring headlights of oncoming cars may be hazardous for most older persons because of slower dark and light adaptation and lessened visual acuity. For the same reasons, moving from light to darker areas in the home (or vice versa) increases the possibility of accidents.

With the dramatic increase in use of digital devices such as computers, tablets, and smartphones, there is increasing concern regarding "digital eye strain" affecting 50% of users (Sheppard & Wolffsohn, 2018). Although symptoms are usually transient, and can range from blurred vision, headaches, dry eyes, and sensitivity to light, digital eye strain can result in frequent and persistent issues (Sheppard & Wolffsohn, 2018).

The visual system is without question one of the most important links to the world in which we live. A variety of gradual changes take place with aging in this very complex sensory system, and awareness of these changes should provide greater motivation for preventive care. Regular eye examinations, proper lighting, and avoidance of excessive eye strain are important in preserving vision. There are effective ways to compensate for age-related visual changes and thereby reduce behavioral limitations associated with visual impairments. This should be a significant area of interest both to the gerontologist and to older adults because it has enormous practical applications for the maintenance of ADL and for the enjoyment of life.

AGE-RELATED DISORDERS OF THE VISUAL SYSTEM

Cataract

A cataract is a cloudy or opaque lens severe enough to interfere with light rays passing through the lens thus causing impaired vision. Cataracts are the most common age-related disorder of the aging eye and some degree of cataract formation is usually found in all people older than age 70, and by age 80 more than half of all Americans have cataracts or have had cataract surgery. Over 25 million people in the United States have cataracts (National Eye Institute, 2019). The exact causes of cataracts associated with aging are not yet clearly identified, but it appears that metabolic changes in the proteins of the lens are significant in cataract formation. Other risk factors are high blood pressure, diabetes, prolonged use of corticosteroid drugs, excessive exposure to sunlight, excessive use of alcohol, and possibly a family history of cataracts.

Early symptoms of cataract include myopia (nearsightedness) and sensitivity to glare as the refractive power of the lens increases when opacities develop. This increase in refractive power temporarily compensates for presbyopia, and some 60- to 70-year-old people can read again without glasses ("second sight"). However, as lens changes progress, vision becomes increasingly impaired, making reading difficult even with glasses. Other classic symptoms experienced eventually are halos around objects, blurred vision, and decreased light and color perception (Touhy, 2016). At one time, older adults were advised to wait until cataracts were "ripe" before having surgery to remove them, but now they are encouraged to have cataracts removed whenever visual acuity changes interfere with their lifestyle. Opacity of the lens can be seen in an ophthalmoscopic examination. Verbal complaints about reduced vision help determine how much the individual is affected by the cataract.

Surgery to remove the lens is the treatment of choice for cataracts. The success rate for cataract surgery is more than 95% (Touhy, 2016). When the lens is removed, there must be some way to make up for its loss. Options are (a) prescription eyeglasses, which are thick, magnify objects by about 25%, and interfere with peripheral vision; (b) contact lenses, which provide more peripheral vision than eyeglasses and do not magnify objects as much but are difficult to manipulate, especially for those with arthritic hands or coordination difficulties; (c) intraocular lens, a plastic lens permanently implanted in the eye that provides good central and peripheral vision and does not magnify objects significantly.

Most cataract surgery is now performed under regional or local anesthesia and often in an outpatient setting. The most common procedure is phacoemulsification, in which a probe is inserted through the lens capsule and high-frequency sound waves break the lens into small pieces, which are removed by suction. The replacement lens is placed inside the capsule. Another surgical procedure to remove the lens and replace it with an intraocular implant is standard extracapsular cataract extraction.

Patient education is an important part of treatment for cataracts. Wearing a hat and sunglasses with UVA and UVB protection when in the sun, not smoking, eating a low-fat diet rich in antioxidants and vitamins C and E, and wearing protective glasses when using power tools are all extremely important. Similarly, possible prevention of cataracts involves these behaviors plus regular eye examinations, taking care of other existing health problems, having a healthy diet, and maintaining a healthy weight.

Glaucoma

Glaucoma, the second most common cause of blindness, is particularly dangerous because it progresses slowly and usually without noticeable symptoms. Approximately 3 million Americans have glaucoma, with 50% not knowing they have the disease, and globally, it is the second leading cause of blindness (Centers for Disease Control and Prevention, 2020). Over 120,000 Americans are blind because of it (Glaucoma Research Foundation, 2021). Age is the most important predictor of glaucoma, with older women affected about twice as often as men. Other contributing factors are African American descent, a family history of glaucoma, diabetes, hypertension, migraines, and possibly long-term use of some medications with anticholinergic effects (Touhy, 2016). Glaucoma occurs when there is an increase in intraocular pressure in the eye. Intraocular pressure is normal when the amount of intraocular fluid (aqueous humor) produced is equal to the amount drained from the eye. When intraocular fluid does not drain as quickly as more is formed, pressure within the eye increases. Increased intraocular pressure, if untreated, damages both the retina and the optic nerve, resulting in irreversible blindness. Glaucoma is classified as primary or secondary. In secondary glaucoma, a pathological process blocks the outflow channels through which aqueous humor drains from the eye. Possible causes are some drugs (corticosteroids, for example), eye diseases, some systemic diseases, and trauma. Treatment of secondary glaucoma is difficult and oriented toward removal or control of whatever prevents the outflow of the aqueous humor.

Primary glaucoma can be primary open angle (POAG) or primary angle closure (PACG). Angle-closure or acute glaucoma is relatively rare and accounts for only 5% to 10% of all glaucomas. Individuals with an anatomically shallow anterior chamber of the eye may develop angle-closure glaucoma as the lens grows and thickens with age, further reducing the size of the chamber and blocking the outflow of fluid. An acute situation then develops, with eye pain, clouded vision, nausea, and vomiting. Prompt treatment is necessary if blindness is to be averted. Laser surgery is often very successful, but some may require medications such as carbon anhydrase inhibitors, mannitol, urea, and glycerin afterward for long-term control of intraocular pressure (Eliopoulos, 2014; Miller, 2009).

Approximately 90% of all primary glaucomas are open angle (chronic). The outflow of aqueous humor becomes impaired gradually as degenerative changes occur in the eye. Symptoms are not usually apparent, and much damage may be done before the condition is ever diagnosed. This type of glaucoma is not curable but can usually be controlled with both topical and systemic medications to help constrict the pupil and increase the outflow of fluid from the eye. Medications may also be prescribed to inhibit aqueous humor production. Typical medications used are beta-blockers, miotics, alpha-adrenergic agonists, and carbon anhydrase inhibitors (Tabloski, 2014). If medications fail, laser treatment, surgery, or microsurgery are indicated to reduce intraocular pressure or to establish an alternate pathway for aqueous humor circulation. Because glaucoma is usually symptomless, individuals older than age 40 should have periodic eye examinations that include glaucoma testing.

The diagnosis of glaucoma is no longer based on measures of intraocular pressure alone, but includes an examination of the optic disc for cupping (atrophy), and visual field evaluations. Another diagnostic procedure is gonioscopy in which a contact lens and a binocular microscope allow for direct examination of the anterior chamber and differentiation between closed-angle and open-angle glaucoma (Eliopoulos, 2014). Glaucoma is a difficult disease to manage because it requires strict adherence to medication regimens and care to prevent further damage to the eyes. Healthcare professionals need to teach patients about the disease and stress the importance of compliance for optimal management.

Diabetic Retinopathy

Diabetic retinopathy is a serious visual problem associated with diabetes and is a leading cause of adult blindness (Eliopoulos, 2014; Touhy, 2016). Prevalence increases with the length of time a person has diabetes and how well the diabetes has been controlled. Many people with diabetes will have some retinopathy after 20 years of diabetes. About 40% to 45% of Americans with diabetes are affected by diabetic retinopathy. Essentially, retinal blood vessels develop small aneurysms, resulting in retinal hemorrhages. Recurring hemorrhages block light from reaching visual receptors and, in addition, damage the receptors themselves. Initially, the macula area is most affected causing distorted central vision, but in time damage occurs over a wider area of the retina. Early-stage diabetic retinopathy is called nonproliferative retinopathy, whereas the more advanced stage is called proliferative retinopathy.

Symptoms usually do not appear until at least 3 to 5 years after the onset of diabetes. Early symptoms are subtle, such as cloudy vision or seeing a shower of spots. Symptoms increase with recurring hemorrhages, retinal detachment, or secondary glaucoma. Early diagnosis is of paramount importance because laser photocoagulation is extremely effective in preventing or slowing visual loss, especially in the early stages of diabetic retinopathy. Those with diabetic retinopathy need to control blood glucose, cholesterol, and blood pressure as well as have regular eye examinations.

Age-Related Macular Degeneration

Age-related macular degeneration (AMD) is a leading cause of legal blindness in older adults and affects both central vision and visual acuity. Types of AMD include dry (nonexudative) and wet (exudative). Approximately 90% of all cases are the dry form, in which cells of the macula start to atrophy, leading to slow but progressive loss of vision as a result of gradual blockage of retinal capillaries. About 2 million older adults in the United States and Canada have dry AMD (Workman, 2013). Only 10% to 15% of those with the dry form will develop the wet form, and those are usually people not under treatment for AMD. The wet form is more severe, and if untreated the majority of people become functionally blind. Abnormal blood vessels grow beneath the retina, which may leak and cause irreversible damage. The macula is the retinal area most densely populated by cones and is responsible for sharp vision. Decreased blood supply to the macula and fovea damages receptors, resulting in central vision loss, usually bilaterally, although peripheral vision is not adversely affected. As central vision declines, tasks involving discrimination of detail or high visual acuity (such as reading or driving) become difficult or impossible. Complete blindness does not occur because peripheral vision remains reasonably intact. Aging is not the only causal factor in macular degeneration; genetics, smoking, cardiovascular disease, family history, and long-term sunlight exposure are also assumed to be significant. Nutrition is recognized as a factor in the progression of AMD, and vitamin C, vitamin E, beta-carotene, zinc, and green leafy vegetables seem to be helpful in the prevention and progression of the disease (Smith & Neely, 2007; Tabloski, 2014). The National Eye Institute's (2020) Age-Related Eye Disease Studies (AREDS) were major clinical trials investigating two vitamin formulas AREDS and AREDS 2, and have found them beneficial for people with intermediate and advanced AMD. The AREDS 2 combination can be found in eye vitamin supplement products such as PreserVision.

Symptoms include evidence of central vision distortion such as objects appearing larger or smaller or straight lines appearing bent, and there is usually other evidence of loss of central vision acuity. Self-checks with an Amsler grid can be a quick and useful way to alert an older adult of the onset of AMD (Boyd, 2020). Ophthalmoscopy is the primary diagnostic procedure. Drusen, yellowish extracellular deposits, occur in the early stages of the dry form. Intravenous (IV) fluorescein angiography and photography are also helpful in determining the extent and type of AMD. Treatment is to reduce the risk of further vision loss and may involve laser photocoagulation, pharmacological control of inflammatory conditions, photodynamic therapy, and use of low-vision aids if laser and pharmacological interventions do not help. Injections directly into the eye with anti-vascular endothelial growth factor (anti-VEGF) can help stabilize wet AMD, and about one third report improved vision. Anti-VEGFs help to inhibit the growth of abnormal blood vessels in the eye. Bevacizumab (Avastin) and Ranibizumab (Lucentis) are the most commonly used, but new anti-VEGFs are being developed along with other treatment procedures.

AUDITION (HEARING)

Hearing is crucial because most of the time we relate to each other primarily through verbal communication. Hearing loss is thought by many to be the most devastating sensory handicap of all, commonly resulting in withdrawal from interactions with family, friends, and society in general. Paranoid ideas and behavior, suspicion, isolation, and loss of contact with reality are phenomena reported to occur in certain individuals as a result of being hearing impaired or deaf (Eliopoulos, 2014; Smith & Neely, 2007). Increasing evidence, however, indicates that people do not necessarily demonstrate such personality changes as a direct result of deafness or hearing impairment, but if these attributes already exist in one's personality, hearing impairment may well exacerbate or intensify them.

The basic structures of the auditory system are as follows:

1. *Outer ear.* The outer ear is composed of the pinna or auricle and the auditory canal. The pinna, the external part of the ear, is useful in directing sounds into the ear. The auditory canal is a short passageway through which sound travels to reach the middle and inner ear. The auditory canal, containing hairs and glands producing cerumen (ear wax), terminates at the eardrum (tympanic membrane).
2. *Middle ear.* The middle ear has substantial functional significance for hearing because mechanical transmission of sound takes place there. The eardrum, or tympanic membrane, separates the outer ear from the middle ear. Pharyngotympanic tubes (formerly called Eustachian tubes) open into the middle ear from the throat and are important in equalizing pressures between the outside and inside of the head. When extreme pressure differences exist between the outer and middle ear, pain results, and the eardrum may rupture unless pressures are equalized. (For example, ears "pop" at high altitudes or when scuba diving as pressures become equalized.)

Structures of importance in the middle ear are three small bones (malleus, incus, and stapes) called the ossicles that transmit sound vibrations from the eardrum through the middle ear to the oval window, a membrane separating the middle ear from the inner ear. Another membrane, the round window, is situated below the oval window. The ossicles are the three smallest bones in the body and are named for their shape: the malleus (hammer), incus (anvil), and stapes (stirrup). The "handle" of the malleus fits against the eardrum, and the base of the stapes fits against the oval window, whereas the incus articulates with the malleus and the stapes. When the eardrum is vibrated by sound waves, the ossicles transmit the vibrations to the oval window, which in turn sets fluids in the inner ear in motion, thus stimulating auditory receptors. Two small muscles attach to the malleus and the stapes, and when unusually loud sounds occur, these muscles pull the ossicles away from the membranes they contact. This is called the tympanic reflex, and it helps protect the auditory receptors from extremely loud sounds. However, the reflex has a lag time long enough to be fairly ineffective in protecting against extremely sudden very loud noises (see Figure 7.2).

3. *Inner ear.* The inner ear, located in the temporal bone, is highly complicated and contains structures for both hearing and equilibrium. It contains the bony (osseous) labyrinth, which consists of the vestibule, cochlea, semicircular canals, and a system of channels through bone, and the membranous labyrinth, which consists of interconnecting membranous ducts in the bony labyrinth. Two fluids, perilymph and endolymph, are contained in the labyrinths to conduct sound vibrations and also to respond to changes in body position and acceleration. The cochlea, contained in the bony labyrinth, is a spiral bony chamber containing the auditory receptors. The cochlea contains the spiral organ of Corti in which hair cells, the actual receptors for hearing, are found. The spiral organ of Corti rests on top of the basilar membrane (Marieb & Hoehn, 2019).

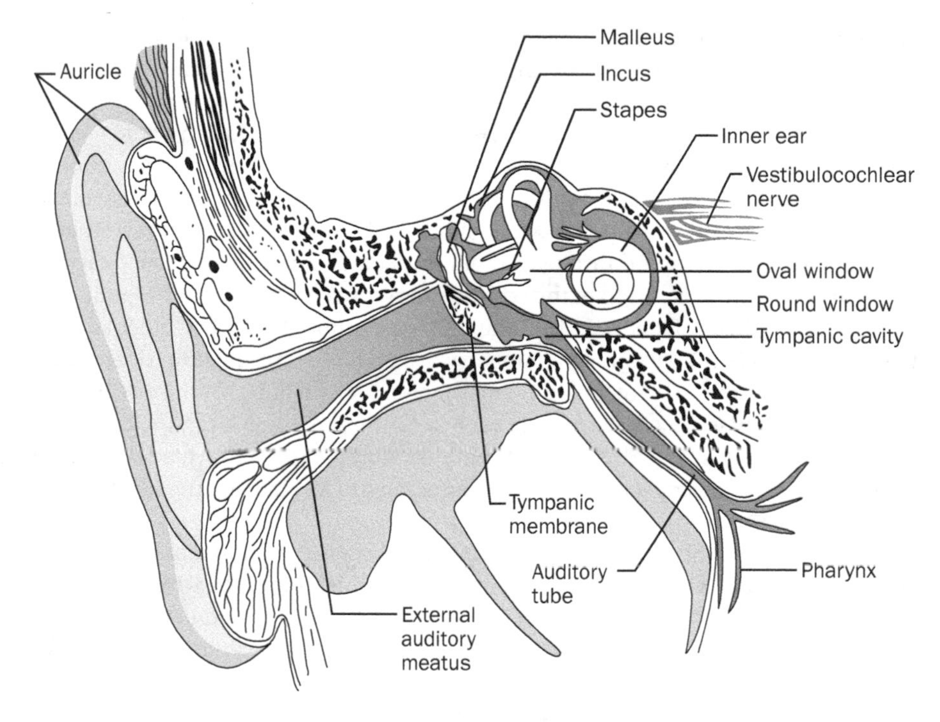

Figure 7.2 The ear.

Sound waves enter the outer ear and initiate vibration of the eardrum, which causes the ossicles in the middle ear to transmit the sound vibration to the oval window. This movement or vibration passes through the oval window, creating a vibration of fluid in the cochlea of the inner ear. In response to a wave of sound pressure, the basilar membrane moves up and down, stimulating the hair cells of the spiral organ of Corti and initiating nerve impulses. Nerve impulses are sent from the ears via the auditory nerve to the auditory center of the brain (located in the temporal lobe of the cerebrum). High-frequency sounds vibrate hair cells on the basal part of the basilar membrane (near the oval window), mid-frequency sounds vibrate hair cells on the middle part of the membrane, and low-frequency sounds vibrate hair cells primarily near the cochlear apex of the membrane. Below the oval window, another membrane called the round window serves to dampen sound waves in the inner ear and restores the system to equilibrium in preparation for the next stimulus entering the ear. The auditory nerve contains 24,000 to 28,000 fibers, and the range of human hearing is approximately 20 to 20,000 Hz (cycles per second). The human ear is thus a very sensitive organ, but if it were more sensitive it would be less efficient. For example, a movement of the eardrum of less than one tenth the diameter of a hydrogen atom can result in an auditory sensation. If the ear were more sensitive, it would respond to the movement of air molecules, and the resulting roaring sounds would all but mask other auditory stimuli.

AGE-RELATED CHANGES IN HEARING

Subtle changes in hearing usually begin in the 40s and progress gradually with age. An estimated 30% of those older than age 65 in the United States are hearing impaired. Hearing impairments occur in 40% to 50% of those older than age 75 and in more than 80% of those older than age 85 (Weinstein, 2010). In our culture, men tend to show hearing loss earlier than women, partly because men have generally been exposed to more prolonged high-level occupational noise than women, and environmental noise factors have a definite effect on auditory integrity. Hearing impairments are generally classified as conductive, sensorineural, central, or mixed. Conductive impairments result from anything interfering with the transmission of sound through the outer or middle ear so that sound stimuli do not reach the inner ear and auditory receptors. Impacted ear wax, arthritic changes in the ossicles, otosclerosis, or stiffening of the tympanic membrane or oval window membrane are possible reasons for conductive hearing impairments. Another common cause of conduction deafness is middle ear inflammations. Sensorineural impairments result from disorders of the inner ear (especially loss of hair cells with age) affecting the transmission of sound to auditory receptors and through the auditory pathways. Other possible causes of sensorineural deafness are degeneration of the cochlear nerve, stroke, tumors in the auditory cortex, and drug toxicity (Marieb & Hoehn, 2019). Many drugs commonly used by older adults such as antibiotics, aspirin, nonsteroidal anti-inflammatory drugs (NSAIDs), and diuretics have potential side effects that affect hearing (ototoxicity). Usually, ototoxicity will be reversed with the discontinuation of the particular drug. Central hearing loss results from lesions in the central auditory pathways. Mixed hearing impairment involves both conductive and sensorineural hearing loss.

Specific changes in hearing associated with the aging process include the following:

1. In the outer ear, the pinna loses some flexibility and becomes a bit longer and wider. Hairs in the external ear canal tend to become stiffer. This, combined with the drying and thinning of tissues in the external ear canal, contributes to a likelihood of cerumen (ear wax) accumulation with age. Cerumen produced in the later years is of thicker consistency (dryer) and is not always easily removed. Accumulation of cerumen in the external ear canal can occlude the auditory canal and is responsible for a substantial amount of conductive hearing impairment (Weinstein, 2010). This can be avoided by keeping the ears free of excessive ear wax buildup. When ear wax accumulates, it should be removed only by a professional because it is not a safe do-it-yourself project and perforation of the eardrum resulting in serious hearing impairment is not uncommon.
2. Aging changes in the middle ear involve lessened flexibility of the tympanic membrane, the oval window membrane, and the round window membrane; an increased possibility of stiffness (calcification) of the ossicles; and probably lessened efficiency of the acoustic reflex. Changes in the membranes of the middle ear and the ossicles somewhat impair transmission of sound vibrations but normally do not impair hearing significantly.
3. Age-related changes in the inner ear are primarily *presbycusis*. With age, auditory loss generally affects perception of high frequencies first. Later, age-related changes may involve middle- and low-frequency ranges as well. Four types of presbycusis have been identified. The most common form is sensory presbycusis, which involves degeneration in the spiral organ of Corti in the cochlea. The other forms involve loss of cochlear and higher auditory pathway neurons (neural presbycusis), atrophy of fibrous vascular cochlear tissue (strial or metabolic presbycusis), and degenerative change in the basilar membrane of the cochlea (mechanical or cochlear presbycusis). Presbycusis has many important influences on behavior because it especially affects the ability to understand speech. Individuals can hear sounds but are not able to discriminate words or comprehend what is being said.

Implications of Age-Related Hearing Impairments

Hearing impairments are not life-threatening but they are disabling because they can seriously disrupt the quality of life. Few people appreciate the extent to which hearing impairments affect all aspects of daily life. Age-related hearing impairment usually occurs gradually, affecting the ability to respond appropriately to auditory signals of danger and, even more important, impairing the ability to understand what is said and to communicate.

Presbycusis interferes substantially with the perception of high-frequency sounds. In the English language, the sounds of consonants have a higher frequency than vowels. Consonants enable us to differentiate one word from another. For example, the words *bear, care,* and *wear* can only be differentiated by consonants. To the individual with presbycusis, these words may all sound alike. Therefore, those with presbycusis hear part of a word or sentence and either have to ask to have it repeated or else guess what was said. Being asked to repeat words or sentences too often produces impatience or irritability in others and frustration for both speaker and listener. When words or sentences are misinterpreted or misunderstood and are responded to inappropriately, those attempting to communicate with the hearing-impaired person often react emotionally and with impatience.

Too often, older hearing-impaired individuals are treated as though they are mentally incompetent or in the early stages of dementia. To further complicate the situation, because hearing impairments are not as obvious as visual impairments, a hearing impairment is often not noticed until communication problems arise. In addition, many older adults are extremely sensitive about being hearing impaired and will deny the problem exists. Because hearing loss is an invisible disability, other people find it difficult to remember the impairment exists and become insensitive to the effect it has on the person affected.

As indicated earlier, certain personality changes have been reported to occur in some, but not all, older individuals with hearing losses. There is widespread variation in the psychosocial effects of hearing impairment in older adults, and each person should be assessed for his or her idiosyncratic reactions to hearing loss. The following suggestions are helpful in communicating with someone who is hearing impaired:

1. Face the person so you can be seen clearly.
2. Speak slowly and enunciate carefully.
3. Lower the pitch of your voice. This is especially important for women, who usually have higher-pitched voices than men.
4. Do not shout because this makes the voice pitch even higher and is embarrassing to the hearing-impaired person.
5. Use touch as an adjunct to communication if appropriate.
6. Remember that it takes a lot of motivation, concentration, and energy for a person with a hearing impairment to function well in communication situations. Situations in which there are multiple individuals and concurrent conversations can become overwhelming. If the person does not feel well or does not have the energy to expend, he or she will not hear as well as at other more optimal times.

Assessment of hearing impairments should include an examination by an otologist or otolaryngologist to assess any medical condition contributing to hearing loss and a complete audiologic evaluation by an audiologist. Audiologic rehabilitation options include an array of assistive listening devices such as amplified telephones, hearing aids, speech reading, and educational/counseling programs. Most people with hearing impairments can now be helped with some type of hearing aid. Hearing aid technology has progressed enormously in the past few years and many options are now available such as programmed and digital aids. Hearing aids available include those in the outer ear, behind the ear, over the ear, and completely in the ear canal. Because of the complexity of each of these, a qualified professional is very necessary to determine the most useful hearing aid for a specific type of hearing impairment (Eliopoulos, 2014). For best results in understanding speech and localization of sound, it is usually necessary to wear hearing aids in both ears, not just in one ear. For those who are profoundly deaf, cochlear implants are becoming more common for older adults. Human voices will sound tinny and not of normal quality, but for those with profound deafness, the perception of these sounds may be helpful. Education, counseling, and referrals to rehabilitation options such as speech reading are also part of an appropriate hearing rehabilitation program. A number of older adults do not wear their hearing aids after purchase because they do not get adequate follow-up and education about expectations, use, and care of these instruments. There are many choices and options available to deal with hearing losses, and older adults need to be encouraged to pursue these to improve their quality of life, but it does take some adjustment to wear hearing aids.

AGE-RELATED DISORDERS OF HEARING

Tinnitus

Tinnitus is the perception of sounds in one or both ears in the absence of an auditory stimulus. It can occur in those with or without hearing loss. Eighty-five percent of those with hearing or ear problems experience tinnitus, and it is more common in older adults. Sounds are described as ringing, buzzing, whistling, or roaring in the ears. Tinnitus sounds are most often subjective (only the affected person hears them) and may be caused by medications, infections, neurological problems, and other disorders related to hearing loss. They may also be objective (the examiner is able to hear the sounds), caused by blood flow in the ear, spastic muscles in the ear, or vibrations of the hair cells. Other possible causes of tinnitus are presbycusis, otosclerosis (bone formation at the oval window), Ménière's disease, certain drugs, and other inner ear problems (Bierschbach, 2013). A thorough history and a complete medical workup are necessary to diagnose the specific cause of tinnitus (Tabloski, 2014).

Once it is clear that tinnitus is chronic and not caused by a surgically correctable medical problem, a long-term program of management can be established. According to Ross et al. (1991), the major components of a management plan include the following:

1. Treating related problems that are correctable; for instance, have the individual avoid loud or continuous noises.
2. Avoiding irritants that may aggravate or perpetuate tinnitus; examples are alcohol, caffeine, chocolate, tea, some of the anti-inflammatory drugs, aspirin, and quinine.
3. Reassuring the older adult that tinnitus is not life-threatening, but a symptom that can be managed; remember that fatigue, worry, and high stress worsen the annoyance.
4. Teaching individuals with tinnitus how to manage it by using various noise-masking techniques; radios and other masking noises may help distract from tinnitus sounds. The American Tinnitus Association (2019) recommends the use of tinnitus maskers, which look like hearing aids and produce sounds that "mask," or cover up, the tinnitus. A proper hearing aid sometimes helps in reducing tinnitus.
5. Teaching biofeedback, relaxation training, and stress management to improve coping strategies.
6. Referring to self-help and support groups for additional information and strategies aiding in more effective management.
7. Treating with medications, which may be of help in certain situations.

VESTIBULAR SYSTEM

The vestibular system is significant for mobility and balance. Receptors providing information on the body's orientation in space are located in the inner ear within the bony labyrinth close to the cochlea. The structures of the vestibular system include two compartments, the utricle and the saccule, plus three tubes (semicircular canals) filled with fluid. The utricle and saccule contain receptors (hair cells) responsive to changes in the position of the head with respect to gravity. Such movements stimulate the hair cell receptors and initiate nerve impulses, which travel the vestibular nerve to the brain. The semicircular canals, placed at right angles to each other, each contain fluid as well as an enlargement (ampulla) at one end containing hair cell receptors. Changes in the rate of motion of the head stimulate hair cells in the ampullae and initiate nerve impulses.

AGE-RELATED CHANGES IN THE VESTIBULAR SYSTEM

The bony labyrinth undergoes degenerative changes similar to those in the cochlea as discussed under age-related changes in the inner ear. In addition, the following age-related changes in the vestibular system can be seen:

1. Sensory receptors (hair cells) decrease in number, and peripheral neural fibers are reduced (Marieb & Hoehn, 2019).
2. Loss of vestibular function becomes more noticeable from increased difficulty walking or standing on uneven or soft surfaces in dark or poorly lit environments. Body sway increases and may be partially responsible for general postural unsteadiness (especially falls) experienced by many older people. Equilibrium and balance become impaired, especially when fast movement is required. Older adults generally adapt by moving slowly and walking with a wide-stance/feet-apart gait to provide greater stability. Pacing one's speed of movement becomes much more important in old age, not only for conserving energy but also for safety.

AGE-RELATED DISORDERS OF THE VESTIBULAR SYSTEM

Disturbances of the vestibular system are commonly implicated in dizziness, vertigo, and other equilibrium problems affecting older adults. Dizziness, a complaint of many older adults, is not easy for the professional to interpret. According to Ross and Robinson (1984), four types of complaints are labeled "dizziness," and these are still used in classifications:

1. Disequilibrium, or imbalance, characterized by difficulties in walking; those with Parkinson's disease or Alzheimer's are illustrative of disequilibrium problems.
2. Faintness, or a feeling of impending loss of consciousness; this type of dizziness is usually caused by circulatory insufficiency. Systemic disorders such as anemias, thyroid disease, hypoglycemia, and medical problems that lessen oxygenation of the brain contribute to faintness.
3. Vague, nonspecific lightheadedness; although usually imprecisely described by those affected, three likely causes are multiple sensory deficits, anxiety with hyperventilation, and chronic systemic disease.
4. Vertigo, a sensation of rotating in space or spinning; the illusion that one is moving, or one's surroundings are moving, differentiates vertigo from other types of dizziness.

Dizziness most often results when several sensory modalities bring contradictory sensory information to the brain. Thus, dizziness often has multiple causes and involves several dysfunctioning systems of the body. Diagnoses of various causal factors are difficult but necessary for effective treatment.

Ménière's Disease/Vestibular Neuritis

Ménière's disease, an inner ear disturbance, results from a dysfunction of the bony labyrinth. Its specific cause is not known. Symptoms characteristic of the disorder include vertigo with nausea and vomiting, tinnitus, neurosensory hearing loss, and a sensation of pressure within the ears. Nystagmus (rapid involuntary movements of the eye) can also be present. Vertigo attacks occur suddenly and may last for several hours. In the early stages of the disease, weeks or months pass between attacks, but as the disorder progresses, attacks can occur every 2 or 3 days. Usually, only one ear is involved.

Diagnostic evaluation may include audiogram, head scan, allergy evaluation, glucose tolerance test, and specific techniques to assess labyrinthine function. Mild episodes may be controlled by antimotion drugs or a low-salt diet and diuretics to decrease endolymph fluid. In more severe cases, draining the excess endolymph from the inner ear may help. As a last resort, removal of the malfunctioning labyrinth is possible (Marieb & Hoehn, 2019). Treatment goals are to eliminate vertigo and stabilize hearing. Treatment approaches thus include medical, surgical, rehabilitative, and dietary strategies, depending on the underlying primary cause. Thus, treatment of vestibular disorders involves alleviating symptoms and correcting underlying causes if they can be accurately identified. Medication choices depend on the individual's tolerance for a particular drug, the efficacy of the drug, and its safety for long-term use. Medications used to treat vertigo often have side effects disturbing older adults. In uncontrolled vertigo, surgery may be the only option.

Vestibular neuritis (sometimes referred to as vestibular neuronitis) has similar symptoms to Ménière's disease except for no or very mild hearing loss. Even though "-itis" after a medical condition usually refers to inflammation, in this case, it is often used as a diagnosis that permanent damage to the vestibular nerve has occurred (Payne, 2016). The most common cause is from a viral infection such as the flu, though damage to the nerve can also result from several other causes, including bacterial infections in the ear, meningitis, injury to the ear, or a tumor developing on the nerve (Payne, 2016). In the vast majority of cases, symptoms subside within 3 months (Payne, 2016). However, in some individuals, the damage caused may be significant, resulting in a chronic condition and persistent lifelong symptoms. Treatment options are also similar to Ménière's disease. Vestibular rehabilitation therapy, a specialized type of physical therapy for balance disorders, may be helpful in some cases.

Balancing the body under the influence of gravity, maintaining equilibrium under a variety of movement conditions, and engaging in coordinated, controlled psychomotor activities all involve the interplay of many intricate mechanisms influenced by the aging process. Falls and other accidents tend to increase with age as these controlling and integrative aspects of movement, balance, and equilibrium decline in efficiency. Promoting a safe environment (nonslip floors, limiting use of rugs or mats that may cause someone to trip, good lighting, uncluttered walkways, secure handrails) and ensuring that well-fitted and nonslip shoes are worn can help prevent falls and their subsequent injuries (Vestibular Disorders Association, 2021).

TASTE (GUSTATION)

Receptors for taste are located in taste buds primarily on the tongue, with each taste bud having 15 to 20 or more sensory cells. These cells replace themselves constantly, but replacement may be slower in older people. There are five different taste sensations: sweet, salt, sour, bitter, and umami (a subtle taste responsible for the beef taste of steak, the tang of aging cheese, and the flavor of monosodium glutamate, a food additive). There is also some evidence for a possible sixth taste modality responding to fatty acids from lipids. This could explain our liking for fatty foods (Marieb & Hoehn, 2019). Taste is one of the chemical senses because substances must be in solution to be tasted; insoluble materials have no taste. The blending of substances produces a variety of taste sensations that contribute to enjoyable eating.

AGE-RELATED CHANGES IN TASTE

Research into age-related changes in both taste and smell is difficult because these senses are so interrelated. Older persons may notice a decrease in the sense of taste around age 60, and especially after age 70 (Linton, 2007). The behavioral significance of changes in taste for eating has not been established. Taste has always been considered to be a relatively minor sensory modality, and its age-related changes occur gradually. Individuals, then, may not be as consciously aware of changes in taste as in vision or hearing. Many older adults do complain that foods taste bland, and they regularly pour on salt, sauces, sugar, or spices to enhance flavor. If, with age, foods begin to taste bland, this may be due partly to changes in taste receptors and partly to other factors contributing to the enjoyment of food. For example, ill-fitting dentures can modify eating patterns; eating alone is a situation often not conducive to the preparation of nutritionally balanced meals, and loss of appetite occurs with inactivity. Dry mouth (xerostomia) occurs more often in older adults because of the reduction of saliva (often a side effect of medications) and also can result in changes in the perception of taste (dysgeusia). An adequate diet is important, especially in older age, and continuing research and education are needed in this area. Nutrition has far-reaching implications for health and vitality in older age.

SMELL (OLFACTION)

Specific receptors for the sense of smell are located in the nasal passages. Various kinds of receptors have been identified, but research is difficult and results are sometimes contradictory.

AGE-RELATED CHANGES IN SMELL

Smell appears to be more affected by age than is taste. As in taste, olfactory receptors are constantly being replaced, but not all receptors may be replaced in older age. Also, environmental factors such as smoking affect olfactory sensitivity and make specific age-related changes difficult to differentiate. There is a higher threshold of smell sensitivity with age, suggesting that odor identification seems to be less efficient and that odors need to be stronger and more intense to be perceived and differentiated by older adults. For example, ethyl mercaptan is an odorous substance once added to propane to enable natural gas consumers to be aware of leaks, but older adults were found to have more difficulty identifying this substance than younger people, posing a serious safety hazard for older persons.

Smell, like taste, is considered a relatively minor sensory modality compared with vision and hearing. Several reasons for the lack of attention to smell in both research and applied settings include the fact that disturbances in the sense of smell are less obvious and have less influence on everyday activities than do the major sensory modalities. Also, this system is considered to be primitive and not as useful to humans, so easy-to-use assessments of olfactory functioning are not readily available. Olfaction is taken for granted, and its significance is not appreciated until disturbances or losses occur.

Changes occurring in the sense of smell have behavioral implications for the proper ingestion of food, for safety, and for personal hygiene. The smell of escaping gas fumes from a stove or heater, electrical wires burning, or spoiled foods are important cues for personal safety. Taste and smell are related senses, and both contribute substantially to the pleasure of eating. Personal cleanliness is important in our society. Older adults with reduced olfactory perception may not be aware of unpleasant body odors or other aspects of personal hygiene. Some older adults use such large amounts of perfume or cologne that it is offensive to those around them, but they are unaware of this reaction.

AGE-RELATED DISORDERS IN TASTE AND SMELL

Pathologies do exist, but diseases in these systems are idiosyncratic and vary greatly from one person to another; no one disorder is particularly associated with aging. The most common causes of loss of smell (and probably taste) are upper respiratory infection, head traumas, and nasal/sinus disease. Olfactory distortions or losses have also been associated with Alzheimer's disease, Parkinson's disease, and Huntington's disease.

SKIN (CUTANEOUS) SENSES

The skin senses are touch, pressure, heat, cold, and pain. Each sense has specific receptors. As with other sensory receptors, differential distribution of cutaneous receptors is found throughout the body. The fingertips, for example, are far more sensitive to touch and pressure than is the forearm.

AGE-RELATED CHANGES IN SKIN SENSES

Research into age-related changes in the skin senses is sparse. Evidence suggests that changes take place gradually, involving some loss of receptors with age and higher thresholds of stimulation in those remaining. Behavioral implications of age-related changes in the skin senses primarily concern personal safety. Burns are likely to occur if an older person does not accurately perceive temperatures. When touch receptors on the soles of the feet do not function effectively, falls occur before the individual even realizes the foot is not on a solid surface. Certain social behaviors may be affected adversely, as when, for example, it is difficult to know how much pressure to exert when holding a glass or fork without dropping it. Some older adults become overly sensitive about such perceived clumsiness and avoid public or social situations. Touch-sensitivity changes can exert subtle influences on various aspects of behavior because touch is necessary to orient ourselves to many aspects of the daily environment and to prevent accidents.

Another significant aspect of touch often neglected in discussions of sensory changes and age is touch as a mode of communication. The use of touch conveys various messages. Appropriate use of affective ("caring") touch improves communication with older adults, whether they are oriented or confused. How touch is perceived, though, depends on cultural and family experiences, gender differences, location of touch, and basic personality preferences. Professionals must be sensitive to the complex dynamics involved in touch because it can serve as a powerful adjunct to verbal communication with older adults when used appropriately.

THE IMPORTANCE OF SENSORY CHANGES IN AGING

Sensory changes with age are some of the most crucial and possibly the most underrated changes associated with the entire aging process. Perhaps it is because these changes usually occur gradually and are not as dramatic as impairments or disabilities that occur suddenly through an accident or health crisis. Perhaps our lack of active concern in this area arises from the "error of familiarity," because most people are at least vaguely aware that sensory changes take place with age but do not dwell on the possible implications of such changes. Perhaps we tend to write these changes off with a "What can you expect from old age?" attitude. Whatever the reasons, most people interested and involved in the study of the aging process do not give sensory changes and their cumulating effect on behavior the significance they deserve.

Changes in each of these systems interfere with a person's ability to gather pertinent information about the environment essential to a high quality of life, and even to the maintenance of life itself. Is it not reasonable, then, that as sensory changes gradually occur, the organism also experiences gradual sensory deprivation that may lead to social isolation as the individual becomes less mobile, increasingly housebound, or more difficult to engage in communication? The next stage might well be "functional senility," a state in which the person generates his or her own world of fantasy or lives in the past because the real world is not interesting enough to provide the variety of stimulation needed to keep psychologically intact. Continued stimulation of sensory modalities is very necessary to maintain adequate functioning in older age. Preventive care and early intervention are extremely important in retaining sensory efficiency in the later years.

REFERENCES

American Tinnitus Association. (2019). *Sound therapies.* https://www.ata.org/managing-your-tinnitus/treatment-options/sound-therapies

Bierschbach, J. (2013). Care of patients with ear and hearing problems. In D. Ignatavicius & L. Workman (Eds.), *Medical–surgical nursing* (7th ed., pp. 1088–1105). Saunders Elsevier.

Boyd, K. (2020). *Have AMD? Save your sight with an Amsler grid.* American Academy of Ophthalmology. https://www.aao.org/eye-health/tips-prevention/facts-about-amsler-grid-daily-vision-test

Centers for Disease Control and Prevention. (2020). *Vision health initiative: Don't let glaucoma steal your sight!* https://www.cdc.gov/visionhealth/resources/features/glaucoma-awareness.html

Eliopoulos, C. (2014). *Gerontological nursing* (8th ed.). Wolters Kluwer Health/Lippincott Williams & Wilkins.

Glaucoma Research Foundation. (2021). *January is glaucoma awareness month.* https://www.glaucoma.org/news/glaucoma-awareness-month.php

Linton, A. D. (2007). Age-related changes in the special senses. In A. D. Linton & H. W. Lach (Eds.), *Matteson & McConnell's gerontological nursing* (3rd ed., pp. 600–627). Saunders Elsevier.

Marieb, E. N., & Hoehn, K. (2019). *Human anatomy and physiology* (11th ed.). Pearson.

Miller, C. A. (2009). *Nursing for wellness in older adults* (5th ed.). Wolters Kluwer/Lippincott Williams & Wilkins.

National Eye Institute. (2019). *Cataract data and statistics.* https://www.nei.nih.gov/learn-about-eye-health/resources-for-health-educators/eye-health-data-and-statistics/cataract-data-and-statistics

National Eye Institute. (2020). *AREDS/AREDS2 frequently asked questions.* https://www.nei.nih.gov/research/clinical-trials/age-related-eye-disease-studies-aredsareds2/aredsareds2-frequently-asked-questions

Payne, J. (2016). *Vestibular neuritis and labyrinthitis.* Patient Info. https://patient.info/signs-symptoms/dizziness/vestibular-neuritis-and-labyrinthitis

Ross, V., Echevarria, K., & Robinson, B. (1991). Geriatric tinnitus: Causes, clinical treatment, and prevention. *Gerontological Nursing, 17*(10), 6–11. https://doi.org/10.3928/0098-9134-19911001-04

Ross, V., & Robinson, B. (1984). Dizziness: Causes, prevention, and management. *Geriatric Nursing, 5*(7), 290–304.

Sheppard, A. L., & Wolffsohn, J. S. (2018). Digital eye strain: Prevalence, measurement and amelioration. *BMJ Open Ophthalmology, 3*(1), e000146. https://doi.org/10.1136/bmjophth-2018-000146

Smith, S. C., & Neely, S. (2007). Visual and auditory problems. In. S. L. Lewis, M. M. Heitkemper, S. R. Dirksen, P. G. O'Brien, & L. Bucher (Eds.), *Medical–surgical nursing* (7th ed., pp. 416–448). Elsevier.

Tabloski, P. (2014). *Gerontological nursing* (3rd ed.). Pearson Education.

Touhy, T. A. (2016). Vision. In T. A Touhy & K. Jett (Eds.), *Toward healthy aging* (9th ed., pp. 130–141). Elsevier.

Vestibular Disorders Association. (2021). *Age-related dizziness and imbalance: Fall prevention.* https://vestibular.org/article/diagnosis-treatment/types-of-vestibular-disorders/age-related-dizziness-and-imbalance/fall-prevention/vestibular

Weinstein, B. (2010). Disorders of hearing. In H. Fillitt, K. Rockman, & K. Woodhouse (Eds.), *Brocklehurst's textbook of geriatric medicine and gerontology* (7th ed., pp. 822–834). Saunders Elsevier.

Workman, M. (2013). Assessment of the eye and vision. In D. Ignatavicius & M. Workman (Eds.), *Medical–surgical nursing* (7th ed., pp. 1039–1051). Saunders Elsevier.

8

The Cardiovascular System

INTRODUCTION

Heart disease remains the major cause of death for both men and women in the United States (Winton, 2019). Health problems in the cardiovascular system resulting from age-related changes and disease are often preventable. Primary modes of prevention include eating a healthy diet and exercising regularly. Data indicate that a consistent exercise program changes both heart functioning and heart size and lowers blood pressure levels as well. Maintaining weight within a normal range and effective management of existing health problems can also do much to decrease the likelihood of heart disease.

Risk factors for cardiovascular disease include those not modifiable by lifestyle changes and those we can modify or change. Nonmodifiable risk factors include family history, gender (men are more likely to have cardiovascular disease than women), ethnic origin, and age. Risk factors that can be modified by lifestyle changes include hypertension (HTN), diabetes, high cholesterol levels, obesity, alcohol use, smoking, a diet high in animal fats and calories, and a sedentary lifestyle (Winton, 2019). Personality characteristics also influence the development of heart disease. Stress, a significant factor in heart disease, may be modified by the use of relaxation techniques, lifestyle changes, and psychosocial therapies.

ANATOMY AND PHYSIOLOGY OF THE CARDIOVASCULAR SYSTEM

The cardiovascular system serves as a pump moving arterial blood containing nutrients and oxygen through the arteries to the cells of the body where metabolism takes place. Waste products from cellular metabolic processes are then returned through the veins to be excreted by the excretory organs.

The Blood

Blood is a sticky, opaque fluid that accounts for about 8% of total body weight and translates into about 5 or 6 L in males and 4 or 5 L in females (Marieb & Hoehn, 2019). The blood is composed of the following:

- Red blood cells (erythrocytes), which carry oxygen to all the cells of the body
- White blood cells (leukocytes), which protect the body from attack by viruses, bacteria, toxins, parasites, and tumor cells
- Platelets (thrombocytes), essential for blood clotting
- Plasma, the fluid component of the blood in which solute (substances dissolved in a solution) and elements are suspended and circulated

Functions of the Blood

Blood, the major medium for the transportation of fluids throughout the body, has four significant functions in the maintenance of life and health. These functions are as follows:

- Respiratory, through the distribution of oxygen from the lungs to the tissues of the body for cell use, and carbon dioxide from the body tissues back to the lungs, where it is expelled
- Nutritive, through the transport of food substances such as glucose, fat, and amino acids, from storage places (the liver and intestines, for example) to body tissues where these materials are needed to produce energy and to maintain life
- Excretory, through the movement of waste products from body cells to the excretory organs
- Regulatory, through the control of body equilibrium (homeostasis) in general, and specifically through hormone distribution, maintenance of water balance, and temperature regulation; for example, excess heat generated in the body is transported continuously by the blood to the lungs and to body surfaces, where it is dissipated

The Lymphatic System

The lymphatic system is composed of the following:

- Lymph, a fluid originating in tissue spaces throughout the body
- A one-way system of lymph vessels transporting lymph from tissue spaces to lymph ducts to the bloodstream
- Lymph nodes

The smallest vessels are lymphatic capillaries, which flow into collecting vessels, lymphatic trunks, and the largest vessels, the lymphatic ducts. The major function of the lymphatic system is to assist in preventing the spread of infection and disease by straining out foreign particles and bacteria as the lymph passes through special lymphoid organs such as lymph nodes, spleen, thymus, tonsils, and Peyer's patches in the small intestine (Marieb & Hoehn, 2019).

The Blood–Vascular System

The human blood–vascular organizational plan is a closed system in which damage to any part will ultimately affect the entire system. The major components of the blood–vascular system are the following:

- The heart, a pumping organ
- The arteries, tubes that conduct blood from the heart to body cells; the smallest artery branches are called arterioles
- The veins, tubes that conduct blood from body tissues back to the heart. Many veins contain one-way valves to prevent blood from flowing backward and thus help return blood to the heart. Valves are most common in veins of the limbs. The smallest vein branches are called venules.
- The capillaries, minute blood vessels connecting arterioles and venules

The Heart

The pump of the blood–vascular system is the heart, a hollow organ with highly muscular walls, situated within the thorax (chest) between the lungs in the space that separates the right and left pleural cavities. In complex organisms such as humans, the heart has four chambers: two *atria* (upper chambers) and two *ventricles* (lower chambers). A thick partition, the *septum*, separates the left side of the heart from the right. The largest artery of the body, the *aorta*, leads out of the left ventricle, and the *pulmonary artery* emerges from the right ventricle. The largest veins of the body (*superior and inferior venae cavae*) enter the right atrium, and the *pulmonary veins* enter the left atrium.

The atria and the ventricles are separated by atrioventricular (A-V) valves that control both the location and the amount of blood in each of the four chambers of the heart. The left valve is called the *mitral or biscuspid*, and the right valve is called the *tricuspid*. Other valves separate each ventricle and its specific artery (aorta or pulmonary); no valves are found between the atria and their respective veins (venae cavae or pulmonary) (see Figure 8.1).

Because the heart is composed of muscle tissue, it needs a rich blood supply in order to maintain proper functioning. *Coronary circulation* involves specific coronary arteries branching from the base of the aorta and distributing blood to the heart muscle. Veins collect the blood to be returned to the right atrium through a large vein called the coronary sinus. If a coronary artery becomes occluded and blocks the supply of oxygen and nutrients to the heart muscle, a heart attack results.

Blood Circulation

In addition to coronary circulation, there are two blood circuits: one called systemic, supplying all body parts with blood, and the other called pulmonary, circulating blood through the lungs to

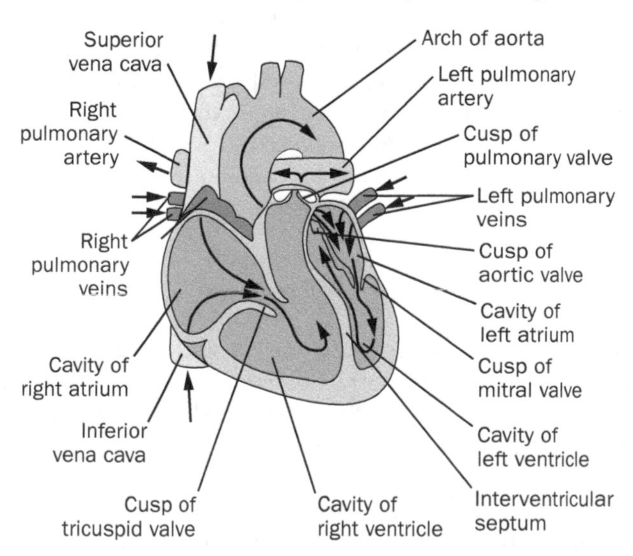

Figure 8.1 Anatomy of the heart and great vessels.

purify it. The right side of the heart receives oxygen-poor blood from body tissues, whereas the left side of the heart receives oxygen-rich blood from the lungs.

Systemic circulation begins as oxygen-rich blood from the lungs enters the left atrium via the pulmonary veins. When the atrium fills with blood, it contracts. The mitral valve opens and blood flows into the left ventricle. Because the valve is a one-way device, blood normally flows only from the left atrium to the left ventricle. When the left ventricle fills with blood, the mitral valve closes, the ventricle contracts, the aortic valve opens, and blood is forced into the aorta, after which the aortic valve closes so blood cannot reenter the ventricle. Blood circulates throughout the body by way of the aorta and other arteries, connects with veins at the level of the capillaries, and returns, depleted of oxygen and carrying carbon dioxide to the heart via various-sized veins ending in the largest veins, the inferior and superior venae cavae, which empty into the right atrium. When the right atrium fills, the tricuspid valve opens and the deoxygenated blood passes into the right ventricle.

Pulmonary circulation begins after blood fills the right ventricle. The tricuspid valve closes, the ventricle contracts, and the pulmonary valve opens, forcing blood into the pulmonary artery to be carried to the lungs to be oxygenated. The pulmonary valve closes so that blood cannot reenter the ventricle. Oxygenated blood returns to the left atrium via the pulmonary veins, and the cycle begins again.

Various estimates suggest the body contains about 70,000 miles of blood vessels, most of which are capillaries. The normal pulse rate for the heart is between 60 and 90 beats a minute, with 72 beats a common average. This equates to more than 100,000 heart beats daily and more than 4,000 gallons of blood pumped through the heart each day.

The normal heart sounds, S1 ("lub") and S2 ("dub"), are produced by the closure of the heart valves. The first heart sound (S1) is produced by the closing of both the mitral (bicuspid) and the tricuspid valves. The second heart sound (S2) is produced by the closure of the aortic and pulmonary valves. These sounds help health professionals assess the functional status of the heart.

Maintenance of Circulation

Circulation is maintained through the continuous rhythmic action of the heart. Although the nervous system affects heart rate, heart muscle is unlike other muscles of the body because it is self-excitatory and has its own built-in pacemaker mechanism to maintain rhythmic and coordinated activity. Specifically, the heart beat is initiated by a segment of tissue in the right atrium designated as the sinoatrial (S-A) node. Excitation begun at the S-A node spreads to similar nodal tissue, the A-V node at the junction of the right atrium and right ventricle, and then through a bundle of fibers (the bundle of His) to the ventricle walls, causing the heart to beat. Normally the atria and ventricles beat in a coordinated rhythm at approximately 72 times a minute. If injury or disease interferes with impulse transmission between the S-A and the A-V nodes, the atria and ventricles beat at different rates and heart block results. Sometimes, if heart rhythm is disrupted, random contractions (fibrillation) occur (see Figures 8.2 and 8.3).

Blood Pressure

The contraction of the left ventricle forces blood into the aorta with a definite force or pressure. The pressure resulting from ventricular contraction is called "systolic" and represents the upper number of a blood pressure reading. During the subsequent brief relaxation of the ventricle, pressure decreases, representing the diastolic pressure (or resting phase), the lower number of a blood pressure reading.

According to most authorities, average blood pressure for a healthy young or middle-aged person at rest should be less than 120/80, although blood pressure fluctuates according to the

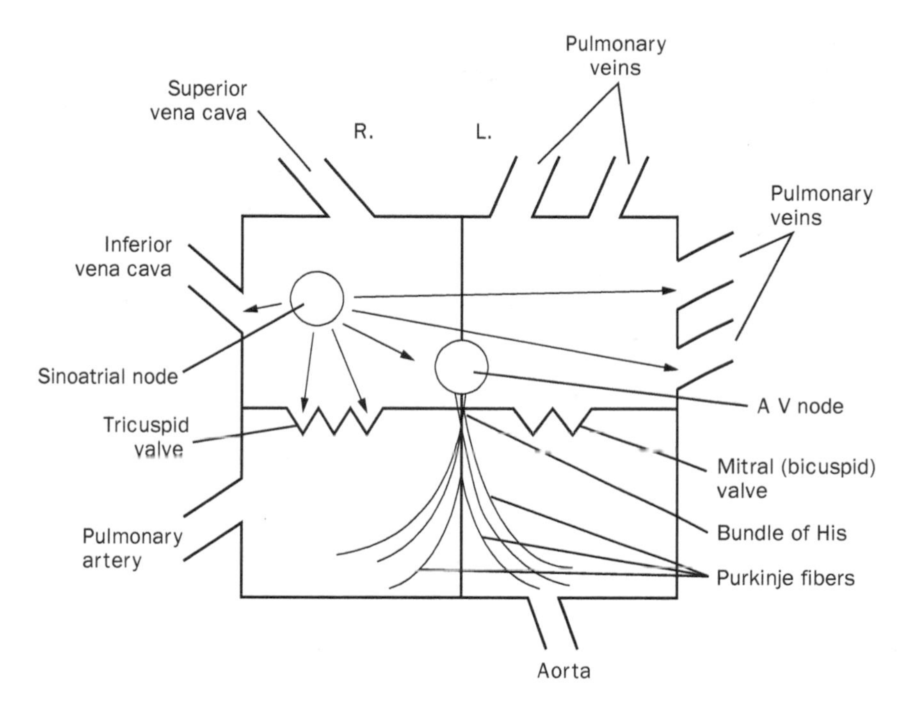

Figure 8.2 Schematic of the electric conduction system of the heart.

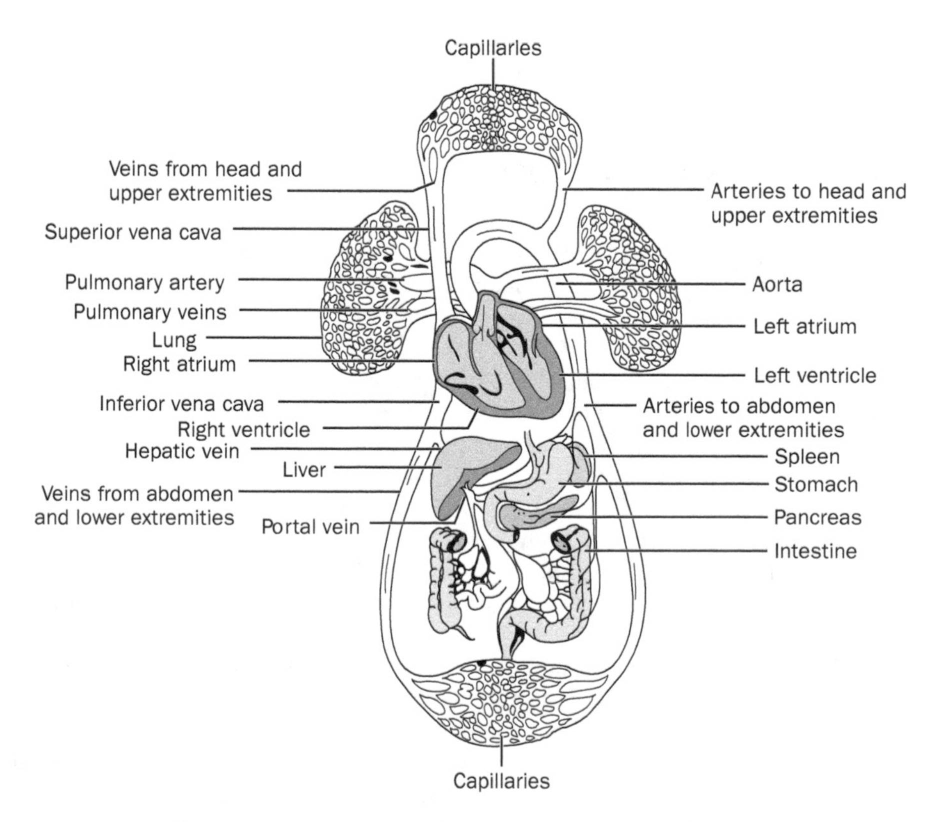

Figure 8.3 Diagram of the pulmonary and systemic circulations.

individual's physiological and psychological status of the moment and a variation of readings occurs throughout the day. High blood pressure is defined as a sustained systolic reading of 140 mmHg or higher over a diastolic reading of 90 mmHg or higher. Hypotension is defined as a systolic reading less than 90 mmHg and diastolic reading less than 60 mmHg. Generally, hypotension is not a concern unless it leads to inadequate blood flow to tissues (Marieb & Hoehn, 2019). Periodically these norms change, thus the importance of keeping abreast of the newest recommendations. Several blood pressure readings need to be taken at different times before HTN or hypotension (too high or too low blood pressure) can be diagnosed. Variables influencing primary HTN are heredity, diet, obesity, diabetes, stress, and smoking. Following are some of the other factors that influence blood pressure:

- Age (blood pressure tends to increase with age)
- Blood volume (the amount of blood pumped)
- Elasticity of arterial walls (which determines how easy or difficult it is for blood to flow)
- Peripheral resistance (especially in the limbs)
- Pumping action of the heart (this varies with age and health)
- Thickness or thinness of the blood (this affects rate of blood flow)

AGE-RELATED CARDIOVASCULAR CHANGES

Although some cells, such as skin or blood cells, are self-replenishing, cardiac (heart) cells are not replaced once damaged or destroyed. However, it is especially difficult to determine which changes in the circulatory system are "normal" aging changes and which are pathological. More research on healthy older adults is needed to clarify this issue. Therefore, we include changes commonly reported in the literature, realizing the distinction between aging and disease is not completely clear.

Age-Related Structural Changes

Substantial evidence now indicates that overall heart size in healthy older adults does not change significantly with age. Contrary to some popular opinion, an enlarged heart is not normal in older age and may instead suggest a pathological condition (Winton, 2019). Age-related changes commonly occurring include the following:

- Increase in fatty tissues in the outermost layer of the heart muscle
- Some increase in the thickness of the left ventricular wall
- Increase in collagen and decreased elastin tissues in the heart and arteries, which causes the vessels to become more rigid and thick
- Decreased efficiency in contractile strength of the heart muscle and decreased maximum heart rate, stroke volume, cardiac output, ejection fraction (the percentage of blood leaving the heart during a contraction), and oxygen uptake (Eliopoulos, 2018)
- Accumulation of lipofuscin, a pigment giving a brown appearance to heart myocardium (middle layer of heart muscle)
- Thickening and sclerosis (hardening) of the valve flaps of the heart, especially the tricuspid and mitral (bicuspid), but also the aortic and pulmonary valves, causing them all to be less efficient and possibly resulting in heart murmurs (Marieb & Hoehn, 2019)
- Significantly decreased number of pacemaker cells (cells that generate impulses and determine the rate of heart activity in the S-A node), with a concomitant decrease in the S-A node rate (Tabloski, 2014)

- Some loss of muscle cells in the A-V node and the bundle of His, an increase in fatty fibrous tissue, and amyloid (starch-like protein) infiltration associated with degeneration
- Increased elastic collagen-type tissue in all parts of the conduction system
- Thickened and stretched veins and less efficiently functioning valves in the veins, slowing return of blood through the veins to the heart
- Calcified, less elastic coronary arteries

Although not substantiated as a definite age-related change, atherosclerosis (fatty deposits on inner walls of arteries) is also associated with age. Additional factors such as lack of exercise, smoking, obesity, and other diseases contribute to vascular changes with age. Usually these changes are experienced gradually and become more obvious when the individual is faced with an infection, increased activity, and enhanced physiological stress (Eliopoulos, 2018).

Age-Related Functional Changes

The following functional age-related changes in the cardiovascular system have been identified:

1. *Longer recovery.* Older heart muscle requires a longer time to recover after each heartbeat; in other words, the heart requires a slightly longer rest period between beats. This fact is not significant in most activity, but it may limit behavior in situations where the heart is stressed and required to beat faster than normal. Maximum attainable heart rate declines, but those who exercise do not show as much decline (Marieb & Hoehn, 2019). Generally, though, older people may be more prone to heart failure than younger adults, who have greater reserve capacity in heart functioning.
2. *Slight arrhythmias.* At rest, heart rate in people of older age is essentially the same as in younger people. However, some evidence suggests arrhythmias such as skipped or extra beats become more common with age. Arrhythmias sometimes produce anxiety in older adults who fail to understand the condition is not necessarily indicative of heart disease.
3. *Decline in cardiac output.* Cardiac output (the amount of blood pumped from the heart in 1 minute) declines somewhat with age, causing less oxygen to be delivered to body tissues and organs. Reduced cardiac output occurs both at rest and with exercise, but the decline usually has little significance for normal everyday behavior (Marieb & Hoehn, 2019). Nevertheless, this fact may help to explain why most older adults tire more quickly than the young and why endurance, especially when doing strenuous work, tends to decline with age.
4. *Increase in atrial fibrillation (irregular, rapid heartbeat) and incidence of heart block (a blockage of the impulse conducted from the atria to the ventricle).* These occur as a result of age-related changes in the conduction system (which controls the rate and coordination of the heartbeat).
5. *Changes in arteries and veins.* The aging process seems to affect the arteries more than the heart itself because the arteries, and to a lesser extent the veins, become more rigid, less elastic, and flexible. Elasticity of arteries is a major factor in regulating blood pressure. For instance, when exerted, the heart beats faster and more blood is pumped through the body at an increased rate. The elastic arterial walls normally expand to accommodate the greater force of blood pushed through and thus arterial resistance is decreased, but if the arterial walls are rigid and cannot expand, the heart must pump harder to move more blood through the system quickly. Blood pressure increases as arterial resistance is increased. Such increases in blood pressure are a common corollary of the aging process. Changes in the veins predispose individuals to a slower return of venous blood to the heart, venous stasis (stagnation of venous blood), varicose veins, and thrombophlebitis (clotting in a vein caused by phlebitis or inflammation of a vein).

6. *Blood components.* Only slight changes are evident in the blood components with age. The volume of blood decreases as it relates to a lower plasma volume. Hemoglobin and hematocrit levels are only minimally diminished and the red and white blood cells and prothrombin (coagulation) time remain unchanged (House-Fancher & Lynch, 2007).

In summary, in nonstressful conditions the normal aging heart functions quite adequately unless there is severe damage to the heart from disease. However, under stress the effects of age become increasingly more obvious and gradually lead to limitations in activity. Research highly recommends regular systematic exercise to promote continued adequate cardiac functioning in the later years. Walking is one of the best and safest ways for elders to exercise and requires no expensive equipment or special locations.

AGE-RELATED DISORDERS OF THE CARDIOVASCULAR SYSTEM

Arteriosclerosis and Atherosclerosis

Arteriosclerosis (hardening of the arteries) is the most common disease of arteries and involves a lessened elasticity and a thickening of the walls of the arteries, especially the small arteries and arterioles, resulting in high blood pressure. *Atherosclerosis*, the most prevalent type of arteriosclerosis, involves small white patches that thicken (atheromas) and protrude into the blood vessel. This is the leading factor for cardiovascular disease. Fatty, fibrous lesions and complicated plaques of scar tissue, calcium salts, and blood-clot formation in the vessel cause HTN, myocardial infarcts (heart attacks), strokes, and vascular diseases (Marieb & Hoehn, 2019). Arteriosclerosis and atherosclerosis usually occur together, and it is rare to find one without the other. Their progress is not continuous but involves a building up and a breaking down of plaques. An increased level of low-density lipoproteins (LDLs) influences the building up of the fatty streaks on the vessel walls causing atherosclerosis, whereas high-density lipoproteins (HDL) transport excess lipids from the peripheral tissues to the liver. Atherosclerosis is often found in the aorta (which arises from the heart and supplies blood to the entire body), the coronary vessels (which supply blood to the heart muscle), and the arteries (which supply blood to the brain, abdomen, and legs).

Risk factors include age, gender (males are more at risk), African American or Hispanic ethnicity, and a family history of the disease. Reversible risk factors include cigarette smoking, obesity, diabetes, elevated lipids, psychological state, HTN, and inactivity. Prevention should begin early in life, but it is never too late to initiate health promotion behaviors.

Diagnosis is made through assessment of cholesterol, triglyceride, and C-reactive protein levels and electron beam computed tomography (EBCT) to identify the presence of plaques. Initial treatment focuses on lifestyle changes such as stopping smoking, regular exercise, weight loss, and a low-fat/low-cholesterol diet. If these are not adequate, various cholesterol-reducing medications (statins are especially helpful but must be used judiciously), a low dosage of aspirin often referred to as "baby aspirin" (81 mg), balloon angioplasty, stents, and bypass surgery are other commonly used options.

Both arteriosclerosis and atherosclerosis lead to ischemic heart disease (lack of adequate blood supply to the heart) as well as cerebral ischemia (lack of adequate blood supply to the brain). They also cause increased blood pressure, produce extra stress on the heart muscle, and set the stage for other diseases of the cardiovascular system. The heart must work harder, but with less overall effectiveness, resulting in insufficient oxygen delivered to body cells and therefore decreased efficiency of body organs in performing their necessary functions.

Hypertension

HTN is defined as a systolic blood pressure (SBP) greater than 140 mmHg or a diastolic blood pressure (DBP) greater than 90 mmHg on two separate occasions. An estimated 30% of Americans older than age 50 are hypertensive, although many do not know it until they experience a heart attack, stroke, or heart failure, or it is discovered during a physical examination. Other complications of HTN are stroke, peripheral vascular disease, chronic kidney disease, and retinal damage; thus early diagnosis and treatment are imperative (Marieb & Hoehn, 2019).

Normal blood pressure is less than 120 mmHg systolic and less than 80 mmHg diastolic; pre-HTN is defined as 120 to 139 mmHg systolic or 80 to 89 mmHg diastolic. Treatment of HTN at an early stage is recommended since it can progress gradually to more serious health issues.

HTN may be differentiated into (a) primary (essential) HTN, the most common, in which there is no obvious or apparent explanation for the sustained elevation in blood pressure; or (b) secondary HTN, which may be caused by stress, smoking, alcohol or drug ingestion, lack of physical exercise, family history, race, high-fat diet and decreased calcium, magnesium, or potassium intake (Winton, 2019).

Persistent abnormally high blood pressure, prevalent in many middle-aged and older adults, can be associated with other factors or systems involved in regulating blood pressure: (a) the circulatory or cardiovascular system, because of its tendency toward sclerosis (hardening of the arteries); (b) the endocrine system, when it acts to retain sodium chloride in the body; (c) the excretory system, when renin (an enzyme involved in raising blood pressure) is released into the blood or when the kidneys do not excrete sodium chloride and water is drawn back from the urinary tubules into the blood; and (d) the nervous system, because it responds to excessive and prolonged emotional tension by increasing peripheral resistance to blood flow, often reflected as high blood pressure. Unfortunately, many individuals have HTN but do not experience symptoms ("*silent*") until body functions become impaired. Even if they experience headaches, dizziness, or fatigue, these may or may not be associated with HTN. Blood pressure should be checked regularly at each healthcare visit or checked at home with personal blood pressure equipment. Opportunities are also available at churches, senior centers, clinics, and other locations.

Risk Factors for Development of Cardiovascular Disease

Nonmodifiable

Age

Blood pressure tends to increase gradually with age until the 60s, when it tends to level off unless other factors cause it to continue to increase.

Gender

Men are more likely to be hypertensive in early and middle age and women in the 55+ years.

Heredity

A family history of HTN and heart disease increases the risk of other family members also developing HTN.

Ethnicity

African Americans have the highest incidence of HTN and suffer from more cardiovascular and major organ complications than Caucasians.

Diabetes

People with diabetes are especially prone to HTN and cardiovascular disease. When these coexist, the risk for complications is greater.

Modifiable

Obesity

Overweight individuals have a greater tendency to develop HTN than those of average weight.

Alcohol

Excessive ingestion of alcohol increases the incidence of HTN. Older persons with HTN should limit their intake of alcohol to 1 ounce daily.

Smoking

Numerous research studies associate cardiovascular disease with smoking and those with HTN are at greater risk for cardiovascular disease.

Elevated Lipids and Cholesterol

A diet high in these increases the incidence of atherosclerosis, which tends to narrow blood vessels and cause both HTN and cardiovascular disease.

High Sodium Intake

Sodium intake promotes fluid retention and increases the likelihood of developing HTN.

Stress

Prolonged high stress may increase blood pressure. The length of time the stress exists, the intensity of the stress, and the individual's response to stress are all modifiable factors.

Sedentary Lifestyle

Regular systematic exercise can reduce blood pressure and weight as well as the risk of developing cardiovascular disease.

Prevention and Treatment of Hypertension

Early diagnosis and treatment are imperative to prevent extensive damage to major organs. Nonpharmacological measures are usually prescribed initially for older adults, depending on the blood pressure reading. They include weight reduction if obese, 30 minutes of regular systematic aerobic exercise daily, moderate alcohol consumption, reduced sodium and caffeine intake, and no smoking. The Dietary Approaches to Stop Hypertension (DASH) diet limits sweets and meats, is rich in calcium and potassium while reducing total and saturated fats, and includes stress reduction and pharmacological treatment.

If lifestyle changes are not effective, medications are prescribed. Initially a thiazide diuretic is usually the drug of choice; however, individuals taking these non–potassium-sparing drugs should be encouraged to eat potassium-rich foods such as bananas, oranges, dried prunes, and raisins or take potassium supplements. Beta-blockers, angiotensin-converting enzyme (ACE) inhibitors, and calcium channel blockers (CCBs) and others may be used alone or in combination to reduce blood pressure levels. Teaching older adults about the particular drug, its action, side effects, how and when to take it, what foods and drinks are not compatible, and the need to check blood pressure levels frequently is necessary. Many of these drugs may not be stopped suddenly, so contact with a primary care provider is very important if one has questions or unexpected drug reactions. Careful prescribing and monitoring is advised along with initiating use of the drug and then proceeding slowly while monitoring for signs of adverse reactions (Winton, 2019). Other nonpharmacological measures to consider for older adults include biofeedback, yoga, meditation, and relaxation exercises, all of which may be effective in reducing blood pressure (Eliopoulos, 2018).

Postural Hypotension

Postural hypotension (orthostatic hypotension) occurs when there is a decease in the SBP of 20 mmHg or more and a decrease in the DBP of 10 mmHg or more and/or an increase in the heart rate of 20 beats per minute when the individual moves from a supine to a standing position (Hutchinson, 2017). Thirty to forty percent of adults older than age 75 experience postural hypotension. It may be caused by age-related changes such as reduced sensitivity of the baroreceptors (receptors that sense blood pressure) and a lessened responsiveness of the sympathetic autonomic nervous system. Other prime causes include medications such as diuretics, antihypertensives, vasodilators, antidepressants, alpha- and beta-blockers, and CCBs. Certain disease states such as HTN, atherosclerosis, Parkinson's disease, arrhythmias, anemia, dehydration, and fluid and electrolyte imbalance might also precipitate the symptoms (House-Fancher & Lynch, 2007). Symptoms include dizziness, light-headedness, fainting, impaired vision, inability to walk properly, fatigue, and confusion, all of which may predispose elders to fall or experience other types of accidents.

Preventive Measures

Care must be taken to prevent blood pressure from falling too low and impairing coronary circulation. Medication regimens need to be continually monitored and evaluated. In addition, older adults should be taught to rise slowly from a lying position and to sit for a short time before standing. Putting on elastic stockings before getting out of bed and reducing physical activity for an hour after a meal are also helpful.

Acute Coronary Syndrome

Even though large amounts of blood continually pass through the heart, this blood does not provide the heart muscle with nutrients and oxygen. Instead, this function is carried out by a group of arteries called coronary arteries that branch off the aorta and envelop the heart. Coronary heart disease results when the blood supply through these arteries is reduced or blocked in any of the following ways:

1. Too high blood pressure in the coronary arteries may result in hemorrhage if a coronary blood vessel should rupture.
2. An aneurysm (a weakened area in the coronary arterial wall) may protrude and rupture, causing a hemorrhage.
3. If blood clots form, they may restrict or block blood flow through the coronary arteries, depriving the heart muscle of blood.
4. Fatty deposits (atherosclerosis) in the inner walls of coronary arteries may interfere with blood flow to the heart muscle and are the most common cause of heart disease in older persons.

Angina Pectoris

The term *angina pectoris* is Latin for "chest pain." It is a symptom of coronary artery disease and occurs when the heart muscle is not receiving an adequate blood supply for effective functioning because of occlusion (closure) or vasospasm (spasm of a blood vessel) of the coronary arteries. Attacks usually last 3 minutes or less and are characterized by pain radiating primarily down the left side of the jaw, neck, between the shoulder blades, and down the arms. Pain, however, may also be on the right side, and individuals may complain of feelings of tightness or pressure in the chest over the sternum or feelings of suffocation. Note that these classic symptoms of angina may not be experienced by older adults and therefore make detection and diagnosis more difficult (Eliopoulos, 2018).

Chronic stable angina (*exertional angina*) is described as episodic (3–5 minutes) chest pain that occurs intermittently over a long period of time. The symptoms, onset, intensity, and duration are similar but increase when causative factors such as cold weather, physical exertion, stress, or after a heavy meal are present (Shaffer & Bucher, 2017). Because of decreased subcutaneous fat, older adults can develop symptoms of angina more rapidly than younger individuals and should therefore wear an extra layer of clothing in cold weather. Chronic stable angina is usually relieved by rest or nitroglycerin (NTG) and is managed with drug therapy.

Unstable angina is characterized by chest pain at rest and occurs with the slightest provocation. It occurs with increasing frequency and represents an emergency requiring immediate attention (Shaffer & Bucher, 2017).

Diagnosis of angina involves a careful history, observing whether the pain is relieved by NTG, plus an electrocardiogram, stress testing, chest x-ray examination, or ambulatory heart monitoring. If medications are not effective, procedures such as cardiac catheterization, angioplasty (in which a catheter is inserted into a coronary vessel and plaque is compressed against the vessel wall by the inflation of a balloon), placing a stent in the coronary artery to open up an occlusion in the vessel, or coronary bypass surgery may be indicated.

The initial treatment is aspirin with NTG, which is considered the cornerstone of anginal therapy because it inhibits spasms of the coronary vessels and improves collateral circulation to the heart muscle. NTG is placed under the tongue or NTG spray is sprayed on the tongue; relief should occur in about 3 minutes after NTG is administered. If not, the dosage is repeated for a total of three times. However, medical attention is advised if the medication is not effective in relieving the discomfort. Other medications that may be prescribed are beta-blockers, CCBs, statins, ACE inhibitors, and ranolazine. Aspirin plus clopidogrel may be used as an ongoing treatment. Stress reduction and avoiding caffeine, heavy meals, and physical exertion are also advised.

Instructing older adults regarding precipitating factors, signs and symptoms of angina, and how and when to take medication or seek medical care is important. If sublingual NTG is prescribed, it should be carried by the person at all times, stored in a capped dark-glass bottle, kept away from heat and light, and a new supply obtained every 6 months.

Myocardial Infarction

A myocardial infarction (MI) results when there is reduced or no blood flow to the heart muscle via the coronary arteries, depriving the heart muscle of oxygen and causing the heart rhythm to become erratic or cease altogether. Most MIs are caused by atherosclerosis of a coronary artery, rupture of the plaque, and blockage of blood flow. Inflammation may also contribute to heart attack if coronary artery walls become inflamed, increasing the buildup of fatty plaques. It may only damage a small area of the heart or it can be extensive. Older adults are more likely to develop complications such as heart failure, arrhythmia, pulmonary edema, and rupture of the heart (House-Fancher & Lynch, 2007). Those with an MI require close observation in an intensive care unit to prevent further complications.

Typical symptoms of an MI include severe, vicelike, continuous, constrictive, chest pain that commonly radiates to the jaw, neck, arm, and back. The pain often occurs in the early morning; is unrelieved by nitrates, position change, or rest; and lasts about 20 minutes or more. Other signs and symptoms include profuse perspiration, moist clammy skin, pallor, and nausea and vomiting. Anxiety, restlessness, drop in blood pressure, arrhythmias, shock, and heart failure may also be present (Shaffer & Bucher, 2017). It is important to note that 25% to 68% of MIs are not identified as an MI; this is especially the case for older adults and women, who may not display the usual clinical symptoms of a heart attack, called a "silent MI." Rather, they may have gastric disturbances, dizziness, nausea and vomiting, fatigue, behavioral or mental changes, or an arrhythmia. The presence of these atypical symptoms may result in the individual not recognizing the seriousness of the situation and not seeking medical attention (Winton, 2019). Greater than one third of women ages 45 to 54 and close to 70% of women age 65 and older are affected (Eliopoulos, 2018). The course of the illness is more complex and the mortality rate is higher among older adults; therefore early, definitive diagnosis and treatment are imperative (House-Fancher & Lynch, 2007).

Early treatment is vital but is sometimes delayed because of the atypical presentation of symptoms in older adults. A chest radiograph, electrocardiogram, various scans (CT or MRI), serum cardiac markers and other blood tests, an angiogram, or stress test may be used to diagnose an MI. The treatment may involve fibrolytic therapy such as tissue plasminogen activator (t-PA) administered within a rigid time frame; however, these drugs have multiple contraindications. Seeking medical attention immediately is crucial because these drugs are administered within a certain period following the time when the symptoms first appear. Other treatments include intravenous NTG, ACE inhibitors, beta blockers, anticoagulants, antiarrhythmics, cholesterol-lowering agents, and possibly CCBs. Aspirin can help to prevent new clots from forming. Procedures such as coronary angioplasty, intracoronary stent placement, or coronary bypass surgery may be necessary treatment choices (Shaffer & Bucher, 2017).

Rehabilitation

Cardiac rehabilitation along with dedication to a healthy lifestyle is paramount in preventing coronary artery disease. Cardiac rehabilitation programs are now widely available throughout the country where individuals are introduced to monitored exercise and educated regarding stress reduction, smoking cessation, modified alcohol ingestion, and weight control. Dietary teaching includes sodium and fat restriction, eating fruits and vegetables, and education about whole-grain, high-fiber foods. Psychosocial issues such as risk modification, psychological support, socialization, and resuming sexual activity are important considerations.

Congestive Heart Failure

Heart failure in older adults increases with age (Eliopoulos, 2018). About 6 million people in the United States have heart failure, and it is the leading cause of hospitalization in both men and women older than age 65. Congestive heart failure (CHF) usually develops gradually and progresses over time, with a high morbidity and mortality rate. It is estimated about 50% of those diagnosed with it will be dead within 5 years (Winton, 2019). Often it is undetected in the early stages because symptoms of fatigue and shortness of breath may be expected and equated with growing older (Tabloski, 2014). Heart failure occurs when the heart is no longer able to pump adequate blood and oxygen to body tissues during exercise or even at rest. There may be left-sided heart failure as a result of the failure of the left ventricle to adequately pump the blood it is receiving from the lungs, resulting in a buildup of fluid in the lungs and less blood pumped through the body. Right-sided heart failure occurs when the right side of the heart cannot adequately empty itself of the blood coming from the venous circulation and less blood enters the heart, producing a backup of fluid (edema) in the extremities or in the abdominal cavity. The signs and symptoms include coughing, shortness of breath, wheezing, edema, loss of appetite, nausea, and tachycardia. Other less common signs include fatigue, coughing, disorientation, and weakness (Mauk, 2018). An individual may have both right and left-sided heart failure (Winton, 2019). Diagnosis involves a history and physical examination; assessing for pulmonary and systemic congestion plus measuring the output of the heart; chest radiographs; urinalysis; an echocardiogram, including calculation of the ejection factor (EF) to measure how well the heart pumps with each beat;

electrocardiogram; Doppler ultrasound; blood chemistries; exercise or stress testing; nuclear imaging studies; and cardiac catheterization (Jett, 2008; Winton, 2019). Classification systems can be helpful in determining therapies for heart failure patients that emphasize the progression of heart failure and the need for different therapies as the disease progresses.

Risk factors for CHF include the following:

1. *Cardiac factors.* Coronary artery disease, myocardial infarction, HTN, and valvular heart disease all reduce the efficiency of the pumping action of the heart.
2. *Noncardiac factors.* Pathological conditions that increase the risk for CHF are chronic obstructive pulmonary disease, pulmonary emboli, kidney disease, liver disease, diabetes, and anemia.
3. *Iatrogenic factors* (caused as a result of treatment). Medications used by elders, such as digoxin, steroids, hormones, and anti-inflammatory drugs, often increase the risk of CHF.
4. Other factors such as malnutrition severe enough to produce fluid and electrolyte imbalances, obesity, alcohol abuse, and prolonged high-stress situations may also cause CHF.

Treatment of CHF includes medications such as ACE inhibitors, vasodilators, beta blockers, digitalis, diuretics, lifestyle and dietary management, (especially reduced sodium) and rest. The goals of treatment are (a) to reduce the body's demand for high cardiac output (through a balanced exercise/rest program and weight reduction); (b) to increase the cardiac output if possible (usually through the use of medication); and (c) to reduce body congestion (water and sodium retention) by sodium restriction and medications such as diuretics. Digoxin (digitalis) is used to increase the force of the heart's contractions and slow down the rate. It is important to teach individuals how to take an accurate pulse and not to take digoxin if the pulse rate is less than 60 unless otherwise instructed by a primary care practitioner. Digitalis toxicity is common among older adults; thus, blood levels need to be monitored regularly.

If prescribed diuretics deplete potassium, foods rich in potassium such as bananas should be encouraged. Sodium restriction helps to decrease the workload on the heart. For heart failure to be adequately managed, older adults and family need to understand as much as possible about the disease, its treatment, medications, and a healthy lifestyle. Possible surgical treatments include coronary artery bypass, heart valve surgery, implantable left ventricular assist device (LVAD), and heart transplant. Ongoing education and monitoring and psychological support are crucial in preventing exacerbations of the disease. The prognosis in older adults is guarded because CHF or heart failure is an end-stage heart disease reflecting the accumulative effects of other serious pathological conditions.

Heart Valve Disease

The incidence of heart valve disease is increasing as larger numbers of people are living longer with degenerative heart diseases. Heart valve disease is thought to be caused by valvular or muscle dysfunction, endocarditis (inflammation of the lining of the heart), or rheumatic diseases. Valvular incompetence or regurgitation occurs when a valve doesn't close properly and blood leaks back into the chamber. Aortic stenosis, the most common valve disorder in older adults, is caused by sclerosis of the aortic cusps or abnormal tissue in the septum or aorta, both of which restrict blood flow (Tabloski, 2014).

Risk factors are high cholesterol, HTN, and diabetes. Diagnosis depends on physical examination, history, echocardiogram, and possibly cardiac catheterization. Treatment often requires mechanical or surgical procedures and possibly valve replacement (Tabloski, 2014).

Cardiac Arrhythmias and Conduction Disorders

With aging, cardiac arrhythmias (irregular heartbeat) and conduction disorders become more common. Arrhythmias and conduction disorders (disorders affecting the heart's ability to regulate a synchronized heartbeat) are more serious in older adults because vital body functions are already less efficient and reduced blood supply to tissues is less well tolerated. Disturbances found more often in the older age group include premature atrial and ventricular contractions, atrial fibrillation (extremely rapid incomplete contractions), and abnormal rhythms of the atrial pacemaker mechanism. Impaired functioning of the S-A node may cause "sick sinus syndrome," resulting in arrhythmias, sinus bradycardia (slow heartbeat), heart block, palpitations (rapid throbbing pulsations), weakness, and dizziness or fainting. These disorders often are first identified during a routine health examination.

Treatment involves antiarrhythmic drugs such as digoxin; sodium, potassium, or CCBs; beta blockers; and other medications. Sometimes the use of a manual or automatic external defibrillator, or synchronized cardioversion, is warranted to interrupt the dysfunctional rhythm and restore normal heart rhythm. Radiofrequency catheter ablation therapy, a permanent pacemaker or cardioverter-defibrillator may be implanted to maintain or restore normal heart rhythm.

Transient Ischemic Attack

A transient ischemic attack (TIA, or ministroke) is an early warning of impairment in the blood supply to the brain and of a possible imminent major stroke. The greatest risk for a stroke is in the first week after a TIA (Tabloski, 2014). TIAs are caused by a sudden interruption in the circulation of blood and oxygen to the brain, usually lasting less than 24 hours, with no permanent brain injury. Symptoms last from a few minutes to 24 hours, with recovery often in 3 hours. TIAs often go unnoticed because symptoms are minimal and of short duration. Typical signs of a TIA include the following (Eliopoulos, 2018):

- Brief vision loss in one or both eyes
- Change in personality or mental abilities such as confusion
- Difficulty understanding speech
- Double vision
- Nausea, vomiting
- Sudden temporary weakness or numbness of the face, arm, or leg
- Temporary dizziness or unsteadiness
- Unexplained headaches or a change in type of headache

If TIAs lead to small strokes, tissue damage will accumulate and eventually produce changes in behavior. Sometimes behavior changes after accumulated ministrokes are so slight that only close family members are aware of the subtle changes taking place, usually in personality or mood. Any unusual or persisting change in normal behavior patterns or unusual symptoms should be evaluated immediately by a professional because early diagnosis and treatment are important.

At risk for TIAs are those age 60 and older with HTN, obesity, or diabetes; those who are smokers or alcoholics; and those with sleep apnea, high cholesterol, and cardiovascular insufficiency. Guidelines of the American Heart Association/American Stroke Association (2011) recommend treating a TIA in the same way as an ischemic stroke. Initially there should be a complete diagnostic workup and the use of the same preventive treatments for a stroke. Various medical options should be considered, including the use of blood thinners and interventional therapies such as endarterectomy (surgical removal of a blood clot from the carotid artery), angioplasty (when a balloon attached to a catheter is blown up to compress the plaque against the vessel wall), or the placement of a stent in the carotid artery. Such interventions carry some risk of a stroke; thus it is important to select a surgeon who has a low rate of complications and who has performed large numbers of these procedures. Individuals taking blood thinners should be carefully monitored for signs of bleeding in the urine, feces, or under the skin. The individual should also be counseled to adopt healthy behaviors such as a low-cholesterol, low-fat diet; not smoking; minimal alcohol intake; and a consistent exercise regimen. These guidelines are updated frequently thus the importance to keep abreast of the newest recommendations.

Cerebrovascular Accident

A cerebrovascular accident (CVA, stroke, brain attack) results when the blood supply to any part of the brain is reduced or completely shut off. Strokes are the third leading cause of death and a significant cause of disability in older adults (Eliopoulos, 2018). The majority of elders experience an *ischemic stroke (occlusive),* which includes either a thrombotic or embolic stroke. A thrombotic stroke occurs when a cerebral artery is narrowed by plaque (fatty deposits) in the artery, causing a clot to form (cerebral thrombosis) that either reduces or closes off the blood flow to an area of the brain. An embolic stroke is caused by air, fat tissue, blood clot, or other foreign matter circulating in the blood, which blocks blood flow in a cerebral vessel. A *hemorrhagic stroke* occurs when a weak spot in a blood vessel of the brain bursts, causing bleeding into the brain tissue.

Significant risk factors associated with a stroke include HTN, heart disease, previous TIAs, smoking, diabetes, atherosclerosis, high cholesterol, sedentary lifestyle, high-fat diet, and family history of strokes. Those of African American ancestry have a higher incidence of strokes, and more men than women have strokes, although postmenopausal women are also at significantly increased risk. Individuals with one or more of these risk factors should be especially attentive to their health and lifestyle because the best treatment for stroke is prevention.

Strokes affect behavior in many different ways depending on location and the amount of brain tissue damaged. Injury to the right half of the brain may result in impaired movement and sensation on the left side of the body, spatial–perceptual difficulties, and memory problems. Conversely, strokes on the left half of the brain affect movement and sensation on the right side of the body. Language or speech aphasia, impaired vision and comprehension, or emotional problems may also result.

After a thorough history and physical examination, careful screening must begin immediately using angiography; a computed tomography (CT), positron emission tomography (PET) scan, MRI or other scans; and laboratory tests. Brain imaging is most important in diagnosing strokes. If the individual has an ischemic stroke and meets the specific criteria, early intervention with fibrinolytic therapy (clot-busting drug) or other drugs may be given intravenously to help reestablish blood flow through the occluded artery. This therapy must be initiated within a 3-hour period or as soon as possible , beginning at the time the individual shows the first clinical signs of a stroke. Thus immediate attention is necessary by calling an emergency number such as 911 and getting the individual to a designated stroke center or an acute care hospital. Initially the person's status is stabilized in the intensive care unit (ICU). Depending on the type, location, and extent of tissue damage, poststroke management can include anticoagulants or antiplatelet therapy and prolonged physical and psychological rehabilitation to help improve functioning and to prevent the many complications of a stroke. A team approach is initiated with medical and nursing care; physical, occupational, and speech therapy; and psychological and social therapy, ideally at a rehabilitation center. Most neurological recovery takes place within the first 3 to 6 months after a stroke. Family functioning is greatly influenced after a stroke in a family member. Because the healing process is often prolonged, family need to assume added responsibilities of caregiving, managing a household, transportation, cooking, and finances. The family experiences many of the same psychological responses as the patient; thus they too need continual social and psychological support. Patient and family teaching are paramount to assist them in understanding the importance of a healthy lifestyle regarding diet, exercise, medication adherence, and stress-free living.

VASCULAR DISORDERS

Peripheral vascular disease is primarily a result of generalized atherosclerosis causing a narrowing of the arteries in extremities as well as the neck, head, abdomen, and legs. Both aneurysmal and occlusive vascular disease result from atherosclerosis.

Aneurysm

Aneurysms tend to occur after age 60, with men more likely candidates than women. Those with HTN and atherosclerosis are more prone to aneurysms. *Aneurysm* is the term for a "pouch" formation in a weakened arterial wall. The pouch fills with blood and may burst, especially if the arterial wall is weak and blood pressure is high; the larger the aneurysm, the more likely it is to rupture. When aortic and cerebral aneurysms rupture, shock and death often result. Other areas commonly affected include the femoral and popliteal arteries in the leg. Although significant pain does not always accompany an aneurysm, some individuals experience chest, back, and abdominal pain or leg weakness and cramping when walking, which is relieved by rest. Treatment involves supportive treatment and dissection of the aneurysm, replacing the area with a synthetic graft material and endovascular coiling.

Arterial Occlusion

The major causes of arterial occlusion are thrombosis, embolism, or trauma. An occlusion develops inside an artery near an area of plaque formation. Most often occlusions occur in the coronary vessels, causing a coronary infarction, or perhaps in the legs. When the legs are involved they become cold, pale, and bluish colored; severe pain is present along with intermittent cramping, especially after walking. Diagnosis requires vascular studies, scans, ultrasound, a thorough history, and physical examination.

Treatment involves increasing the flow of blood to the area through exercise, anticoagulant therapy, or surgery. Older adults should be advised not to cross their legs or wear tight clothing, and shoes should be well fitted and comfortable. Feet should be kept clean and dry, and foot care should be provided by a podiatrist. Any break in the skin, trauma, or blister needs to be reported immediately to a primary care practitioner.

Varicose Veins

Varicose veins (varicosities) are caused by inefficiency of one-way valves in peripheral veins that return blood from the peripheral to central circulation. Blood then pools, especially in the lower extremities, vein walls become weak, and swollen "knotted" veins result from the slowed circulation. Varicose veins are more prevalent in women and in those who are obese, with a predisposition to varicosities occurring in families. Treatment may involve keeping the affected limb elevated; avoiding trauma to the leg; wearing support hose; and the use of sclerotherapy, laser, high-intensity pulsed-light therapy, endovenous occlusion, or saphenofemoral ligation. Surgical removal of the vein is possible if other treatment modalities are not effective.

Varicose veins in the lower part of the rectum and anus are called hemorrhoids.

Phlebitis and Deep Vein Thrombosis

Phlebitis is an inflammation of a vein, often in the leg, producing conditions favorable for the formation of blood clots (thrombi) that can break loose and occlude a major vessel in the lungs (pulmonary embolism), heart, or brain and they may be life-threatening. Particularly at risk are postoperative knee or hip surgery patients who are older, obese, or smokers, and those who are dehydrated or have major heart or circulatory problems. Signs and symptoms include a bluish-red color in the leg and a leg that is warm to touch, tender, swollen, and painful. Appropriate leg exercises, antiembolic stockings, intermittent compression devices (ICDs), and anticoagulant therapy are among the appropriate treatment choices.

SUMMARY

Heart disease remains the most common cause of death in individuals older than age 65 but is often preventable, especially if health promotion behaviors are initiated early in life. However, positive results can be attained even if they are begun in later life. Maintaining an active exercise regimen, weight control, and managing stress are all necessary for optimal cardiac health.

Cardiovascular disease often leads to fear and anxiety, increasing self-preoccupation and impatience with those who are healthy and active. It is important that efforts be directed toward assuming a normal lifestyle after each episode of cardiac dysfunction. Many individuals with cardiovascular disease live normal, well-balanced lives under medical supervision. Participation in a cardiac rehabilitation program is especially recommended to promote a longer, healthier life.

REFERENCES

American Heart Association/American Stroke Association. (2011). Guidelines for the prevention of stroke in patients with stroke or transient ischemic attack: A guideline for healthcare professionals from the American Heart Association/American Stroke Association. *Stroke, 42*, 227–276. https://doi.org/10.1161/STR.0b013e3181f7d043

Eliopoulos, C. (2018). *Gerontological nursing* (9th ed.). Wolters Kluwer.

House-Fancher, M. A., & Lynch, R. J. (2007). Cardiovascular system. In A. D. Linton & H. W. Lach (Eds.), *Matteson & McConnell's gerontological nursing* (3rd ed., pp. 313–352). Saunders Elsevier.

Hutchinson, M. L. (2017). Hypertension. In S. L. Lewis, L. Buchner, M. M. Heitkemper, & M. M. Harding (Eds.), *Medical-surgical nursing* (10th ed., pp. 681–701). Elsevier.

Jett, K. (2008). Chronic disease in later life. In K. Jett, P. Hess, T. A. Touhy, K. Jett & A.S. Luggen (Eds.), *Toward healthy aging* (7th ed., pp. 222–268). Mosby Elsevier.

Marieb, E. N., & Hoehn, K. (2019). *Human anatomy and physiology* (11th ed.). Pearson Education Limited.

Mauk, K. (2018). Management of common illnesses, diseases, and health conditions. In K. Mauk (Ed.), *Gerontological nursing* (pp. 303–385). Jones & Bartlett.

Shaffer, R., & Bucher, L. (2017). Coronary artery disease and acute coronary syndrome. In S. L. Lewis, L. Bucher, M. M. Heitkemper, & M. M. Harding (Eds.), *Medical-surgical nursing* (10th ed., pp. 702–736). Elsevier.

Tabloski, P. (2014). *Gerontological nursing* (3rd ed.). Pearson Education.

Winton, M. B. (2019). Cardiovascular function. In S. E. Meiner & J. J. Yeager (Eds.), *Gerontologic nursing* (6th ed., pp. 331–363). Elsevier.

9

The Respiratory System

INTRODUCTION

Of all the body systems, the respiratory system is most exposed to damage from the environment. Here the oxygen taken into the lungs is transferred to the blood, and carbon dioxide is released from the lungs. It is extremely difficult to separate age-related changes in the lungs from environmental or outside insults such as pollution, smoking, diseases, or infections. Symptoms such as breathlessness or fatigue may be attributed to getting older when in reality they are sometimes caused by unrecognized diseases. Assessment, treatment, and outcome of respiratory diseases are closely linked to the health practitioner's knowledge of age-related changes, the effects of the environment, and relevant lifestyle issues.

STRUCTURES OF THE RESPIRATORY SYSTEM

Structures involved in the respiratory system are as follows:

- The various air passageways, including the nasal cavities, mouth, pharynx, larynx, trachea, bronchi, and bronchioles
- The lungs, which contain the tiny air sacs (alveoli), and alveolar ducts

Air Passageways

Air enters the body primarily through the nasal cavities, where it is warmed, moistened, and filtered by the mucous membranes in the nose. Air may also enter through the mouth. Either way, incoming air enters the pharynx, a funnel-shaped passageway connected to the larynx. Seven cavities or tubes open into the pharynx: the mouth, trachea, esophagus, two nostrils, and two pharyngotympanic tubes (formerly called the Eustachian tubes) from the ears. The pharynx has three divisions: nasopharynx, oropharynx, and laryngopharynx. The tonsils are located near the pharynx and serve to help protect these cavities against bacterial infection.

Passing through the pharynx, air enters the larynx (or voice box). The larynx is about 2 in. long and contains vocal cords, which produce the voice. When food is being propelled through the pharynx, the opening of the larynx (the glottis) is closed by the reflex action of the epiglottis (a thin lid of fibrocartilage) to prevent food or liquids from entering the trachea (the windpipe).

Anatomically, the larynx is composed of nine cartilages bound together by an elastic-like membrane. One of the cartilages, the thyroid cartilage, is ordinarily more prominent in men than in women and is referred to as the "Adam's apple." During puberty, the larynx becomes larger in males and the vocal cords become longer and thicker, causing men to have deeper voices than women. Human voice quality, with all its variations and complexities, involves not only the larynx but also the pharynx, nasal cavities, mouth, teeth, and tongue; the resonating chambers in the head (sinuses); and the learned ability to control the inhalation and exhalation of air.

Extending downward from the larynx is the trachea, about 4 in. long and 1 in. in diameter. Situated in front of the esophagus, it is composed of elastic tissues and from 16 to 20 C-shaped cartilaginous rings. Hair-like projections called cilia line the trachea and help to push mucus containing debris and dust particles up toward the pharynx. The trachea, which is elastic and flexible, stretches when one breathes in and recoils when one breathes out, but the cartilage rings prevent it from collapsing and cutting off the air supply to the lungs.

Upon entering the chest region, the trachea divides into left and right bronchi (smaller tubes) leading into the lungs. The bronchi continue to divide into smaller and smaller tubes until, at about 1 mm in diameter, they become tiny elastic tubes called bronchioles. Bronchioles branch into even smaller alveolar ducts leading to many alveoli (air sacs). The tiny alveoli in the lungs are covered with many pulmonary blood capillaries where exchange of gases between alveoli and blood takes place. It is here that carbon dioxide, a waste product, is removed from the blood and a fresh supply of oxygen is picked up by the hemoglobin in the blood to be delivered to the heart and then to body tissues for immediate use. To summarize, the bronchial tree is composed of a series of tubes that become progressively smaller until they end in a network of alveoli surrounded by blood capillaries. The life-sustaining carbon dioxide–oxygen exchange occurs in the alveoli.

The Lungs

The two lungs, the major organs of respiration, are soft, spongy, elastic tissue able to change shape during respiratory movements. Located in the thoracic cavity (the chest), they are somewhat cone shaped. The top, or apex, of each lung extends into the base of the neck, and the lower part rests on the diaphragm, a large muscle forming the partition between the thoracic and abdominal cavities. The left lung is divided into two lobes because the heart is located on the left side, whereas the right lung has three lobes.

Each lung is enclosed in a thin, double-layered membrane called the pleura. Pleural fluid is found between the layers and creates an adhesive force that holds the lungs close to the thorax wall. Negative pressure in the pleura, along with positive pressure equal to atmospheric pressure in the lungs themselves, allows the lungs to expand and recoil as the size of the chest cavity increases and decreases (see Figure 9.1).

FUNCTIONS OF THE RESPIRATORY SYSTEM

Breathing

Movements of the respiratory muscles allow for changes in the size of the chest or thoracic cavity and make breathing possible. During inspiration (inhalation), for example, the size of the chest is increased by the contraction and flattening of the diaphragm and contraction of the ribcage muscles, which causes the ribs to move upward and forward. As a result, chest capacity or volume increases, pressure within the lungs decreases, and air is sucked in. As the respiratory muscles relax, the diaphragm resumes its normal dome shape, the ribs move back to resting position, and chest volume (size) decreases. As the size of the chest cavity becomes smaller, pressure in the lungs increases and air is forced out (expiration or exhalation). Breathing thus is not a function

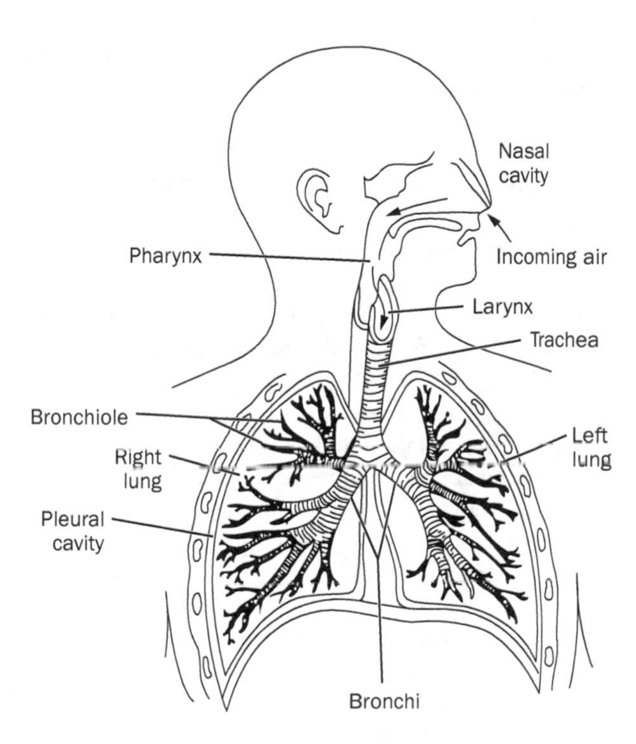

Figure 9.1 Respiratory system.

of the lungs alone but is also due to the action of diaphragm and ribcage muscles. The lungs are not muscular tissues but act more like balloons subject to pressure differences between the lung cavity in the body and atmospheric pressures outside the body. They are sometimes compared with a bellows in action.

Respiration rate is essentially under involuntary control by brain centers in the medulla and pons (at the base of the brain), although it is also subject to substantial voluntary control mediated by the cerebral cortex. We cannot voluntarily breathe while swallowing, though, because of a powerful reflex that prevents food or liquids from passing down the trachea to the lungs instead of down the esophagus to the stomach. When this reflex is interfered with in any way, we choke. Another strong reflex regulating breathing is controlled essentially by carbon dioxide levels in the blood. When the amount of carbon dioxide exceeds a certain level, we are forced to breathe. The carbon dioxide level in the blood, then, controls breathing more than the level of oxygen in the blood. Other nonbreathing processes such as coughing, sneezing, hiccupping, crying, laughing, and yawning also move air in and out of the lungs (Marieb & Hoehn, 2019).

External and Internal Respiration

Each day the intact respiratory system exchanges about 2,600 gallons of oxygen from the air with carbon dioxide from the blood. This exchange, which takes place in the lungs, is called external respiration. The exchange in the body cells of oxygen from the blood and carbon dioxide from the tissues constitutes internal respiration. Body cells are highly dependent on a constant supply of oxygen for metabolism and on the regular pickup and excretion of carbon dioxide, a major waste product of the body's metabolic processes. Because body cells and tissues are unable to store any significant amount of oxygen over time, a new supply must be delivered continuously to all tissues of the body via the hemoglobin in the bloodstream. Cells die rapidly without oxygen. This process is infinitely more complicated than this brief description implies.

AGE-RELATED CHANGES IN THE RESPIRATORY SYSTEM

Age-related changes in the respiratory system are often indistinguishable from changes in the system arising from such factors as air pollution, occupational hazards, smoking, and other lifestyle and environmental factors. Respiratory efficiency, though, does decrease with age.

Specific age-related changes identified in this system include the following:

1. Calcification of the laryngeal and tracheal cartilage occurs, resulting in a stiffening of those structures. The number of cilia and their activity in the bronchial mucosa is reduced. Glandular cells in the large airway are reduced, resulting in decreased production of protective mucus to help ward off respiratory infections (Tabloski, 2014). The cough reflex is blunted, causing decreased effective coughing. Because coughing is beneficial in clearing the upper airway of small or large particles, there is a greater risk of choking or aspirating materials into the lungs, possibly resulting in aspiration pneumonia. Because there are fewer nerve endings in the larynx, the gag reflex may also be less efficient and older adults may then be prone to develop respiratory tract infections.
2. The actual number of alveoli does not change significantly with age, but their structure is altered. With age, the number of functional alveoli decreases as alveolar walls thin, alveoli enlarge, and fewer blood capillaries are available for gas exchange. Overall, there is a decrease in the surface area available for oxygen–carbon dioxide exchange (Sanders, 2019).
3. *Presbylaryngis* refers to aging changes in voice pitch caused by thinning or aging of the vocal cords. This results in a higher, reedy voice that is difficult to hear and requires more effort for an older adult to produce. If needed, voice therapy may help older adults learn to speak with greater breath support. Greater variability in the pitch of the voice occurs more often in older individuals in poor health. Breathiness in speech can also result from less air reaching the throat, incomplete closure of the glottis, or decreased joint mobility of the jaw (Jett, 2012).
4. Skeletal changes such as calcification of the costal (rib) cartilages, osteoporosis, and weakened respiratory muscles all affect respiratory functioning. Kyphosis (hunchback), scoliosis (lateral spinal curvature, S-shaped), a shortened thorax, and chest wall stiffness all contribute to limiting chest expansion and reducing effective ventilation. The resulting greater dependence on the use of the diaphragm in breathing requires increased energy expenditure and promotes fatigue.
5. The skeletal muscles of the thorax and diaphragm responsible for inhalation and exhalation lose strength and endurance as a part of the aging process, there is increased stiffness of the chest wall and kyphosis, and a barrel chest may result (Eliopoulos, 2018). Because these muscles are primarily responsible for increasing and decreasing the size of the thoracic cavity, age-related muscular changes are extremely important in regulating the amount of air actually in the lungs (Davies & Bolton, 2010).
6. The lungs decrease in size, become flabbier, and decrease in weight (Miller, 2009). They have less elastic recoil because the elastic fibers decrease, and there is an increase in cross-linked collagen. Forced vital capacity, or the maximum amount of air that can be expelled from the lungs after a full inspiration, is diminished somewhat in older age. In addition, residual volume, the amount of air left in the lungs after a forced expiration, increases (Eliopoulos, 2018). The alveolar ducts and alveoli become enlarged, leading to decreased efficiency of the oxygen–carbon dioxide exchange.

7. The decreased effectiveness of the oxygen–carbon dioxide exchange causes increased levels of carbon dioxide and decreased levels of oxygen in the blood, predisposing the older adult to a lower oxygen supply available to vital organs in acute respiratory conditions and perhaps also to increased incidence of sleep disorders.

Overall, with age, less oxygen is delivered to body cells and consequently there is a lessened reserve capacity in the respiratory system when dealing with high-demand situations. These changes contribute to the greater fatigability of most older persons. Many factors contribute to lung degeneration and disease. Cigarette smoke, both active and secondhand, causes reduced ciliary action, inflammation of the respiratory tract, constriction of the bronchioles, and reduced breathing capacity. The air breathed into the lungs contains irritating and toxic gases from cigarette smoke, as well as gases such as carbon monoxide, nitrogen dioxide, and hydrogen cyanide. A smoker's forced expiratory volume (the amount of air that can be expelled forcefully) declines at double and triple the rate of nonsmokers. Smoking increases the risk of developing chronic debilitating or fatal diseases such as cancer of the larynx or lung, chronic obstructive pulmonary disease, and other serious lung infections as well as cardiovascular, urinary, and other organ system dysfunctions (Eliopoulos, 2018; Miller, 2009).

Other risk factors for respiratory difficulties include environmental hazards such as pollution in the environment and in the workplace, immobility, and the presence of chronic diseases that predispose an individual to lung problems. The use of feeding tubes and the aspiration of food or liquids could also lead to pneumonia. Age-related changes in the skeletal and muscular systems, obesity, and immobility decrease breathing effectiveness and increase the likelihood of developing respiratory infections and impaired lung function. Some medications may also increase the risk of respiratory impairment.

AGE-RELATED DISORDERS OF THE RESPIRATORY SYSTEM

Chronic Obstructive Pulmonary Disease

Chronic obstructive pulmonary disease (COPD) is a classic disease of older individuals because it may very likely result from age-related changes, disease processes, and lifestyle. *COPD* refers to a group of diseases in which there is (a) reduced flow of air in and out of the respiratory system; (b) excessive secretions of mucus within the airways; (c) chronic infection; (d) an increase in the air spaces beyond the terminal bronchioles, with an accompanying loss of the alveolar walls; (e) decrease in recoil ability (elasticity) of the lungs; and (f) narrowing of the bronchi as a result of factors such as allergies. Included in COPD are chronic bronchitis and pulmonary emphysema. Older individuals often have a combination of COPD diseases. Asthma is sometimes included in COPD, and it occurs more often in older adults than previously recognized.

COPD ranks as the third leading cause of death in the United States, with the death rates for women exceeding the death rates for men (Tabloski, 2014). According to the American Lung Association (2021) women are often misdiagnosed, and women are more vulnerable than men to sustain lung damage from cigarette smoke and pollutants because their lungs are smaller. COPD leads not only to respiratory impairment but also to work disability and hospitalization, placing a tremendous economic burden on individuals, families, and society. The major risk factor in the development of COPD is smoking (more than 80% are smokers) and exposure to secondhand smoke. It should be noted that some of the damage to the respiratory system is reversible in the individual who stops smoking, no matter what the person's age. Actual lung damage caused by emphysema, though, cannot be remedied significantly. Other factors contributing to COPD are a genetic risk factor, air pollution, lower socioeconomic status, infections, and occupational exposure to toxic materials.

An early symptom is an intermittent cough, usually in the mornings. Later, labored or difficult breathing (dyspnea) is characteristic, mostly during exertion, but dyspnea is progressive and eventually occurs at rest as well. Wheezing and increased sputum production are also common in COPD. Often those with advanced COPD experience anorexia and weight loss. Unfortunately, commonly about 50% of lung function is lost before definite symptoms appear and the individual seeks medical care (Jett, 2012).

Diagnosis is made with a thorough medical history, physical examination, sputum studies, pulmonary functioning studies, scans, x-ray examination, measurement of arterial blood gases, and other blood tests. Treatment begins by removing the individual from irritants such as cigarette smoke and pollution. Bronchodilators also may help to relieve breathing difficulties so prevalent in these individuals. Corticosteroids in aerosol form (inhalers), anticholinergics, beta-adrenergic drugs, antibiotics, respiratory therapy, diaphragmatic or pursed-lip breathing, oxygen, proper nutrition, and appropriate exercise are also often prescribed. For some patients, surgical options such as lung volume reduction surgery may help, but it is only used in severe cases of COPD (Workman, 2013a). It is recommended that those with COPD take the vaccine for pneumonia and be given a yearly influenza vaccination. Most necessary is the complete cessation of smoking by those who have this disease. Psychological support is of utmost importance because COPD is usually present for the rest of one's life. Pulmonary rehabilitation centers have comprehensive programs to meet the physical and psychosocial needs of the individual and their significant others. Most involve exercise training, breathing retraining, education about the disease, medication information, nutrition information, and smoking cessation (Hendrix, 2011). Such interventions have proven to be very helpful in managing and coping with these difficult and frightening diseases.

Chronic Bronchitis

Chronic bronchitis is defined as a recurrent, persistent cough with excessive mucus secretions for at least 3 months per year for 2 consecutive years, or for 6 months during 1 year. Chronic bronchitis is an inflammation of the bronchi with small airway obstruction caused by hypersecretion of mucus and the presence of edema. This interferes with the flow of air in and out of the lungs. The more years a person has smoked, the greater the likelihood of having chronic bronchitis. However, it may also be caused by other irritants. Treatment includes rest, use of cough suppressants, acetaminophen for pain, and humidification of the air. Antibiotics may be used to prevent the bronchitis from progressing to pneumonia (Tabloski, 2014). Chronic bronchitis often coexists with emphysema and in many instances it is difficult to distinguish between them.

Emphysema

Emphysema is defined as an enlargement and destruction of alveolar walls in the lungs. The lungs lose their ability to stretch and relax, thus remaining partially filled with stale, oxygen-poor air. The flow of air is impeded during breathing as the alveolar walls are gradually destroyed and air spaces expand. Often mucus and infected material pool in these structures, causing coughing, sputum production, infection, and greatly increased effort to breathe. The disease usually progresses, especially if individuals continue to smoke, and they experience lethargy, weight loss, weakness, and respiratory acidosis (excessive acidity of body fluids) as a result of decreased oxygen in the body. The incidence of emphysema increases with age, and by age 90 most individuals are likely to have some signs of the disease. Smoking is a major initiating factor; other factors of importance are secondhand smoke, air pollution, allergies, poor nutrition, and alcohol consumption. Patient education to learn to deal as effectively as possible with this disease is very important. Treatment involves postural drainage, breathing exercises, bronchial dilators, inhalation therapy, and avoiding stress. It is essential to stop smoking and avoid respiratory infections (Eliopoulos, 2018).

Pulmonary Tuberculosis

Tuberculosis (TB) is an infectious disease transmitted by inhalation of the microorganisms causing TB that may attack any organ of the body but usually develops in one or both lungs. Twenty-five percent of all those identified with TB are aged 65 or older. In the United States, people at greatest risk for TB are those in frequent contact with an untreated person; those with decreased immune function or HIV; those who live in crowded areas such as long-term care facilities (the incidence rate among nursing home residents is two to seven times higher than in those living in other settings), mental health facilities, or prisons; older homeless people; those who abuse injection drugs or alcohol; lower socioeconomic groups; and some foreign immigrants (Workman, 2013b). Often TB in older adults is not a new infection but a reactivation of a long-dormant TB infection. Predisposing factors for reactivation include smoking, alcohol abuse, reduced immune system efficiency, long-term corticosteroid therapy, type 1 (insulin-dependent) diabetes, and malnutrition (Miller, 2009). As the immune system of those previously infected with tuberculosis declines with aging, inactive TB often becomes an active disease.

The older person with TB may merely feel tired without exhibiting a cough or other respiratory signs. However, other common signs of the disease are weight loss, malaise, night sweats, low-grade fever that lasts for months, cough, and depression. The sputum is often green or yellow and may be blood streaked. Shortness of breath and dull chest pain may also be present (Marieb & Hoehn, 2019). Symptoms develop slowly in older adults, and the disease is often misdiagnosed and far advanced before it is identified. Skin testing is the most common diagnostic tool, followed by other procedures such as chest radiographs, CT scans, and isolating the organism from sputum or tissue analysis. The usual skin test used to check for TB is not reliable to use with older adults since it tends to have false negatives due to immune system reduction. Other variations of it are available for possible use (Sanders, 2019). As for treatment, special care should be taken to identify and treat tuberculosis in nursing home settings before the disease spreads to other residents. Various drug regimens, especially antibiotics, infection control, and patient education, are recommended to treat tuberculosis.

Pneumonia

Pneumonia is an inflammation of the lungs caused primarily by bacteria, viruses, or chemical irritants. Bacterial pneumonia is most common in older adults and leads all the infectious diseases causing death in this age group. Those most susceptible have chronic diseases such as COPD, cardiovascular disease, diabetes mellitus, and alcoholism. Most bacterial pneumonias develop from breathing in disease-causing bacteria that have survived for some time in the oropharynx.

Rather than the typical symptoms of difficulty breathing, chest pain, fever, chills, and productive cough, older adults with pneumonia often have weakness, confusion, lethargy, poor appetite, rapid breathing, mild fever, and lower heart rate (Miller, 2009; Workman, 2013b). The most common type of nosocomial pneumonia (pneumonia acquired in a hospital) in older adults is streptococcal pneumonia, with a 20% to 30% death rate. Individuals older than 60 have a two to three times higher risk of developing nosocomial pneumonia, especially if they have had surgery, diagnostic procedures, mechanical ventilation therapy, or tube feedings. Viral influenza pneumonia, with a fatality rate of about 50%, is common in older adults as well. Pneumonia can also be caused by the aspiration of fluid or food particles into the lungs, and older adults are more susceptible because of less efficient gag and cough reflexes. Pneumonia typically is diagnosed through x-ray examinations and sputum specimen studies. Abnormal lung sounds may also be present. Treatment includes the use of antibiotics, respiratory therapy, and rest. Vaccinations for pneumonia are recommended for everyone older than age 65, especially if the individual has an accompanying chronic illness. In addition, yearly influenza vaccinations are highly recommended for older adults.

Lung Cancer

Lung cancer is the primary cause of cancer death for both men and women in North America. Today in the United States more people die of lung cancer than any other cancer (Marieb & Hoehn, 2019). About 20% of those diagnosed with lung cancer are older than age 70. Aroud 85% to 90% of all lung cancers are correlated with smoking, but other causative factors include asbestos exposure, occupational hazards, dust, and chronic lung damage.

Early symptoms include cough, difficulty breathing, blood in the sputum, weight loss, and frequent lung infections. More bothersome symptoms often occur later, when the possibility of cure is drastically reduced. Too often lung cancer remains undetected until it has spread and hope for cure is minimal; thus the cure rate for this type of cancer is low, with most people dying within 1 year of diagnosis (Marieb & Hoehn, 2019). Removal of the affected lung is the most effective treatment, but often the cancer has spread beyond the lungs before it is diagnosed. Chemotherapy and radiation, then, are the other available treatment choices, and they have only moderate success rates.

PREVENTION OF RESPIRATORY DISEASE

The first line of defense in preventing respiratory disease is to maintain good health by drinking 1.5 to 2 quarts of fluid each day and eating a well-balanced diet. Annual physical examinations can help to detect health problems early and thus initiate treatment in time to be maximally effective. It is imperative that individuals stop smoking because smoking (and being exposed to secondhand smoke) is a major cause of lung diseases. Smoking-cessation programs and the use of the nicotine patch, pill, or newer medications are quite effective but do require motivation and social support. Avoiding respiratory infections is especially important for those with respiratory disease. The pneumonia vaccine is recommended to help prevent individuals from contracting various types of pneumonia. Because influenza is often accompanied by bacterial pneumonia, a yearly influenza vaccination is also advisable. A regular exercise regimen will help maintain and improve lung functioning and is extremely important in older age.

SUMMARY

Even though age-related changes do occur in the respiratory system, they usually do not handicap older adults significantly in the performance of normal daily activities unless disease, strain, or stress is imposed on the system. As is true in so many of the body systems, most people have enough reserve capacity to be able to tolerate some degree of reduced organ efficiency without producing substantial or even noticeable limitations.

Those who have a respiratory disease, however, may be weak and short of breath, resulting in fear, anxiety, and an inability to participate in activities of daily living. Because chronic lung conditions are often increasingly disabling as they progress, the need for continued health monitoring and for caregivers places considerable stress on the family and finances. The American Lung Association offers substantial assistance with physical needs and also provides individual and group therapy. For most people, adequate respiratory efficiency may be retained well into older age by regular systematic exercise and general physical fitness. Stress situations accentuate the reduced efficiency of the aging body, whereas exercise and physical fitness at least partially offset the effects of aging.

REFERENCES

American Lung Association. (2021). *Learn about COPD.* https://www.lung.org/lung-health-diseases/lung-disease-lookup/copd/learn-about-copd

Davies, G., & Bolton, C. (2010). Age-related changes in the respiratory system. In H. Fillit, K. Rockwood, & K. Woodhouse (Eds.), *Brocklehurst's textbook of geriatric medicine and gerontology* (7th ed., pp. 97–100). Saunders Elsevier.

Eliopoulos, C. (2018). *Gerontological nursing* (9th ed.). Wolters Kluwer.

Hendrix, T. (2011). Respiratory function. In S. Meiner (Ed.), *Gerontologic nursing* (4th ed., pp. 432–461). Mosby Elsevier.

Jett, K. (2012). Physiological changes. In T. A. Touhy & K. Jett (Eds.), *Ebersole & Hess' toward healthy aging* (8th ed., pp. 44–61). Mosby Elsevier.

Marieb, E., & Hoehn, K. (2019). *Human anatomy and physiology* (11th ed.). Pearson.

Miller, C. (2009). *Nursing for wellness in older adults* (5th ed.). Wolters Kluwer/Lippincott Williams & Wilkins.

Sanders, D. L. (2019). Respiratory function in aging. In S. L Meiner & J. J. Yeager (Eds.), *Gerontologic nursing* (6th ed., pp. 364–394). Elsevier.

Tabloski, P. (2014). *Gerontological nursing* (3rd ed.). Pearson Education.

Workman, L. (2013a). Care of patients with noninfectious lower respiratory problems. In D. Ignatavicius & L. Workman (Eds.), *Medical-surgical nursing* (7th ed., pp. 600–639). Saunders Elsevier.

Workman, L. (2013b). Care of patients with infectious respiratory problems. In D. Ignatavicius & L. Workman (Eds.), *Medical-surgical nursing* (7th ed., pp. 640–661). Saunders Elsevier.

10

The Gastrointestinal System

INTRODUCTION

The gastrointestinal (GI) system takes in food, transforms it into nutrient molecules, absorbs these into the blood, and excretes whatever is indigestible. Many GI complaints in the ambulatory older population are not caused by organic disease but are functionally based. It should be kept in mind, however, that serious disruptions in this system can and do occur. Hiatal hernia, gallstones, diverticulosis, and cancer of the colon are especially prevalent in older adults. Thus, any complaint involving the GI system should be thoroughly investigated.

The status of the GI system has significant influence on an individual's nutritional state because various disease conditions influence the absorption of nutrients and the ability to ingest a well-balanced diet. Furthermore, age-related changes in other body systems, the presence of diseases and their treatment, plus various psychosocial issues affect the ability of older adults to purchase and prepare food necessary to maintain adequate health. Whether or not individuals ingest the necessary carbohydrates, fats, proteins, vitamins, minerals, and water, and the manner in which they are digested once they are eaten, are also important variables. Because eating is a major source of satisfaction for older adults, we must be alert to specific age-related changes in the GI system and to major pathological conditions and medical treatments that may prevent or reduce the enjoyment of eating.

Symptoms of GI disturbances are often not those typically present in younger age groups. For example, older persons seem to complain less of pain when they have appendicitis or a peptic ulcer. GI bleeding might go unnoticed for days; even irritation and early perforation may not be as evident as in the young. Fever and an increase in white blood cell count are not always present when there is an infection. Furthermore, symptoms of diseases in other organ systems such as the cardiovascular and neurological systems are often displayed as GI connected.

COMPONENTS AND FUNCTIONS OF THE GASTROINTESTINAL SYSTEM

The components of the GI system are organized into two divisions:

1. The alimentary canal (the GI tract), a coiled, hollow, muscular tube with an opening at each end, digests food and absorbs digested particles into the blood. It consists of (a) mouth, (b) pharynx, (c) esophagus, (d) stomach, (e) small intestine (duodenum, jejunum, ileum), and (f) large intestine (cecum, appendix, colon, rectum, anal canal).

2. Accessory digestive organs and digestive glands consisting of (a) organs (teeth, tongue, gallbladder), (b) glands (liver, pancreas), and (c) salivary glands (parotid, submandibular, tubuloalveolar). These glands produce various secretions contributing to digestion (see Figure 10.1).

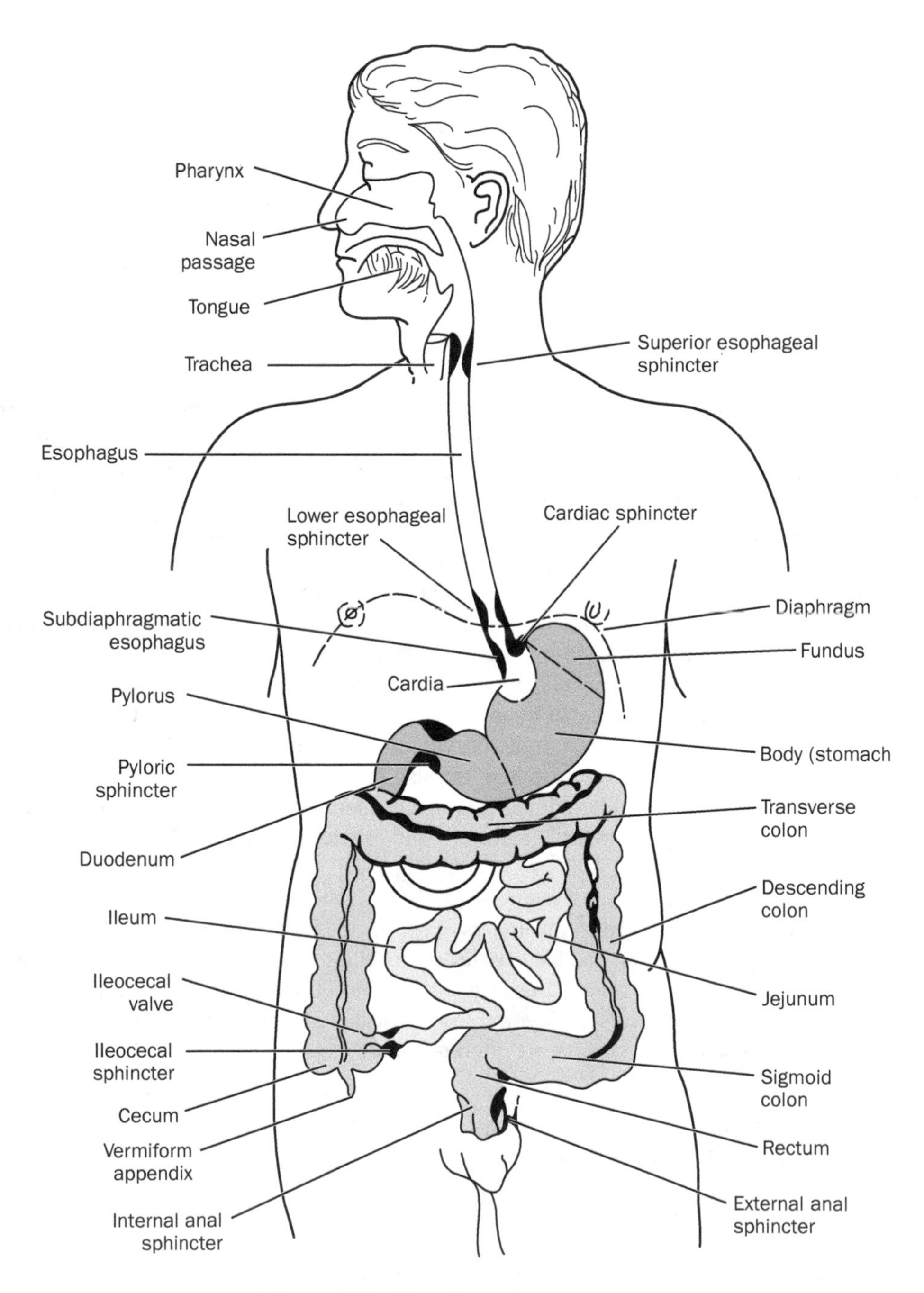

Figure 10.1 The digestive system.

Digestion

Food enters the mouth (ingestion) and is propelled through the GI tract by swallowing and by involuntary peristalsis (contraction and relaxation) of the muscles in the walls of the esophagus, stomach, and small and large intestines. During this passage, food material is also acted on by mechanical and chemical digestion. *Mechanical digestion (breakdown of food material)* is accomplished by chewing movements in the mouth, churning movements in the stomach, and rhythmic contractions and relaxation in segments of the small intestine. *Chemical digestion* is a process in which enzymes secreted by the various digestive glands break down large food molecules into substances more readily absorbed through the lining of the GI tract. Carbohydrates are broken down into galactose, glucose, and fructose; fats into fatty acids and glycerol; and proteins into amino acids. These products are transported from the GI tract (mostly through the walls of the small intestine) to the blood and lymph systems. The large intestine then propels the undigestible material toward the anus, where it is expelled as feces (Marieb & Hoehn, 2019).

The Mouth

When food enters the mouth, saliva is secreted by the salivary glands. The secretion of saliva is primarily a reflexive activity, although it is conditioned to some extent by learned experiences. For example, think of some food you like and your mouth will probably "water"; the thought alone is sufficient to trigger a salivation response. Each day 2 or 3 pints of saliva are produced. The functions of saliva are (a) to moisten and lubricate mouth surfaces and thus aid in both speech and swallowing; (b) to partially dissolve food, which can then better stimulate the taste buds; (c) to lubricate food so it can be swallowed more easily; (d) to initiate the digestive process through the direct action of salivary enzymes; and (e) to act as a cleanser for the mouth cavity and teeth.

Thirty-two permanent teeth are located on the upper and lower jaw, and the tongue is attached to the floor of the mouth. When the jaws open and close, the tongue moves the food about, mixing it with saliva while the teeth chew it into smaller pieces to facilitate swallowing.

The Pharynx and Esophagus

During the act of swallowing, food of a suitable size is pushed back toward the throat by the tongue into the pharynx, the common passageway for both food and air. Once food reaches the pharynx, the act of swallowing becomes involuntary and is no longer under voluntary control. The swallowing center in the medulla and pons of the brainstem takes over swallowing action at this point. The pharynx contracts when food enters it, forcing food substances into the esophagus. The position of the tongue during the first stage of swallowing prevents food from returning to the mouth. Food does not normally pass into the nasal cavity because the soft palate moves up to block off the cavity, nor does food enter the larynx (the respiratory passage) because the muscles of the larynx seal off the laryngeal opening. At this time the vocal cords draw tightly together, and during the act of swallowing respiration is inhibited. Thus, food has but one place to go into the esophagus and then into the stomach.

The rhythmic alternating muscular contractions and relaxations of peristalsis literally push food down the esophagus. Most food normally passes from the mouth to the stomach in 6 to 7 seconds. At the junction of the esophagus and stomach is a circular muscle called the cardiac or gastroesophageal sphincter that opens, allowing food to enter the stomach, and then closes to prevent regurgitation.

The Stomach

Located in the upper left side of the abdominal cavity, the stomach temporarily stores food while it is further broken down by chemical and mechanical action. Proteins are particularly acted on in the stomach. In its empty state, the stomach resembles a deflated balloon with longitudinal rugae (folds or creases). Its concave inner surface is called the lesser curvature, and the convex outer surface is called the greater curvature. Fortunately for those who enjoy eating, the stomach can expand (within limits) depending on the amount of food delivered to it.

About 30 to 40 million gastric glands are located in the stomach. These produce gastric secretions such as pepsin, a protein-digesting enzyme; lipase, a fat-digesting enzyme; hydrochloric acid, which causes the stomach to be very acidic, thus killing bacteria in the food; intrinsic factor, necessary for vitamin B_{12} absorption in the small intestine; and mucin, which produces an alkaline mucus that protects the walls of the stomach from the stomach's highly caustic gastric juices.

Food in the stomach is continually pushed by peristaltic movements that churn, pulverize, and thoroughly mix it with gastric juices, breaking it down into a mush-like liquid called chyme. At the lower end of the stomach is the pyloric sphincter, a circular muscle separating the stomach from the small intestine and functioning as an effective strainer. It allows liquids to pass through first, carbohydrates next, then proteins, and lastly fats (the hardest and slowest to digest). The emptying of the stomach's contents into the duodenum, the first section of the small intestine, is a gradual process usually completed in about 3 to 5 hours. Emotional states such as excitement, fear, anger, or grief can inhibit gastric motility and alter glandular secretions.

The Small Intestine, Liver, Gallbladder, and Pancreas

Considered to be the major digestive organ of the body, the small intestine is a coiled tube connecting the stomach to the large intestine. It is the longest part of the alimentary system. Here the majority of absorption occurs and digestion is completed. It includes three sections: the duodenum, the jejunum, and the ileum. On the inner surface of the small intestine are fingerlike projections called villi, where digested food is absorbed into the blood and lymph systems. Food moves through the small intestine by peristaltic and segmentation action (contractions of segments of the small intestine), while rhythmic movements churn and mix it with bile, pancreatic, and intestinal juices to facilitate the absorption of carbohydrates, proteins, fats, vitamins, minerals, and water.

Glandular digestive secretions come from the liver, gallbladder, and pancreas. The liver, the largest gland in the body, is located primarily in the right upper section of the abdomen and contains four lobes. Arterial blood from the aorta enters the liver through the hepatic portal vein and empties into the inferior vena cava.

The gallbladder is located in a depression on the undersurface of the liver. Bile, a fat emulsifier, is produced by the hepatocytes (liver cells) in the liver and is concentrated and stored in the gallbladder. Bile leaves the liver through the hepatic duct, which fuses with the cystic duct from the gallbladder to form the common bile duct entering the duodenum. Hepatocytes also store the fat-soluble vitamins A, D, E, and K; pick up nutrients from the blood; store glucose as glycogen; make plasma proteins from amino acids; and break down heme (iron-containing pigment) from hemoglobin to create bilirubin. They also play an important role in detoxifying substances such as drugs and alcohol.

The pancreas extends across the upper abdomen behind the stomach and is the largest enzyme producer in the body. It functions in two ways: through exocrine secretions that are carried through ducts and through endocrine secretions that empty hormones directly into the bloodstream. *Exocrine cells* secrete pancreatic juice-containing enzymes such as trypsin, amylase (which breaks down starches), and lipases (protein- and fat-digesting enzymes) via the pancreatic duct into the common bile duct, which enters the duodenum. *Endocrine glands* produce insulin and glucagon, which regulate the metabolism of carbohydrates in the islets of Langerhans of the pancreas.

The Large Intestine

Anatomically, the large intestine is about 5 feet long, is not arranged in folds as is the small intestine, and does not have villi on its interior surface. The first part of the large intestine, the cecum, is a pouch from which projects a narrow tube, the appendix, on the right side of the body. Rupture of the appendix is dangerous because body waste material is then expelled into the body cavity, resulting in peritonitis (inflammation of the abdominal cavity). The colon part of the large intestine includes ascending, transverse, descending, and sigmoid sections. The sigmoid colon joins the rectum, which merges with the anal canal and terminates in the anus. The anal canal has an internal involuntary anal sphincter and an external anal sphincter, which is under voluntary control.

Food residue enters the large intestine from the small intestine through a valve (ileocecal valve), which prevents backflow. In the large intestine, residue is subjected to strong muscle action that carries the remaining undigestible substances (feces) to the lower part of the large intestine (the colon), where periodically it passes into the rectum to be expelled through the anus. Defecation is a reflexive act initiated by the accumulation of feces in the rectum. Because voluntary control of the anal sphincter muscle is absolutely required for social acceptability, one of the most devastating assaults on self-image is to experience partial or total loss of bowel control.

Defecation habits vary greatly among individuals, and it is of practical importance in gerontological education to recognize that substantial variation is both common and normal. Relying on a laxative every day may do more harm than good because this easily fosters medicinal dependency. Dependence on laxatives leads to sluggishness of the intestinal musculature and a need for continued artificial stimulation. Such unnecessary dependence may set the stage for the development of serious GI problems. Exercise, proper diet, and reduced stress promote regularity in normal GI systems at any age. These factors are especially important in older adults.

AGE-RELATED CHANGES IN THE GASTROINTESTINAL SYSTEM

Although age-related changes have been identified in the various components of the GI system, they evidently exert relatively little effect on overall functioning of the GI tract. The process of digestion slows with age and may become somewhat less efficient, but marked changes are not common and the system usually remains adequate to meet most reasonable demands imposed on it. In spite of this resilience, many older adults' complaints center around various GI problems. Over the years, folklore and misconceptions about digestive functioning and age have played a considerable part in encouraging older adults to attribute various signs and symptoms of digestive malfunctions to age alone. It is much more likely that lifestyle factors such as poor dietary choices, lack of adequate fluid intake, lack of bulk in the diet, excessive straining when defecating, and lack of exercise are more responsible for many of the common GI complaints than age.

Specific age-related changes in the GI system are difficult to identify, but tend to include the following:

Mouth

Tooth enamel and dentin gradually wear down, making teeth more susceptible to cavities. Older teeth often appear darker or stained. In recent years the number of older adults without teeth has declined in the United States, probably because of better preventive dental care, although geographic and socioeconomic factors are important determinants of adequate dental care. Still, about 25% of adults 65 and older are without teeth (edentulous) (Tabloski, 2014). Also, many older adults have dentures that often create discomfort. There is a decrease in saliva production, although the commonly reported dry mouth is more likely caused by nutritional disturbances, medication side effects, and pathological conditions rather than normal aging. Neuromuscular changes can affect chewing and swallowing and both tend to slow with age. Taste buds decrease in number with age, possibly affecting taste acuity and enjoyment of food.

Throughout life the mouth is exposed to continual trauma from substances ingested, such as alcohol, drugs, and nicotine, which may eventually accentuate age-related changes or produce disease. For these reasons it is often difficult to differentiate between pathological conditions and normal age-related changes.

Esophagus

Weaker smooth muscle of the esophagus and lessened sphincter motility (presbyesophagus) may cause delayed emptying of food into the stomach. If it does occur, it increases the risk of gastroesophageal reflux disease (GERD), which is quite common in older people. These changes may cause the older adult to have a feeling of fullness, difficulty in swallowing, pain beneath the sternum, and regurgitation of stomach contents back into the esophagus. Nevertheless, barring the effect of disease, only minimal changes in the esophagus have been found to be associated with age.

Stomach

There appears to be some slowing of gastric emptying with age. Decreased hydrochloric acid has long been assumed to be an age-related change, but research now suggests infection (especially with *Helicobacter pylori*) is also a likely cause of decreased hydrochloric acid. The stomach wall also becomes less elastic, which affects gastric emptying.

Small Intestine

With age a slight decrease in most digestive enzyme secretions is seen, which may affect the absorption of some nutrients. Also, decreased muscle tone may result in slower peristalsis, especially in those who are sedentary.

Large Intestine

Both decreased anal sphincter tone and muscle tone can delay movement and expulsion of material from the large intestine, possibly contributing to constipation. Diverticuli or weak spots in the wall of the colon also occur in many people over age 60.

Liver, Gallbladder, and Pancreas

The liver decreases in size and weight but usually continues to function normally. Blood flow in the liver declines somewhat with age. Drug metabolism by the liver may become less efficient, especially in drug overload. With age, gallstones may tend to increase in the gallbladder. Fewer pancreatic enzymes are produced, but there seems to be no appreciable decline in fat, carbohydrate, and protein digestion. Thus, although activities of these structures tend to slow with age, age-related effects appear to be relatively minimal.

For the most part, aging has little significant effect on GI functioning, although age-related changes may make older adults more susceptible to functional changes and GI disease. GI complaints should always be evaluated for possible pathological conditions rather than the effects of aging.

AGE-RELATED DISORDERS OF THE GASTROINTESTINAL SYSTEM

Attention to teeth and mouth is often neglected in the latter years of life and in many nursing homes it is even considered a low priority for care (Dirks, 2016). Periodontitis impacts 60% of older persons and in 10% to 15% it is considered to be severe (Chapple & Genco, 2013). The health of the teeth, mouth, and perioral tissue greatly influences the intake of foods so needed to maintain health and sustain life. Lesions and inflammation in the mouth, around the teeth, and oral mucosa impede the eating of certain foods such as those that are fibrous, acidic, or with seeds and husks, causing them to be painful or unpleasant to eat. Diabetics have a higher incidence of periodontitis, and certain medications such as antibiotics, pain and antianxiety medications may contribute to perioral disease. Cognitive deficits and physical limitations may prevent many individuals from receiving necessary mouth and teeth care. Advanced periodontal procedures, oral surgery, and dental implants, so common today, pose a greater risk for older adults with chronic diseases (Chernoff, 2016; Saunders, 2016).

The importance of preventive oral care cannot be overestimated. If the mouth and teeth are not maintained and treated, systemic diseases along with comorbid illnesses that are chronic may result, resulting in increased expenditures that are difficult to manage. Unless teeth with dental restoration are maintained they will likely result in the loss of teeth, oral infections, and pain, all of which may influence the intake of adequate nutrition, result in weight loss, influence the health of major organs, and even cause cancer or aspiration pneumonia (Dirks, 2016; Saunders, 2016).

Age-Related Disorders of the Mouth

Multiple mouth problems may occur in older adults. These can develop from age-related changes in the mouth, poor nutrition, inadequate dental hygiene, decayed teeth, and bacterial invasion of the tissues. Various health disorders caused by endocrine, cardiovascular, and mental impairments also contribute to mouth problems. A number of medications, such as chemotherapy drugs or drugs prescribed for seizures, often irritate mouth tissues. Taste sensations can be diminished by loss of taste buds, mouth or gum inflammation, smoking, and the ingestion of alcohol or other toxic substances.

Xerostomia (Dry Mouth)

Some older persons suffer from dry mouth resulting from a variety of causes, such as decreased salivary secretions, dehydration, diabetes, hormonal or vitamin deficiencies, anemia, and radiation therapy. Many drugs cause xerostomia, primarily the antihistamines, antipsychotics, antidepressants, antianxiety drugs, antihypertensives, anticholinergics, decongestants, and diuretics (Tabloski, 2014). Alcohol and tobacco use add to mouth problems. To compensate for dry mouth, older adults sometimes "dunk" their food in liquids to promote ease of chewing. Synthetic saliva or oral swab sticks help rehydrate the mouth. A careful review of an individual's drug regimen should be undertaken with a view to discontinuing drugs causing dry mouth if they are not absolutely needed. Increasing fluid intake, chewing gum, or using sugar-free mints may also be helpful.

Dysphagia (Difficulty Swallowing)

Dysphagia results from a variety of causes, such as weakened muscles of the esophagus, less efficient sphincter functioning, loss of teeth, poorly fitting dentures, atrophy of chewing muscles, and decreased salivation. A stroke or accident may suddenly impair swallowing, whereas diseases such as Parkinson's, myasthenia gravis, and multiple sclerosis or pharyngeal tumors gradually interfere with swallowing (Eliopoulos, 2018).

Swallowing usually takes about 20 seconds. A normal swallow has three stages: the oral stage, in which food is mixed with saliva to form a bolus ready to be swallowed; the pharyngeal stage, when the bolus is propelled into the esophagus; and the esophageal stage, in which the bolus is sent to the stomach by peristaltic activity (Tabloski, 2014). Impairment in swallowing is best diagnosed by a multidisciplinary team of therapists, dietitians, physicians, and nurses. Assessment involves observing the swallowing reflex, the pocketing of food in the mouth, choking or coughing, spitting or leaking food from the mouth, and regurgitating food through the mouth or nose. A swallowing video fluoroscopy or modified barium swallow helps to analyze the stages of swallowing and diagnose the difficulty more accurately.

Efforts to promote adequate swallowing and food intake include reducing distractions and increasing the time allowed for swallowing. Some individuals need to swallow twice to empty the pharyngeal tract. Those with difficulty swallowing should sit up straight, have the head flexed forward when they swallow, and remain upright for at least 15 minutes after eating. After a stroke involving the mouth and throat, it helps to place the food on the unaffected side of the mouth with the spoon placed on the tongue and rocked back and forth. Do not touch the teeth or place the spoon too far down the mouth. At times it may be helpful to hold the mouth shut during swallowing or retrieve food from the cheek if it is stored there. If liquids are not readily swallowed, they may need to be thickened. Never wash down food with liquids. Teaching the individual and family methods to facilitate swallowing is very important for recovery of function, for safety, and to reduce fear. Speech therapists are professionals trained to rehabilitate individuals with swallowing disorders.

Dental Caries

Dental caries involves progressive loss of tooth surfaces linked with bacterial plaque. There are two types of dental caries: (a) coronal, involving the tooth enamel and (b) root caries, located around the tooth root. The latter is more prevalent in older adults. Some factors considered as causes of caries formation in the general population are smoking, increased carbohydrate in the diet, susceptible teeth, plaque formation, and bacteria in the mouth. Older adults are at higher risk for caries because of decreased saliva production, which reduces adequate cleaning of the teeth and gums; recessed gums that expose root surfaces; chronic disease; dry mouth; and lack of effective mouth and tooth care. Prevention of caries depends on daily brushing and flossing, as well as at least a biyearly dental assessment and professional teeth cleaning. Other preventive measures include topical applications of fluoride, chewing fibrous foods, and reducing carbohydrate intake, especially soft, sticky, sugary snacks.

Periodontal Disease

Bacteria in gum crevices surrounding the roots of the teeth cause gingivitis (inflammation of the gums) and, if not treated, periodontitis with bone involvement. Risk factors include smoking, diabetes, nutrition, some medications, stress, and illnesses. Calcified deposits around teeth also contribute to periodontal disease. Because periodontitis is asymptomatic until it has progressed significantly, it is considered a leading cause of tooth extractions in middle and late adulthood. It also contributes to systemic infections in older adults (Eliopoulos, 2018). Signs of periodontal disease are bleeding gums, bad breath, permanent teeth that are loose, any change in the bite, and any change in the way partial dentures fit.

Treatment programs include regular dental cleaning and mouth assessments every 6 months, plaque removal, brushing, flossing, gum surgery, and bone grafts if needed. Older disabled individuals may benefit by using assistive devices that help them hold a toothbrush and dental floss and by learning how to do good preventive care themselves. Healthcare providers are responsible for thousands of older adults in short- and long-term healthcare settings, so concerted efforts are necessary to teach nurses and nursing assistants effective mouth care techniques especially for those individuals unable to care for themselves. An even greater challenge involves continual monitoring of mouth care provided to older residents of nursing homes to ensure it is adequate.

Oral Cancer

Oral cancer appears in many forms in older adults. Lumps, swelling, a sore throat that does not heal, white scaly patches (leukoplakia), persistent pain, numbness, and bleeding of mucous membranes of the tongue and mouth are all warning signs. These symptoms may be mostly prevented by an adequate diet, appropriate vitamin intake, and regular observation of mouth and tongue tissues, especially if dentures or jagged teeth cause continual irritation or if the individual is a smoker or drinks alcohol, two major risk factors for oral cancers.

Leukoplakia can be a precursor to cancer, and any white scaly patch or reddened area in the mouth should be assessed by a dentist or primary care practitioner for diagnosis, biopsy, and treatment. Many cancerous lesions are squamous cell cancers and need to be identified early. Treatment depends on type of lesion, size, and location. Surgery, radiation, and chemotherapy are the most common treatment options (Tabloski, 2014). Regular dental checkups and teaching older adults and caregivers how to inspect the mouth and tongue tissue are imperative for prevention or early detection.

Age-Related Disorders of the Esophagus

Chronic complaints of discomfort by older adults, such as heartburn, substernal pain, belching, and general discomfort in the region above the stomach, often relate to the esophagus. Often esophageal pain is thought to be cardiac pain. Whether it is esophageal reflux (backflow), spasm of the sphincter, or esophagitis (inflammation of the esophagus) caused by acid reflux, a diagnosis of the complaint is essential. Whereas anginal pain (caused by spasms of the coronary arteries that supply blood to the heart muscle) presents as pressure-like pain on exertion with accompanying blood pressure or pulse changes, esophageal pain is associated with eating; lying down; stooping; or drinking tea, coffee, or acidic juices. Obesity, smoking, and overeating contribute to esophageal reflux. Anginal pain usually responds to nitroglycerin, whereas esophageal pain is relieved by medications such as antacids, sitting upright, or taking antispasmodic drugs (Tabloski, 2014). "Pill esophagitis" is esophageal injury caused by pills that do not pass to the stomach but remain in the esophagus. To prevent this, medications should be taken sitting upright and with at least 8 ounces of water.

Cancer of the Esophagus

The incidence of cancer of the esophagus has decreased somewhat, but most people who develop esophageal cancer are older. Ninety-five percent of esophageal cancers are squamous cell cancers, which are caused by poor oral hygiene, alcohol abuse, or smoking. Initially, individuals with this cancer may have difficulty swallowing solids; later, the difficulty extends to liquids. Diagnosis is confirmed by viewing the esophagus with a scope, CT and MRI scans, barium swallow, and biopsy. Sections of the esophagus may be surgically removed, followed by chemotherapy or radiation therapy. Laser therapy and photodynamic therapy are also possible treatment options. Often, however, diagnosis is made too late for a positive prognosis.

Gastroesophageal Reflux Disease

GERD occurs in individuals who have had a long history of reflux esophagitis, especially those older than age 65 (Tabloski, 2014). It is the most common upper GI problem in adults. In GERD, stomach contents flow up into the esophagus and irritate it. Symptoms may include mild or more severe heartburn, indigestion, severe chest pain, regurgitation, and dysphagia. GERD may be exacerbated by medications; certain foods or liquids such as coffee, alcohol, citrus, and spicy foods; eating meals high in fat; and eating 2 to 3 hours before bedtime. In addition, smoking, obesity, and anxiety also contribute to GERD. Sometimes cancer in the form of Barrett's esophagus occurs in those with GERD when symptoms are not well controlled.

Diagnosis is made by taking a personal history, barium studies, upper gastroscopy, other diagnostic procedures, and biopsy. Treatment involves lifestyle modifications and drug therapy. Elevating the head of the bed; reducing the size of the evening meal; avoiding caffeine, alcohol, fat, and chocolate; and reducing weight are all appropriate. Antacids, histamine receptor antagonists (also called histamine blockers), and proton pump inhibitors (PPIs) are useful in controlling symptoms. Surgery is available if other treatments do not reverse esophageal erosion (Tabloski, 2014).

Hiatal Hernia

In hiatal hernia, a portion of the stomach slides or rolls up through the opening where the esophagus passes through the diaphragm. Hiatal hernias are classified as sliding, in which a part of the stomach slides into the chest cavity, especially when lying down, or rolling or paraesophageal, in which part of the stomach rolls up through the diaphragm, forming a pocket alongside the esophagus. Probable causes of hiatal hernia include muscle weakness around the diaphragmatic opening, kyphosis (hunchback), scoliosis (lateral curvature of the spine), and straining during bowel movements. Other major risk factors are obesity, smoking, and age. Hiatal hernia may be present for years but go unnoticed until symptoms such as heartburn, regurgitation, belching, difficulty swallowing, indigestion, or chest pain occur especially when lying down. The presence of pain may be misdiagnosed as a heart attack. Symptoms should always be assessed by a primary care practitioner. Diagnosis usually involves a barium swallow and esophagoscopy (Eliopoulos, 2018). Individuals with hiatal hernia are advised to eat three small meals each day, refrain from eating 3 to 4 hours before going to bed, elevate the head of the bed 3 to 6 inches, maintain weight within a normal range, avoid straining and stooping, take antacids 1 to 3 hours after meals and before bedtime, and avoid alcohol, chocolate, fats, peppermint, and smoking. In some cases surgery is required to repair the hernia.

Age-Related Disorders of the Stomach

Gastritis (Inflammation of the Stomach)

Gastritis is an inflammation of the stomach mucosa. It may be acute or chronic, diffuse or localized. Acute gastritis is often caused by the ingestion of aspirin, nonsteroidal anti-inflammatory (NSAID) drugs, alcohol intake, smoking, and highly stressful situations.

Symptoms include belching, abdominal pain, nausea and vomiting, heartburn after meals, and a poor appetite. For some individuals chronic gastritis presents no symptoms (Yeager, 2019).

Diagnosis generally includes endoscopy studies with possible biopsy, stool samples, and serum antibody tests. Treatment for acute gastritis involves finding the specific cause and eliminating it or avoiding it. Drugs provide symptom relief and reduce the irritation of the gastric mucosa. Treatment for chronic gastritis focuses not only on finding and eliminating the cause but also on appropriate lifestyle modifications and adherence to strict drug regimens. Gastritis often precedes gastric ulcers.

Gastric (Peptic) Ulcer

A gastric ulcer is an erosion of the mucous membrane of the lower esophagus, stomach, pylorus (the lower portion of the stomach opening into the duodenum), or duodenum. This type of ulcer is common in older people, and they are more likely to have complications than younger adults. It often begins as an acute condition and gradually becomes chronic over the years. Its symptoms are usually milder and less specific than in those who are younger, so diagnosis is more difficult. However, morbidity and mortality rates are higher for older adults than for the general population. *Helicobacter pylori* infections are found in approximately 90% to 95% of those with duodenal ulcers.

Typical symptoms are pain in the upper abdomen, weight loss, nausea, vomiting, and thirst, although many older adults do not report these symptoms until major complications occur. Older adults showing unusual weight loss, general debility, anemia, or any abdominal distress should be suspected as possibly having a gastric ulcer. A review of medications is advised because many medications such as NSAIDs and anticoagulant drugs often cause stomach irritation. Smoking and chronic alcohol abuse are also significant factors. Bleeding is the most common complication and accounts for one-half to two-thirds of all fatalities related to ulcers. Perforation is another major complication of gastric ulcer disease and involves spilling gastric or duodenal contents into the peritoneal cavity. A third major complication is gastric outlet obstruction, in which emptying stomach contents is slowed and eventually becomes virtually impossible.

Diagnosis relies primarily on endoscopy, as well as serum antibody tests, gastric analysis, laboratory test for the presence of *H. pylori*, blood tests, and stool examination. Other diagnostic procedures may be added as needed. Treatment includes drugs such as histamine-2 (H_2) receptor antagonists or PPIs (the drugs of choice) to decrease gastric secretion, reduce spasms, or reduce hydrochloric acid. Antacids and antibiotics are also useful therapeutic regimens and can be effective in reducing symptoms and preventing reoccurrences. Individuals are encouraged to stop smoking; reduce stress; rest; eat three meals a day of well-tolerated foods; and avoid coffee (both caffeinated and decaffeinated), cola, tea, meat extracts, very hot or cold foods, and irritating drugs. Surgery may be necessary for those who do not respond to medical and lifestyle management.

Cancer of the Stomach

Stomach cancer is common in individuals who have little or no hydrochloric acid in the stomach or who have chronic gastritis. Dietary factors, smoking, obesity, *H. pylori* infections, and gastritis are risk factors. Early symptoms are few or absent, but in time weight loss, anorexia (loss of appetite), abdominal pain, nausea, vomiting, and anemia occur. Viewing the stomach with a scope and obtaining a biopsy of tissue is the usual diagnostic procedure. Ultrasound, CT scanning, and upper GI barium studies help in more precise diagnosis. Blood studies and stool examination are also useful diagnostic tools. Surgery is the treatment of choice, but this cancer is often discovered only in the later stages, when metastasis (spreading) has already occurred. Chemotherapy or radiation therapy may then be used to control the disease or alleviate symptoms (Yeager, 2019).

Age-Related Disorders of the Small Intestine

Although the small intestine is less often diseased than other parts of the GI system, obstructions characterized as mechanical or paralytic may occur. Both antibiotics and surgery are used for treatment, depending on the specific cause. Ischemia (decrease of blood supply to the intestines) or even infarcts (occlusion of the blood vessels serving the intestine) may occur as a result of cardiac failure or thromboembolisms (blood clots). Other causes of obstruction include strangulation (a constriction or shutting off of the blood supply to the bowel), radiation injury, and local irritants. Cancer is rare in the small intestine but it can occur, even in the appendix. Surgery is usually the treatment of choice.

Age-Related Disorders of the Large Intestine

Appendicitis

Appendicitis is the most common inflammatory lesion of the bowel. It is not rare in older adults, but its classic signs and symptoms, such as fever and severe abdominal pain, are often absent. The individual becomes acutely ill and may even die if the cause is not identified quickly enough. Those who are obese, have diabetes, or whose appendix is retroperitoneal (behind the peritoneum) are most likely to be symptomless. Often appendicitis is not diagnosed until the attack begins to subside. Confirming diagnostic tests are ultrasound or CT scans. Surgery is not performed until the condition is stabilized. Antibiotics are often used to reduce inflammation and infection.

Diarrhea

Diarrhea is generally more serious in older adults than in younger persons because the homeostatic equilibrium of the elderly is more precarious. Rapid loss of fluid in older age can quickly lead to dehydration and electrolyte imbalance, both potentially life-threatening conditions. Factors contributing to diarrhea include fecal impaction, laxative abuse, intestinal infections, medications, diverticulitis, malignancy, and food or water impurities. It is not uncommon for older adults to have alternating bouts of constipation and diarrhea, although usually one will be more dominant. Any change in bowel habits, especially if persistent over a few weeks or accompanied by pain, fever, or weight loss, should be thoroughly investigated by a primary care practitioner.

Diagnosis involves a thorough history, laboratory tests, stool analysis, and a colonoscopy or radiological studies with barium contrast. Treatment involves adequate fluids and electrolytes such as sodium, chloride, potassium, calcium, bicarbonate, and magnesium. Antidiarrheal medications to slow peristalsis in the intestines are also commonly prescribed, but only for a short period.

Constipation

Constipation is defined as difficulty in passing hard, dry stools or as a decrease in the frequency of an individual's normal pattern of elimination. Normal patterns of bowel elimination vary considerably from one person to another. For instance, a normal pattern for one person might be one or more bowel movements a day, but another person may have three or fewer a week. Some older adults believe "regularity" means a bowel movement every day; if this does not happen, they believe they are constipated. Therefore, complaints about constipation must be evaluated carefully and always with reference to normal bowel activity for that particular person.

Factors contributing to constipation include (a) slowed intestinal motility with age, although generally this is not a highly significant factor for most older adults; (b) too little bulk in the diet and reduced fluid intake; (c) certain medications; (d) depression; (e) lack of exercise; and (f) cancer of the colon and a variety of other medical conditions. Some of these factors involve lifestyle choices and can therefore be modified quite easily.

Guidelines to help avoid constipation include the following:

1. Drink eight or more glasses of water a day.
2. Eat high-fiber foods regularly.
3. Avoid refined carbohydrates.
4. Exercise regularly.
5. Allow adequate time for bowel movements when the urge to defecate occurs.
6. Do not use laxatives for prolonged periods.
7. If constipation persists, get a complete medical evaluation.

In most cases constipation in older adults can be resolved with nonmedical approaches. Only a few require the use of selected medicines. Fiber-rich foods, increased fluid intake, and regular exercise are highly recommended. Other possibilities are synthetic bulk agents and some medications that also include stool softeners. Laxatives that are irritating should be avoided. To reiterate, the use of laxatives regularly is not recommended because they may injure the myenteric plexus and actually reduce the ability to relieve constipation.

Because there are many reasons for constipation, ranging from functional reasons to serious pathological conditions, any change in usual bowel habits over several weeks, especially if accompanied by weight loss, fever, or pain, should be thoroughly investigated by a qualified primary care practitioner.

Diverticulosis and Diverticulitis

Diverticulosis is the presence of pouches in the intestinal wall and is very common in older adults. It may occur in either the small or large intestine but typically in the large intestine. If inflammation occurs in a pouch, diverticulitis results. Complications of diverticulitis include possible perforation, abscess, and bleeding. Diverticular disease affects a large percent of the U.S. population by the age of 80. Weakened muscle mass, diets high in refined carbohydrates, and lack of dietary fiber are often assumed to be primary causes. Most individuals with diverticulosis experience no pain, but if diverticulitis develops, pain in the left lower quadrant, fever, and abdominal tenderness are common.

Diagnosis uses blood studies, urinalysis, radiographs, ultrasound, and CT scans as well as a thorough physical examination. A barium enema or colonoscopy may be desirable if the individual does not have acute diverticulitis since there is danger of possible perforation if the situation is acute. Treatment revolves around dietary modifications to include high-fiber intake, weight reduction, and a regular exercise regimen (but not exercises that increase intra-abdominal pressure). For those with complications, bowel resection or colostomy may be necessary.

Cancer of the Colon and Rectum

Colorectal cancer (cancer of the colon and rectum) is one of the most common malignancies and is a leading cause of cancer death in the United States. Incidence increases throughout life and almost doubles for each decade older than age 50. Some of these cancers develop from polyps (tumors on a stem) in the colon that grow slowly over time, and symptoms do not occur until the disease is quite advanced. However, other flat growths in the colon are now known to be more likely to be cancerous than polyps (Marieb & Hoehn, 2019). Risk factors include (a) age older than 50, (b) history of rectal and colon polyps, (c) personal and family history of rectal and colon polyps, (d) history of chronic inflammatory bowel disease, (e) diet high in fat and low in fiber, (f) lack of exercise, (g) smoking, and (h) drinking more than two alcoholic drinks a day.

The signs and symptoms of colon cancer include changes in bowel habits, rectal bleeding, weakness, and weight loss. Symptoms may not occur until after the cancer has been present for some time, and pain is often not experienced until late in the disease.

Diagnostic procedures generally involve a thorough history and physical examination, fecal occult blood test, sigmoidoscopy, colonoscopy, ultrasound, CT and MRI scans, and complete blood work. Treatment is dependent on the stage of cancer involvement. Malignant tumors may be removed with a colonoscope or laser, but when the tumor has invaded the walls of the intestine, a colon resection and possibly a colostomy (opening of the bowel through the abdomen) or lymph node removal may be necessary. Radiation and chemotherapy are also used, as is biologic or targeted therapy as deemed necessary.

The American Cancer Society (www.cancer.org) 2020 recommends the following screening guidelines.

For people of average risk:

- Initiate regular screening at age 45 using a stool-based test or a visual examination of the rectum and colon.
- Individuals in good health with a life expectancy greater than 10 years continue colorectal screening through age 75.
- In individuals ages 76 to 85, the decision for screening is based on personal preferences, overall health, life expectancy, and prior screening history.
- Individuals over age 85 should no longer receive colorectal cancer screening.

Individuals are considered to be average risk if they DO NOT have:

- A personal history of colorectal cancer or certain types of polyps
- A family history of colorectal cancer
- A personal history of inflammatory bowel disease
- A confirmed or suspected hereditary colorectal cancer syndrome
- A personal history of radiation to the abdomen (belly) or pelvic area to treat a prior cancer

Additional recommendations include: increasing fruit and vegetable intake, lowering dietary fat to 20% to 25% of total calories (mostly unhydrogenated types from plant origins or unsaturated fats), and getting the majority of carbohydrates from whole grains; limiting red meat; ensuring adequate intake of calcium, vitamin D, antioxidants, and folic acid; exercising at least 30 minutes a day; avoiding smoking; maintaining a healthy weight; and limiting alcohol use.

Hemorrhoids

Hemorrhoids are varicose veins in the anal canal. External hemorrhoids appear outside the external anal sphincter and internal hemorrhoids appear above the internal anal sphincter. Chronic constipation, straining during bowel movements, and prolonged sitting contribute to hemorrhoids. Symptoms include rectal itching, protrusion of the internal hemorrhoids into the anal canal, bleeding during bowel movements, and pain related to the external hemorrhoids. Hemorrhoids can be diagnosed by digital examination and sigmoidoscopy or colonoscopy. Treatment depends on size but generally involves a high-fiber diet and adequate fluids, along with stool softeners, sitz baths, and suppositories of hydrocortisone cream, topical anesthetics, or analgesics. If hemorrhoids continue to bleed, itch, or cause pain, nonsurgical procedures or surgical removal may be the treatments of choice. Nonsurgical methods include rubber band ligation or cryotherapy (rapid freezing of the hemorrhoid). Surgical procedures involve surgical excision of the hemorrhoid (hemorrhoidectomy).

Age-Related Disorders of the Pancreas

Cancer of the Pancreas

Cancer of the pancreas is a leading cause of cancer death in the United States. It is difficult to diagnose early and is usually first discovered in the late stages of development. In 90% of newly diagnosed individuals, the cancer has spread either locally or metastasized to other parts of the body often to the lungs or bone (Yeager, 2019). Known risk factors are age, diabetes, smoking, alcohol use, family history, high-fat diet, and exposure to certain chemicals. Initially symptoms may be vague, but eventually anorexia, weight loss, fatigue, chills or fever, possibly pain, and jaundice occur. Transabdominal ultrasound, CT, and MRI scans are commonly used in diagnosis. Tumor markers are also used, but these techniques generally only detect advanced stages of the disease. Chemotherapy or radiation may relieve pain and shrink the tumor somewhat but have minimal success in increasing survival time. Surgery is the most effective treatment, but prognosis for pancreatic cancer is poor because it is often not diagnosed early enough.

Age-Related Disorders of the Liver

Cirrhosis

Cirrhosis of the liver involves inflammation and degeneration of the liver. Its highest incidence is in men, and most who acquire it are between the ages of 40 and 60. Four types of cirrhosis have been identified: (a) alcoholic, caused by alcohol abuse; (b) postnecrotic, a complication of hepatitis; (c) biliary, associated with biliary infection or obstruction; and (d) cardiac, from severe right-sided heart failure. Early symptoms include GI disturbances, fever, and enlargement of the liver as a result of cells filled with fat. The liver takes on a "hobnail" appearance. Weakness, weight loss, jaundice, and later chronic liver failure and circulation obstruction are common.

Diagnosis of cirrhosis involves liver function studies. Treatment is long term and tedious, focusing primarily on a diet high in carbohydrates, proteins, and vitamins. Fat intake must be limited. Antacids are used to relieve gastric distress, often vitamins are prescribed, and potassium-sparing diuretics are used to relieve fluid buildup in the abdomen. Rest is necessary and alcohol consumption is prohibited (Wu, 2017).

Age-Related Disorders of the Gallbladder

Gallstones (Cholelithiasis)

Cholelithiasis is the presence of stones in the gallbladder and cholecystitis is the inflammation of the gallbladder, usually associated with cholelithiasis. Incidence is higher for women and individuals older than 40 years. Other factors significant in the occurrence of gallbladder disease are familial tendency, obesity, and sedentary lifestyle. Cholelithiasis can occur with infections and disturbance in cholesterol metabolism or any other circumstance in which the balance of cholesterol, bile salts, and calcium is changed so that precipitation of these substances occurs. Stones often stay in the gallbladder with no symptoms, but if they migrate to the cystic duct or common bile duct they can be extremely painful. Symptoms are upper right quadrant pain, nausea, vomiting, jaundice, and inability to tolerate fatty foods. If the duct is obstructed, dark urine and light-colored stools are typical.

Diagnosis involves ultrasonography and is 90% accurate in detecting the presence of stones. Laboratory tests are useful in determining inflammation. If the gallbladder still functions normally and the gallstones are small, then extracorporeal shock wave lithotripsy (ESWL) is a nonsurgical alternative to manage gallstones. Laparoscopic surgical removal is the most common treatment, and it is usually quite effective. Medicines that dissolve the stones may also be used if surgical procedures are not deemed appropriate.

SUMMARY

Although the GI tract is the focus of numerous complaints by older adults, it stands the test of time better than some of the other organ systems in the human body. Many GI complaints in older age result from inappropriate or unhealthy lifestyle behaviors rather than from the aging process per se. It is believed that many functional disorders and diseases can very likely be avoided by more careful attention and adherence to healthy diets, regular exercise, stress reduction, and other positive health regimens.

REFERENCES

Chapple, I. L., & Genco, R. (2013). Diabetes and periodontal diseases: Consumer report of the joint EFP/AAP workshop on periodontitis and systemic diseases. *Journal of Periodontology*, *84*(4), S106–S112. https://doi.org/10.1902/jop.2013.1340011

Chernoff, R. (2016). The symbiotic relationship between oral health, nutrition and aging. *Generations*, *40*(3), 32–38. https://www.jstor.org/stable/26556216

Dirks, S. J. (2016). Nursing facility dentistry. *Generations*, *40*(3), 52–59. https://www.jstor.org/stable/26556226

Eliopoulos, C. (2018). *Gerontological nursing* (8th ed.). Wolters Kluwer.

Marieb, E., & Hoehn, K. (2019). *Human anatomy and physiology* (9th ed.). Pearson.

Saunders, M. (2016). Oral health and older adults: A history of public neglect. *Generations*, *40*(3), 6–18. https://www.jstor.org/stable/26556208

Tabloski, P. (2014). *Gerontological nursing* (3rd ed.). Pearson Education.

Wu, K. H. (2017). Liver, pancreas and biliary tract problems. In S. L. Lewis, L. Bucher, M. McLean Heitkemper, & M. M. Harding (Eds.), *Medical-surgical nursing* (10th ed., pp. 974–1013). Elsevier.

Yeager, J. J. (2019). Gastrointestinal function. In S. E. Meiner, & J. J. Yeager (Eds.), *Gerontologic nursing* (6th ed., pp. 395–426). Elsevier.

11

The Urinary System

INTRODUCTION

The urinary system includes two kidneys, two ureters, the bladder, and the urethra. This body system interacts with other organs of excretion—the lungs, skin, and intestines—to maintain the homeostatic equilibrium necessary for maintenance of life. The primary functions of the urinary system are to do the following:

- Excrete toxic substances and waste products of metabolism.
- Regulate water balance in the body.
- Help maintain acid–base balance (when the pH of the blood is maintained between 7.35 and 7.45) in body fluids.
- Aid in controlling concentration of salts and other necessary substances in the blood.

The kidneys filter about 200 L of fluid from the blood each day and excrete from it toxins, metabolic waste products, drugs, and excess ions (particles carrying an electrical charge), which become part of the urine. At the same time, reusable substances needed by the body, such as glucose, amino acids, vitamins, sodium, calcium, chloride, potassium, and phosphate ions, are returned to the bloodstream. The kidneys also regulate the volume and chemical composition of blood and regulate the balance between water and salts and between acids and bases. They produce an enzyme called renin, which assists in regulating blood pressure and kidney function, and also a hormone called erythropoietin, which stimulates red blood cell production in the bone marrow (Marieb & Hoehn, 2019).

STRUCTURE OF THE KIDNEYS

The kidneys are paired, bean-shaped organs situated behind the abdominal cavity and slightly below the diaphragm, but outside the peritoneum, the membrane enclosing the abdominal cavity. In the average adult, each kidney is about 5 in. long. The kidneys have enormous reserve capacity, and it is estimated we can lose about 60% of the 1 million nephrons in the kidneys before blood chemistry is significantly impaired. Humans are able to live successfully with only one functioning kidney. Fibrous and fatty tissues anchor the kidneys to surrounding structures. Each kidney has three distinct areas: the cortex, or outer area; the medulla, below the cortex; and the pelvis, continuous with the ureter, a tube connecting each kidney with the bladder.

Kidneys have a rich blood supply because the renal arteries deliver approximately one-fourth of the total cardiac output to the kidneys every minute. Each renal artery comes from the abdominal aorta and subdivides into smaller arteries in the kidneys. Veins leaving the kidneys trace the arterial blood pathway in reverse and exit the kidneys as the renal veins, which empty into the inferior vena cava (see Figure 11.1).

The basic unit of the kidney is the nephron, in which urine formation and other life-maintaining activities of the kidneys take place. Each kidney contains more than 1 million nephrons. A nephron consists of (a) a capsule enclosing a glomerulus (a coiled series of small blood capillaries) and (b) an attached renal tubule. Blood is carried from the renal artery to the capillaries of the glomerulus, an anatomical arrangement providing a rich blood supply to the nephron. Many substances (except blood cells and most plasma protein) pass freely from the blood entering the kidney into the glomerular capsule. Fluid filtering from the bloodstream into the capsule is called glomerular filtrate and is processed as it passes through the renal tubules. Some filtrate is converted to urine and reusable substances are sent back into the bloodstream. Material to be expelled from the body as urine passes successively through renal tubules, collecting tubules, renal pelvis, ureters, bladder, and urethra. Nephrons are located primarily in the cortical area of the kidney, whereas collecting tubules are located primarily in the medulla of the kidney. As collecting tubules approach the renal pelvis, they fuse to form ducts that deliver urine to the ureter (see Figure 11.2).

FUNCTIONS OF THE URINARY SYSTEM

The kidneys process about 47 gallons of fluid daily. Of this, only about 1% leaves the body as urine; the rest is returned to the blood circulation to be reused, an amazing conservation mechanism

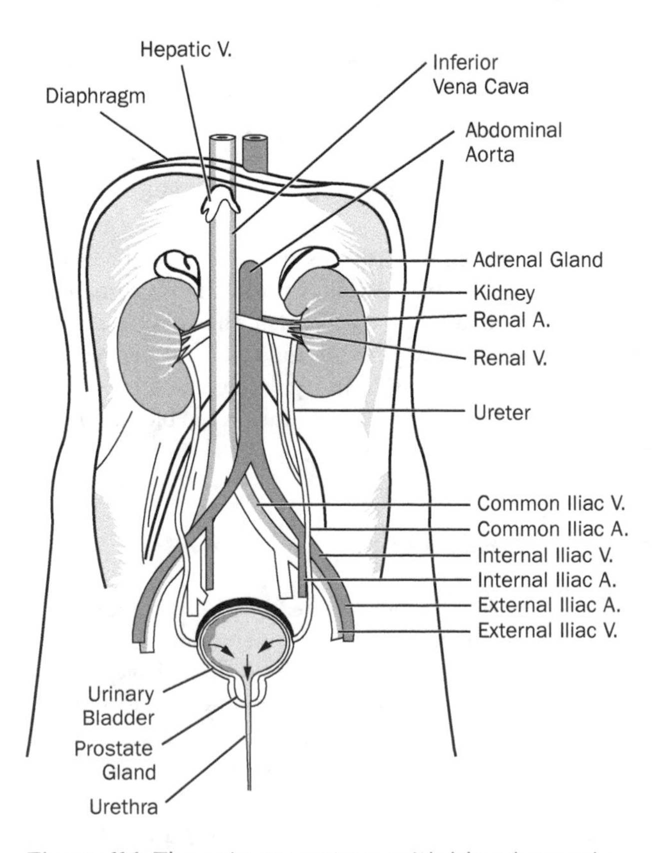

Figure 11.1 The urinary system, with blood vessels.

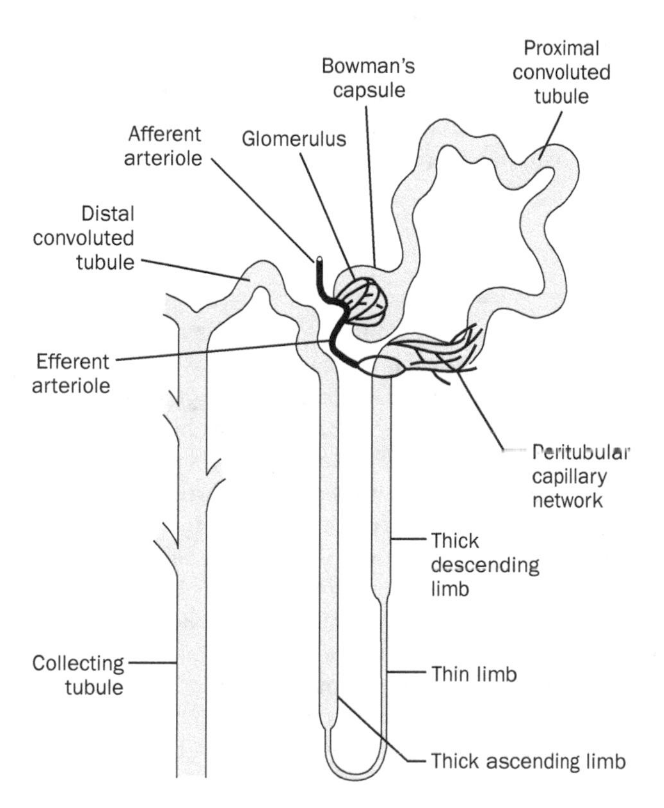

Figure 11.2 Diagram of the nephron, its blood supply, and attached collecting tubule.

in the body. Urine consists of about 95% water and 5% solids (organic and inorganic materials). The nephrons are totally responsible for urine formation. Urine formation and the adjustment of blood composition involve three processes: glomerular filtration, a function of the glomerulus; tubular reabsorption; and tubular secretion, both functions of the renal tubules.

Glomerular Filtration

The glomerulus is a highly efficient filter. As blood passes through the glomerulus, materials such as water, glucose, amino acids, and nitrogenous wastes pass easily from the glomerulus into the glomerular capsule and renal tubule, whereas other materials such as proteins and blood cells are usually not able to pass through the glomerular filtration membrane. Blood pressure provides the force for this filtration process. The amount of blood filtered by the glomeruli in a given time is the glomerular filtration rate (GFR).

Tubular Reabsorption

Urine is formed as the filtrate moves through renal tubules to the collecting ducts, the renal pelvis, and then into the ureters. Most of the contents of the renal tubules are passed through the tubule walls and returned to the bloodstream by way of blood capillaries surrounding the tubules. Material left in the tubules will become urine. This process is called tubular reabsorption. Creatinine is one substance not reabsorbed, making it useful in measuring GFR.

Tubular Secretion

Some selected substances are added to the filtrate through the tubular walls, or tubular reabsorption in reverse. Hydrogen and potassium ions, urea, creatinine, ammonia, and some organic acids enter the tubular fluid through the process of secretion. Urine is thus composed of both filtered and secreted substances.

System Dynamics

Renal clearance refers to the ability of the kidneys to clear (or cleanse) a given volume of plasma (fluid portion of the blood) of a particular substance in a given time, usually 1 minute. Renal clearance tests are used in the evaluation of GFR to determine glomerular dysfunction or damage.

Filtration, reabsorption, and tubular secretion constitute highly efficient processes for water conservation in the body. Concentration of urine is regulated according to the body's supply of water; for example, if there is excess water in the body, urine will be diluted because it will contain large quantities of excess fluid to be excreted. On the other hand, if body fluid level is low, the kidneys will concentrate urine, and more water will be reabsorbed (returned to body tissues). Water loss and retention in the kidneys are complex processes under both neural (nervous system) control and local control. When water concentration in the body is low, hypothalamic receptors stimulate the pituitary gland to produce an antidiuretic hormone. This hormone increases the reabsorption of water through tubule walls, so more water is returned to body tissues for their use. Urine in the tubules, then, is more concentrated. Local control of water loss and retention depends on the amount of sodium in the glomerular filtrate. Certain cells near the glomerulus are stimulated by the sodium content of the filtrate to produce renin, an enzyme, reducing the amount of water in the urine. Renin also stimulates secretion of an adrenal gland hormone called aldosterone. Aldosterone increases water loss from tubules by increasing the movement of sodium out of the tubular filtrate. Water is then drawn from urine to be made available for body use.

The analysis of urine (urinalysis) is a simple but valuable diagnostic tool that indicates disease by the presence of substances not normally found in urine or by alterations in the proportions of substances normally present. When the kidneys stop functioning properly, three potentially dangerous situations occur.

1. The level of waste products in the blood increases.
2. Acidity of the blood increases because excess acid is no longer removed by the kidneys.
3. Sodium and water balance, crucial for life, is disrupted, producing serious disequilibrium of the internal environment.

Contractions of the smooth muscle walls of the ureters connecting the kidneys and the bladder force urine into the bladder. The bladder, where urine is collected and temporarily stored, is a muscular sac situated in the pelvic cavity. When approximately 300 mL (a little more than a cupful) of urine collects in the bladder, sensory receptors in the bladder walls are stimulated and the conscious desire to urinate results. As the bladder's sphincter muscles relax and open, urine is forced through the urethra and outside of the body (the act of urination or micturition). The urethra is short and exclusively excretory in females, whereas in males it extends from the bladder through the penis, carries both urine and semen, and thus has both excretory and reproductive functions.

Bladder function may be either voluntary or involuntary, depending on a variety of factors. For example, children must be taught bladder control, but this cannot be accomplished until the child is physically mature enough to be able to voluntarily withhold urine. Although normally controllable, in situations of extreme emotional stress or excitement, loss of voluntary bladder control (called stress incontinence) may occur, even in healthy adults. Urination is essentially a voluntary act, one of tremendous significance in our culture, and any loss or decreased efficiency of bladder functioning (as sometimes happens in older age) is particularly embarrassing and psychologically difficult.

AGE-RELATED CHANGES IN THE URINARY SYSTEM

As with other body systems, the aging process results in gradual reduced efficiency of the urinary system. There may be greater sensitivity in older adults to admit to changes in the urinary system, and those who work with older adults must recognize this possibility when evaluating possible urinary difficulties.

Anatomical changes in the kidney associated with age include the following:

1. Kidney size decreases by age 80. This loss occurs primarily in the cortex of the kidney where the glomeruli are located. The overall number of glomeruli decreases 30% to 50% by age 70 (Touhy, 2016). Functioning nephrons decrease both in number and size, although the kidneys generally continue to maintain homeostasis by regulating body fluid.
2. A steady increase in glomerular sclerosis (hardening of the glomeruli) occurs.
3. There is a decline in the number of cells of the renal tubules, an increase in tubular diverticula (outpouchings or pockets in walls), and a thickening of the tubular walls. These changes affect the kidney's ability to concentrate urine and reduce its ability to remove drugs from the body.
4. Blood vessels in the kidneys become smaller and thicker, and atherosclerotic changes occur that reduce blood flow through the kidneys and decrease GFR (Touhy, 2016).
5. The ureters, bladder, and urethra are all muscular structures and tend to lose tone and elasticity. The bladder especially may be affected as a decrease in muscle tone leads to incomplete emptying of the bladder, with a consequent greater risk of infections. Prostate enlargement in men can impair the emptying of the bladder and contribute to bladder contractions.
6. Some decline in bladder capacity often occurs, and less urine can be stored in the bladder. In addition, more urine is retained in the bladder after voiding (residual urine). Total bladder capacity declines from approximately 350 to 450 mL to about 200 to 300 mL in older age (Yeager, 2019). Lessened reserve capacity, weaker muscle tone, and increased bladder contractions contribute to more frequent and also more urgent urination in many older persons, especially at night. In addition, the need or signal to urinate may be delayed until the bladder is almost full, resulting in even greater urgency and possibly incontinence (Miller, 2009).

Functional changes in the kidneys associated with the aging process include the following:

1. Blood flow decreases. Renal blood flow is reduced from approximately 600 to 300 mL/min between ages 40 and 80 (Yeager, 2019). Both a decrease in cardiac output and fewer blood vessels in the kidney contribute to decreased blood flow in the aging kidney.
2. GFR declines. The GFR as measured by serum creatinine levels is stable during young adulthood, but begins to decline after approximately age 40, although a decline is not universal. A decrease in GFR makes the removal of potentially toxic substances such as medications less efficient (Eliopoulos, 2018).
3. Ability to concentrate urine is decreased. With age, the kidney is not able to concentrate urine as well as it formerly did or dilute urine as needed. Older persons then are not as able to adapt as efficiently to dehydration or water overload. Crises may occur when water intake is reduced because of confusion, fear of incontinence, or any other reason, and especially if diuretics are being used.
4. Other functional changes identified in the older kidney indicate that maintaining the acid–base balance in the blood and regulating sodium and potassium levels may become more difficult (Tabloski, 2014).

AGE-RELATED DISORDERS OF THE URINARY SYSTEM

Medical problems involving the urinary system generally arise from the progressive decrease in renal function and renal blood supply with age, a greater likelihood of obstruction in the lower urinary tract, and increased susceptibility to urinary infections. Urinary tract and kidney disorders range from those easily treated to those that are life threatening, requiring long-term dialysis and organ replacement (Eliopoulos, 2018).

Urinary Tract Infections

An increased incidence of urinary tract infections (UTIs) occurs with age, especially after age 80 among both men and women. Older adults have higher incidences of UTIs if institutionalized, more likely to be the result of catheterization, soiling, and not emptying the bladder completely (Yeager, 2019).

UTIs are commonly classified as uncomplicated or complicated. Uncomplicated UTIs occur in individuals who are essentially healthy and have normal voiding. They typically respond well to antibiotics and usually do not result in permanent damage. Complicated UTIs usually involve either functional or structural abnormalities in voiding, are difficult to treat, and readily result in permanent damage.

UTIs are caused by pathogenic microorganisms present in some part of the urinary tract with or without accompanying signs and symptoms. Sites of infection (and the names of their infections) include the bladder (cystitis) and the kidney (pyelonephritis). Women are more likely to develop UTIs because of the short female urethra and its proximity to the vagina, urethral glands, and rectum, leading to possible bacterial contamination (Winkelman, 2013; Yeager, 2019).

Changes in bladder functioning associated with strokes or diabetes increase the risk of UTIs caused by incomplete emptying of the bladder, as do decreased estrogen levels in women, neurogenic bladder, and prostatic hyperplasia. Indwelling catheters also increase the risk of a UTI. UTIs are diagnosed by the presence of bacteria in the urine detected by urine culture and sensitivity tests. Renal ultrasonography may be used to determine structural problems. Treatment of UTIs involves judicious use of antibiotics, and men usually require longer periods of treatment than women (Tabloski, 2014).

Cystitis (Lower Urinary Tract Infection)

Cystitis is an inflammation of the bladder often found in older adults, especially women. It commonly results from urine flowing back into the bladder from the urethra, as a result of fecal contamination or from a catheter or cystoscope. Typical symptoms include urgency and frequency of urination, voiding small amounts, burning or pain when urinating (dysuria), lower abdominal pain, nocturia (getting up at night to urinate), sometimes blood in the urine, and an overall sense of feeling unwell. A healthcare professional should be seen when symptoms are first noticed, because infections can spread to the kidneys. However, in some older adults these typical symptoms may not be present. Diagnosis depends on the history of the symptoms, physical examination, urinalysis, and urine culture. Short-term antibiotic therapy is the preferred treatment, often with various over-the-counter (OTC) drugs to relieve symptoms, but recurring infections often require long-term, low-dose therapy. Liberal amounts of fluids are encouraged to cause voiding every 2 to 3 hours to flush bacteria from the urinary tract.

Pyelonephritis (Upper Urinary Tract Infection)

Pyelonephritis is an inflammation of one or both kidneys. Bacteria may enter the bladder via the urethra and ascend to the kidney, or they may enter the kidney through the bloodstream. Left untreated, pyelonephritis can lead to an accumulation of toxic materials in the blood, structural damage to the kidneys, uremia (retention in the blood of nitrogenous substances usually excreted by the kidneys), and progressive renal failure. Many older adults are asymptomatic, and pyelonephritis often goes undetected until it is at an advanced stage such as renal failure. If symptoms do occur, they include pain in the kidney area, tenderness in the back over the kidneys, chills and fever, nausea, vomiting, and a possible change in mental status. Chronic pyelonephritis symptoms include fatigue, weight loss, frequent urination, and anorexia. Diagnosis is by CT scan, ultrasound, urine culture, blood cultures, and measurement of serum creatinine levels. Treatment is essentially the same as for cystitis, but the course of treatment is longer (Yeager, 2019).

Acute Glomerulonephritis

Although once considered to be primarily a disease of the young, data now indicate glomerulonephritis is more prevalent in older adults than previously thought but may often be so subtly presented that it is not diagnosed (Eliopoulos, 2018). It is actually a group of kidney disorders characterized by an inflammatory response in the glomeruli resulting from immune system reactions; it is not an infection of the kidney per se. Symptoms include fatigue, anorexia (loss of appetite), hypertension, anemia, facial edema (swelling) preceding pharyngitis (inflammation of the pharynx), proteinuria (proteins in the urine), and oliguria (decreased amount of urine excreted). Diagnosis involves urinalysis, serum creatinine, and blood urea nitrogen (BUN) studies. Renal biopsy is usually considered to be the best diagnostic procedure. Treatment considerations are to protect kidney functioning and treat complications quickly. Bedrest is helpful during the acute phase. Antibiotics are used to eliminate causative factors, sodium and protein are restricted, and fluid intake and excretion are monitored. Control of high blood pressure with various hypertensive medications is an extremely important part of successful treatment. Chronic glomerulonephritis may occur if the acute phase lasts longer than 1 year, although actually it does not usually follow an acute phase. Most individuals with chronic glomerulonephritis progress to renal failure. Treatment with peritoneal or hemodialysis allows individuals to live for some years, but eventually, unless a kidney transplant is performed, death ensues.

Benign Prostatic Hyperplasia

Considered part of the male reproductive system, the prostate gland often interferes with urinary function in older males. Anatomically it is located near the base of the bladder and surrounds the urethra. Benign prostatic hyperplasia (BPH) is a nonmalignant prostate enlargement that often produces symptoms of urethral obstruction after age 50, affecting approximately 50% of men between the ages of 51 and 60 and increasing to 90% of men older than age 80 (Tabloski, 2014). Its etiology is still somewhat unclear, but hormones and aging both play some role. As the prostate gland enlarges, it compresses the urethra, which leads to urinary obstruction and ultimately to urinary retention. Urinary obstruction results in incomplete emptying of the bladder, urinary stasis (stagnation of normal flow of urine), UTI, bladder stones, and bladder diverticula.

Although many middle-aged and older men have some BPH, not all experience symptoms. Early symptoms include straining to urinate, difficulty in starting the urinary stream, longer time necessary to urinate, and a feeling of incomplete bladder emptying. These symptoms are referred to as "lower urinary tract symptoms," or LUTS. As prostate enlargement continues, urinary urgency and frequency occur, as does nocturia (getting up at night to urinate), which often contributes to sleep disorders (Quallich, 2017; Tabloski, 2014).

Diagnosis usually includes a thorough history, physical examination of the prostate, neurological examination to rule out neurological problems, urinalysis, and a blood test for prostate-specific antigen (PSA) to exclude prostate cancer. Imaging studies such as transabdominal ultrasound and transrectal ultrasound (TRUS) are also used. Diagnosis is essentially one of exclusion after other possible diagnoses have been eliminated from consideration.

Treatment depends on how much the symptoms are a problem for the person. The International Prostate Symptom Score is a questionnaire ranking severity of symptoms on a numerical scale to indicate how bothersome symptoms are to the individual (Ignatavicius, 2013). Current treatment options are weight loss if needed, increased exercise, behavior modification, and various anticholinergic medications. If symptoms worsen, other medications are available. Minimally invasive surgical therapies (MISTs) may be used for moderate to severe symptoms, and transurethral resection of the prostate (TURP) is often used in more severe situations (Tabloski, 2014). Some newer types of treatment modalities include laser therapy, stent therapy, and various types of thermotherapy (Yeager, 2019).

Many men do not report symptoms until they are extreme and immediate intervention is necessary. Sometimes embarrassment is the primary reason, or fear about impotence if surgery is necessary. Healthcare professionals need to be sensitive to individual concerns about possible impotence, and this issue should be addressed before any surgical intervention.

Urolithiasis (Kidney Stone Disease)

Urolithiasis refers to the presence of stones (calculi) in the urinary tract, usually either in the kidney or ureter. Stones are formed by the deposition of crystalline substances excreted in urine, but the exact mechanism of stone formation is not clear. Three specific conditions are necessary for stone formation (Winkelman, 2013):

1. A slow flow of urine resulting in supersaturation of urine with some element (such as calcium) that crystallizes and later becomes a stone.
2. The lining of the urinary tract becomes damaged by irritation from crystals.
3. Insufficient amounts of substances in the urine that tend to inhibit and prevent supersaturation of urine and formation of crystals.

Factors involved in the incidence and type of stone formation include metabolic (at least 90% of those who form stones have a metabolic risk factor), dietary, genetic, geographic location, and family history. If stones block urine flow, a urinary infection usually occurs. Some stones cause very little pain, whereas others, especially those lodged in the ureter, result in excruciating pain. Spontaneous passage of stones occurs in approximately 80% of cases. However, stones too large to be passed through the system have to be removed or broken up so they can pass through and be excreted.

Diagnosis involves medical history, including family history of urological stones, physical examination, urinalysis, and blood work. Stones can be visualized on radiographs and CT scans, and noncontrast CT is especially sensitive to identify stones. Renal ultrasonography is also useful to identify size and placement of stones. Treatment is directed toward pain control, removal of stones, preserving nephrons, and controlling infection. If pain is too severe, or if another infection is present, or if stones block the urinary tract, medical or surgical treatment is warranted. Lithotripsy is a procedure used to break up stones in the kidney so they can be voided. Extracorporeal shock wave lithotripsy (ESWL) is a very common procedure for treating kidney stones.

If obstruction occurs or the stone is too large to be excreted, various minimally invasive surgical procedures may be necessary. If they are not effective, major surgery may be performed. Although there are different types of stones, general prevention includes increasing fluid intake; modifying diet to increase fiber; eating more calcium-rich foods; eating less meat (beef, pork, poultry); avoiding foods high in oxalate such as nuts, chocolate, and dark green vegetables; restricting purines; and reducing salt intake. If dietary modification is not effective, certain medications may help dissolve stones or prevent other occurrences.

Cancer of the Bladder

Cancer of the bladder is more common in those older than 50 and affects men more than women. Risk factors for bladder cancer include cigarette smoking and prolonged exposure to carcinogens (such as dyes, rubber, leather, ink, paint) in the workplace. There also may be a causal relationship between bladder cancer and excessive use of analgesics. Cancers of the prostate, colon, and rectum in men and lower gynecological tract cancers in women commonly metastasize to the bladder. Symptoms include blood in the urine, painful or difficult urination (dysuria), urinary frequency or urgency, and lower back pain. Diagnosis involves cystoscopic examinations and biopsy, CT scan, and MRI. Other scans may also be deemed appropriate depending on the specific situation. Treatment options include surgery, chemotherapy, radiation, and immunotherapy (Eliopoulos, 2018; Yeager, 2019). Metastases and recurrences of bladder cancer are common.

Urinary Incontinence

Urinary incontinence refers to the involuntary passing of urine in quantities that constitute a social or health problem. Institutionalized individuals experience more incontinence than those living in the community, but incontinence is often a major reason for institutionalization because of difficulty managing it in the home. Although not an inevitable part of the aging process, many older adults do experience varying degrees of incontinence, with women more often affected than men. Estimates are that approximately one in three older adults living in the community have some urinary incontinence (Tabloski, 2014). Incontinence is a significant cause of disability and dependency plays a significant role in the decision to place older adults in long-term care. Incontinency though is not a part of normal aging (Yeager, 2019).

Age-related changes contributing to urinary incontinence in women may often be related to lessened estrogen levels, which cause a weakening of the pelvic floor and bladder outlet as well as a decrease in urethral muscle tone that contributes to vaginal inflammation. In men, age-related enlargement of the prostate gland can lead to decreased urinary flow, increased possibility of urine retention, and weakness of the detrusor muscle of the bladder, all contributing to urinary incontinence. Other possible causes of incontinence in both women and men include delirium, drugs, diuretics, infections, and diabetes.

Incontinence is usually classified as either transient or established. Transient incontinence develops suddenly (acute) and is related to an accompanying health (medical or surgical) condition, including medications as a possible cause. When the medical condition is resolved, incontinence is relieved. Established incontinence, however, is chronic and persists over time, becoming progressively worse (Eliopoulos, 2018). Several types of incontinence have been identified:

1. Stress incontinence, primarily caused by weakened muscles in the pelvic floor, is common in women who have had many children. Involuntary passage of urine occurs when intra-abdominal pressure is increased such as in laughing, coughing, sneezing, or during exercise.

2. Urge incontinence, or an inability to delay voiding after the perception that the bladder is full, may be due to a UTI, prostate enlargement, bladder or pelvic tumors, or central nervous system (CNS) impairment after a stroke.
3. Reflex incontinence, a variation of urge incontinence, is the sudden leaking of large amounts of urine without a sensation of urgency or full bladder. It is caused by lesions in the cerebral cortex, multiple sclerosis, or other neurological disturbances.
4. Overflow incontinence is caused by prostate enlargement and obstruction, or by some medications, or nervous system disturbances affecting the bladder. Small amounts of urine leak from the distended bladder, frequently or even continuously.

Other types of incontinence reported in the literature are mixed incontinence, a combination of stress and urge incontinence; functional incontinence, or loss of urine because of inability or unwillingness to get to the toilet in time or because of cognitive impairment; and iatrogenic incontinence, primarily caused by medications (Eliopoulos, 2018). Nocturia, or frequency of urination at night, is often very disruptive for older adults, with potentially serious consequences for health because, aside from the danger of falling when getting up at night for a trip to the bathroom, sleep interruption may result in fatigue, forgetfulness, disorientation, depression, and sleep disorders (Yeager, 2019).

Incontinence can be a devastating behavior problem for older adults, with enormous psychological and social implications. Control of both bowel and bladder functions is required and closely related to socialized behavior in our society; loss of these functions is viewed as personal incompetence and usually has a decidedly negative effect on self-esteem. Older adults often severely restrict their activities, social interactions, and interpersonal relationships because of concerns about incontinence. Many older adults will not admit to incontinence and therefore do not seek assistance in managing it. Others do not think incontinence can be treated because they believe it is part of growing old (Winkelman, 2013). Those who do seek aid may be virtually ignored by many healthcare professionals, who may also erroneously believe it is an inescapable part of the aging process and that effective intervention is not possible or necessary. Although there are age-related changes in the urinary tract that predispose older adults to varying degrees of urinary incontinence, it is not an inevitable occurrence in the normal aging process. Helpful treatment and management techniques are available and should be made easily accessible to older adults.

Urinary incontinence has many possible causes, and careful evaluation of the individual is necessary in order to plan individualized treatment. Assessment must include a thorough history of the problem, physical examination, relevant environmental and social factors, a functional assessment, consideration of associated medical conditions possibly influencing urinary patterns, previous surgeries, and current medications. The physical examination should assess mental status; mobility; dexterity; and neurological, abdominal, rectal, and pelvic status. Urodynamic tests such as ultrasound of the kidneys and bladder and a provoked full-bladder stress test can provide additional information. Specific tests useful in assessment are urinalysis, serum creatinine or BUN levels, measurement of postvoid residual urine volume, possibly urine culture, and measurement of blood glucose levels.

The selection of an intervention depends on the cause of the incontinence and the individual's personality and motivation. Treatment options include (a) timed voiding and bladder training; (b) Kegel, or pelvic floor, exercises; (c) judicious use of medicines; and (d) incontinence pads if absolutely necessary and used as a last resort.

Other techniques that may be useful include the following:

- *Biofeedback.* Use of visual or auditory instruments to provide moment-to-moment information on how effectively an individual is controlling muscles associated with urination. With practice, many people can learn to control the relevant muscles.

- *Urinary control methods.* Newer devices for controlling the flow of urine include a small balloon that rests at the neck of the bladder and can be inflated or deflated to control urination; urethral plugs and intraurethral catheters with valves to control the flow of urine in women; and an external device that provides a watertight seal to prevent leakage. Such devices are constantly being developed for both women and men with varying degrees of success, but they offer other possible options for incontinence control. Minimally invasive surgical procedures are also available for both women and men to treat incontinence.
- *Catheters.* In some situations catheters are used to control and manage incontinence, but other treatment options should be evaluated and tried first. If other forms of treatment are deemed not appropriate and do not help, catheters may be a remaining option. Clean intermittent catheterization or indwelling catheters may be used, but both must be used carefully with older adults (Eliopoulos, 2018).

In actuality, most cases of urinary incontinence can be cured or improved, and every individual with urinary incontinence is entitled to proper evaluation and appropriate treatment. As indicated previously, too often older adults, family, and healthcare professionals do not seek appropriate interventions for incontinence. A persistent, prevalent myth is that incontinence is normal and inevitable in older age, a most inappropriate point of view leading to unnecessary embarrassment and discomfort. One further concern is that fluid intake should never be restricted in an effort to reduce incontinence episodes in older adults because it can quickly cause dangerous dehydration. Fluids can be taken earlier in the day to minimize getting up at night to urinate, but adequate fluid intake must be maintained in this age group.

Renal Failure

When the kidneys are unable to perform their regulatory functions or remove metabolic waste products from the body, renal failure results. Substances normally excreted in urine accumulate in body fluids, disrupt endocrine and metabolic functions, and cause serious disturbances of fluid, electrolyte, and acid–base levels.

In acute renal failure (sometimes identified as acute kidney injury) there is a sudden loss of kidney function caused by failure of renal circulation or glomerular or tubular dysfunction. Some leading causes of acute renal failure are a sudden decrease in blood flow to kidneys; damage from certain medications, poisons, or infection; and sudden blockage that prevents urine from leaving the kidneys. Specifically, chronic diseases such as hypertension and diabetes can cause renal failure, as can prostatic hypertrophy. In older adults, renal failure may reflect lessened reserve capacity for maintaining homeostasis (a) when there has been a loss of body fluids and electrolytes because of diarrhea or vomiting, (b) in situations involving inadequate intake of fluids, (c) when infections are present, or (d) after stress induced by surgery or a heart attack. Symptoms of acute renal failure are lethargy, feeling restless, nausea, vomiting, diarrhea, dry skin and mucous membranes, headache, pain in the back (flank pain), and low urinary output. Diagnosis is based on urinalysis, urine chemistry examinations, radiography, and renal ultrasonography. CT and MRI scans may be useful as well. Treatment involves correcting any reversible cause, restoring fluid levels, correcting biochemical imbalances, and maintaining good nutrition (LaCharity, 2013; Yeager, 2019).

Chronic renal failure (CRF) is progressive deterioration of renal function resulting in uremia (an excess of urea and other nitrogenous waste products in the blood). Stages of CRF are decreased renal reserve leading to renal insufficiency, which then progresses to renal failure and, finally, to uremia. Death will occur unless dialysis or a kidney transplant is performed. Diabetes, prostatic hyperplasia, hypertension, and long-term use of nonsteroidal anti-inflammatory drugs

(NSAIDs) can contribute to CRF (Tabloski, 2014). Symptoms include fatigue and lethargy, headache, general weakness, and gastrointestinal disturbances. However, symptoms may go unnoticed until the disease is in advanced stages. If untreated, symptoms increase in severity, followed by deep coma, often accompanied by convulsions, and finally death.

Diagnosis involves evaluation of the following: anemia, elevated serum creatinine or BUN, elevated serum phosphorus, decreased serum calcium, low serum proteins, and usually low carbon dioxide and acidosis (low blood pH). Treatment is concerned with maintaining homeostasis and kidney function for as long as possible. Attention to diet is important, especially protein intake, fluid intake, sodium intake, adequate calories, and vitamins. Specific symptoms are treated when they occur to improve renal function. When conservative management techniques fail, maintenance dialysis or kidney transplantation is the remaining treatment choice (Tabloski, 2014). One note of caution is that symptoms of renal disease in older adults are often nonspecific, and other disorders present may mask symptoms of renal disease.

SUMMARY

The urinary system becomes less efficient with age, but barring accident, disease, or unusually high-demand situations, it will function adequately into extreme old age. Exercise and proper diet—including adequate fluid intake, limited use of medications, and not smoking—help the urinary system maintain adequate functioning.

REFERENCES

Eliopoulos, C. (2018). *Gerontological nursing* (9th ed.). Wolters Kluwer.

Ignatavicius, D. (2013). Care of male patients with reproductive problems. In D. Ignatavicius & L. Workman (Eds.), *Medical-surgical nursing* (7th ed., pp. 629–1651). Saunders Elsevier.

LaCharity, L. (2013). Care of patients with acute kidney injury and chronic kidney disease. In D. Ignatavicius & L. Workman (Eds.), *Medical-surgical nursing* (7th ed., pp. 1537–1573). Saunders Elsevier.

Marieb, E., & Hoehn, K. (2019). *Human anatomy and physiology* (11th ed.). Pearson Education Limited.

Miller, C. (2009). *Nursing for wellness in older adults* (5th ed.). Wolters Kluwer/Lippincott Williams & Wilkins.

Quallich, D. (2017). Male reproductive and genital problems. In S. L. Lewis, L. Bucher, M. McMclean Heitkemper, & M. M. Harding (Eds.), *Medical-surgical nursing* (19th ed., pp. 1268–1291). Jones and Bartlett.

Tabloski, P. (2014). *Gerontological nursing* (3rd ed.). Upper Pearson Education.

Touhy, T. A. (2016). Elimination. In T. A. Touhy & K. Jett (Eds.), *Ebersole and Hess' toward healthy aging* (9th ed., pp. 200–220). Elsevier.

Winkelman, C. (2013). Care of patients with urinary problems. In D. Ignatavicius & L. Workman (Eds.), *Medical-surgical nursing* (7th ed., pp. 1489–1517). Saunders Elsevier.

Yeager, J. J. (2019). Urinary function. In S. E. Meiner & J. J. Yeager (Eds.), *Gerontological nursing* (6th ed., pp. 427–448). Elsevier.

12

The Reproductive System

INTRODUCTION

The male and female reproductive systems are composed of both internal and external organs.

COMPONENTS AND FUNCTIONS OF THE FEMALE REPRODUCTIVE SYSTEM

The external female reproductive organs include (a) the external genitalia, or *vulva,* made up of the labia majora, labia minora, clitoris, vestibule, and hymen; (b) the *mammary glands* (breasts), each composed of a nipple, lobes (15–25 arranged radially around the nipples within fat), connective tissue, and excretory ducts. These glands respond to estrogen, progesterone, and prolactin by increasing and decreasing in size and by secreting milk.

The female internal reproductive organs include the following:

- The *vagina,* a tubular canal forming the birth passageway. It extends from the cervix (neck of the uterus) to the exterior of the body and serves as the female organ of copulation.
- The *uterus,* a muscular, pear-shaped organ located between the bladder and the rectum. Four sets of ligaments hold the uterus in place. The functions of the uterus are to receive, retain, and nourish a fertilized egg (Marieb & Hoehn, 2019).
- The *ovaries,* two almond-shaped organs about 1.5 in. long on either side of the uterus near the fallopian tubes. The ovaries are held in place by several ligaments and are considered to be the primary organs of reproduction in the female. Their functions are to develop and release mature ova and to produce the hormones estrogen and progesterone. The fallopian, or uterine, tubes transport mature ova from the ovaries to the uterus (see Figure 12.1).

AGE-RELATED CHANGES IN THE FEMALE REPRODUCTIVE SYSTEM

The female climacteric includes the transitional period in which reproductive capacity diminishes and finally ceases. Climacteric changes often begin in the 40s, and menopause, the cessation of

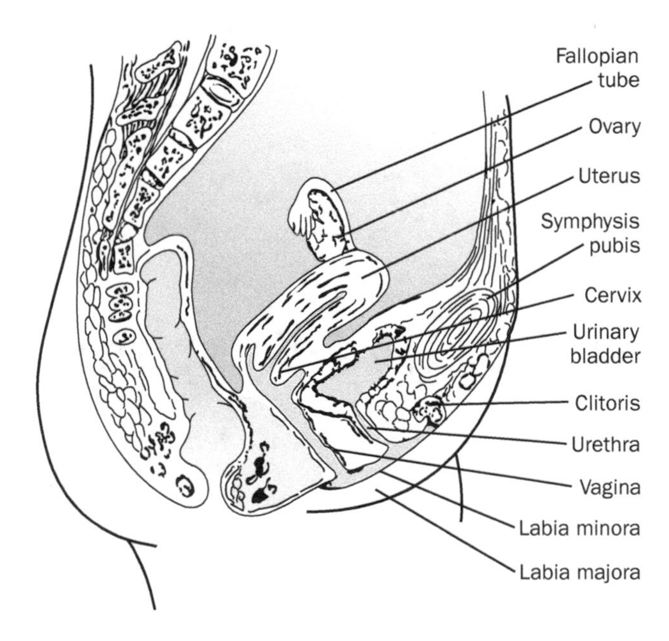

Figure 12.1 Female urogenital system.

the menstrual cycle, is completed by the early to mid-50s for most women. During the climacteric period menstrual cycles become less regular. The length of the menstrual period tends to be shorter, although both shorter and longer menstrual periods may occur during the premenopausal months. Menopause is defined as having occurred when 1 year has passed without menstruation. Climacteric changes result from decreased production of hormones associated with menstruation, especially estrogen and progesterone produced by the ovaries.

Specific age-related changes in the female reproductive system include the following:

- The occurrence of the climacteric, culminating in menopause, the cessation of menstrual flow.
- Thinning and graying of pubic hair.
- A loss of subcutaneous fat and elastic tissue in the external genitalia, which shrink slightly.
- Thinning of the vaginal walls, which also become drier and less elastic. They, too, shrink slightly. Decreases in blood flow and in the amount of vaginal lubrication produced may cause sexual intercourse to be uncomfortable or even painful.
- Decreases in size and weight of the ovaries and uterus with age; the latter becomes more fibrous. Ovulation gradually ceases.
- Decreased secretion of estrogen.
- Loss of some elasticity of the ligaments supporting the ovaries and uterus.
- Diminished muscle and glandular tone. Skin is less elastic, resulting in a loss of firmness and sagging of the breasts and other body tissues (Eliopoulos, 2018; Miller, 2009).

The psychological implications of these changes depend on a woman's body image, the significance of reproductive ability to her self-image, her sense of personal competency, and how much she believes the various myths about menopause. In addition to the physical changes cited here, other symptoms commonly include hot flashes and sweats caused by vasodilation of the skin's blood vessels (especially in the head, upper chest, and back), irritability and other mood changes (including depression in some women), headaches and muscle pain, and sleep deprivation if hot flashes occur excessively at night. Accelerated loss of bone mass predisposes some women to osteoporosis, and rising cholesterol levels increase the risk of heart disease. For some women, sexual desire decreases; for others, it increases; and for still others, there is no perceptible change in sexual desire before, during, or after menopause. Not all women experience bothersome symptoms associated with menopause, but for others symptoms may be severe and require medical intervention. Those who have a hysterectomy (surgical removal of the uterus and often the ovaries) experience immediate menopause rather than a gradual menopause and may develop physical and psychological symptoms much more quickly.

It is also necessary to keep in mind that many other life transitions or age markers occur during the middle years and that menopause is not the only event taxing women's coping and adaptive skills. For instance, many women at this time of life are also confronting issues regarding children leaving home, the care of aging parents, a peak in career and financial earnings, plus other physical changes indicating that they are aging. Women thus face many challenges that require major adjustments in the middle years, not just menopause.

Hormone replacement therapy (HRT, or hormone therapy) is still a rather controversial subject. Administration of estrogen and progesterone combined has been the leading treatment for menopausal symptoms for many years, although it has many side effects and associated health risks. The Women's Initiative Study found that hormone therapy can increase the risk for cardiovascular disease, venous thrombosis, and pulmonary embolism, especially in women older than age 60. The amount of risk depends on the specific hormone used and the length of time taken (Crowley, 2011).

Current practice is to consider the advantages and disadvantages of hormone therapy on an individual basis. Specific factors to consider are age of the woman, severity of symptoms, other health problems present, and family history. If hormone therapy is recommended, it should be a part of a program that includes a healthy diet, exercise, smoking cessation, health promotion, and as low a dosage of hormones for the shortest time possible (Eliopoulos, 2018; Tabloski, 2014). Generally, most research now agrees that long-term usage of hormone therapy is not clinically appropriate because of its additional health risks. However, shorter term use may be beneficial and safer for younger women who have severe postmenopausal symptoms that interfere with their quality of life. Prescribing (HRT) will change over time as research findings impact clinical practice (Winton, 2019).

AGE-RELATED DISORDERS OF THE FEMALE REPRODUCTIVE SYSTEM

Atrophic Vaginitis

In women older than age 55, atrophic vaginitis, or inflammation of the vagina, most often results from lowered estrogen levels. The vaginal walls become thinner, drier, and more fragile and are easily irritated by sexual intercourse or by nonsexually related irritants or chafing. Symptoms may include itching, a change in normal vaginal discharges, vaginal odor, burning during urination, and bleeding.

Vaginitis, especially if accompanied by bleeding, should be thoroughly investigated by a physician to rule out more serious conditions such as a malignancy. Oral medications or topical applications of estrogen in suppository or cream form or antibacterial or antifungal creams inserted directly into the vagina generally relieve inflammation. Douches may be prescribed, which require teaching the individual the proper procedure to prevent injury. Good hygiene helps treat and prevent vaginitis (Eliopoulos, 2018).

Pelvic Organ Prolapse (Cystocele, Rectocele, and Prolapsed Uterus)

With age, ligaments and fibrous tissue supporting pelvic organs (bladder, uterus, rectum, and urethra) lose elasticity, allowing various types of hernias to occur. A herniation of the bladder into the vagina is called a *cystocele.* A herniation of a part of the large intestine into the vagina is called a *rectocele. Prolapse of the uterus* occurs when supporting ligaments stretch and allow the uterus to drop into the vaginal cavity. Other prolapses are urethral, vaginal, and small bowel, but they are not common.

The three most common of these conditions are influenced by estrogen depletion and by having many children (because childbearing tends to stretch the ligaments and other supporting tissues). Other possible causes include obesity, constipation, and respiratory problems accompanied by a chronic cough. Common symptoms include lower back pain, a feeling of fullness or heaviness because of a mass protruding into the vaginal cavity, difficulty in urinating, stress incontinence, and difficulty in defecation. For mild to moderate prolapse, various approaches and therapies are recommended. These include avoiding high-impact aerobics, repetitive bending and heavy lifting, chronic coughing, and obesity. Daily Kegel exercises often are used to tighten the vaginal supporting muscles.

Diagnosis involves a pelvic examination, possibly a urinary tract x-ray examination, measurement of residual urine by ultrasound, urine culture, and perhaps a CT scan or MRI. Nonsurgical management may include Kegel exercises to increase pelvic support and pessaries in the vagina to assist in pelvic support. Bladder training and high-fiber diets, stool softeners, and laxatives may also be useful. Surgery may be necessary for severe symptoms. Various minimally invasive techniques as well as more complicated open surgical procedures are available (Ignatavicius, 2013a). Preventive measures include Kegel exercises to keep muscle tone, maintaining a healthy weight, avoiding lifting heavy objects, not smoking, avoiding constipation or other straining, and living a healthy lifestyle.

Cancer

Although cancers of the vulva and vagina are relatively rare, women need to be alert to any lesion, unusual itching, discharge, or change in appearance of body tissues. Early diagnosis and immediate treatment are necessary to prevent disfigurement or death.

Cancer of the cervix (the neck of the uterus) is the third most common cancer of the female genital system, peaks in women in their 30s and 40s, and then declines in incidence. Risk factors for cervical cancer include low socioeconomic status, early onset of sexual activity, multiple sexual partners, lowered immune system, smoking, and infections with human papillomavirus (HPV), the most common cause of cervical cancer. There are currently two HPV vaccines recommended for young women to protect against cervical cancer. Ideally these should be given before the first sexual contact (Ignatavicius, 2013a). Pain usually is not present, but vaginal bleeding and leukorrhea (yellow or white mucous discharge) may occur, especially in older women.

The American Cancer Society (ACS, 2020) highly recommends that women aged 30 to 65 have a cotest of both a primary HPV test and Papanicolaou (Pap) test every 5 years, or at least a Pap test alone every 3 years. For women aged 65+, those who have had ≥3 consecutive negative Pap tests or ≥2 consecutive negative HPV and Pap tests within the past 10 years, with the most recent test occurring in the past 5 years, should discontinue cervical cancer screening (ACS, 2020). Women who have had a total hysterectomy will also no longer require screening. As is true for other forms of cancer, early detection and prompt treatment give the best chance for cure. Treatment usually involves a hysterectomy and removal of pelvic lymph nodes plus radiation and/or chemotherapy.

Cancer of the uterus (endometrial) occurs most often after menopause because of hormonal imbalances. Risk factors for uterine cancer include taking estrogen alone as hormone therapy, obesity, starting menstruation at an early age, beginning menopause later, never being pregnant, and getting older (Tabloski, 2014). The cure rate is high with early detection and treatment, but too often it is not detected and treated early enough to be cured. Vaginal bleeding, spotting, and staining are the most noticeable symptoms. Although bleeding may be caused by other diseases or physical problems, it should never be ignored because it is a cardinal symptom of uterine cancer.

Diagnosis generally involves a biopsy of cells in the lining of the uterus and ultrasound. Treatment of uterine cancer depends on its severity and the stage in which it is discovered. Typical treatment modes include surgical removal of the uterus, cervix, ovaries, and fallopian tubes, often followed by chemotherapy or radiation.

Cancer of the ovary tends to increase with age and be a disease primarily of women older than 50. It has the highest mortality rate of all gynecological malignancies even though it represents only 5% of the gynecological malignancies. Unfortunately, there are usually only vague symptoms specific to ovarian cancer, so early detection is difficult. In the early stages women are often asymptomatic (Eliopoulos, 2018). As it progresses, there may be abdominal swelling and bloating, feeling the need to urinate urgently or frequently, a sense of pelvic heaviness, back pain, appetite loss, or a change in bowel habits and fatigue. Late in the disease there is persistent abdominal pain and ascites (accumulation of serous fluid in the peritoneal cavity). A biopsy is the only way to definitely diagnose ovarian cancer. Surgical removal of the tumor, chemotherapy, and radiation have been reasonably successful if the tumor is not far advanced. When experiencing any of these symptoms screening by a healthcare professional is highly recommended.

Breast cancer is the second leading diagnosis of cancer for women following skin cancer (Mauk & Silva-Smith, 2018). Men can get breast cancer also but only account for less than 0.5% of all cases. Most common sites are in the area near the armpit and the upper, outer part of the breast. Changes in skin color, dimpling, or discharge from the nipple are potential signs of breast cancer. It should be noted, however, that many lumps in the breast are not malignant but instead may be fibrocystic changes in breast tissue. However, all changes in breast tissue should be medically evaluated immediately because early detection is crucial for treatment and survival. The ACS (2020) recommends women age 55 and over should have a mammogram every 2 years or continue screening yearly. Screening should continue as long as a woman is in good health and expected to live 10 years or more. All women should examine their breasts monthly.

Risk factors include older age, family history, early menstruation and late menopause, estrogen hormone therapy, obesity, alcohol use, no pregnancies, and hereditary factors. Women with an abnormal BRCA1 or BRCA2 gene have a much higher risk for developing breast cancer. These gene mutations can be detected by DNA testing. However, more than 70% of those who develop breast cancer have no identifiable risk factors for the disease (Marieb & Hoehn, 2019).

Two categories of breast cancer are noninvasive, in which the cancer remains in the duct, and invasive, when it spreads to tissue around the ducts. Metastasis occurs when cancer cells leave the breast and spread to other sites such as bone, lungs, brain, and liver.

Diagnosis usually involves mammography, especially digital mammography. Ultrasonography and MRI are also useful diagnostic techniques. However, breast biopsy is more definitive in diagnosis. Treatment for breast cancer usually includes a biopsy of breast tissue for accurate diagnosis, chemotherapy and/or radiation therapy, drug therapies, and surgical removal of the tumor (lumpectomy) or the entire breast and surrounding tissue (mastectomy). A wide range of treatment options are now available to treat breast cancer. The treatment of choice in breast cancers depends on the type of cancer, the areas involved, and the stage of the cancer. The most common staging method for breast cancer involves tumor size, nodal involvement, and presence of metastasis. Stages range from 0 to IV. With Stage 0 the cancer is limited inside the milk ducts and is noninvasive. Stage IV involves metastasis where the cancer has spread outside the breast and lymph nodes (Choma, 2017). A combination of treatment modes should be selected in consultation with medical specialists. Cure rates for breast malignancy are significantly higher with early detection before the cancer has spread into the surrounding lymph tissue.

COMPONENTS AND FUNCTIONS OF THE MALE REPRODUCTIVE SYSTEM

The external organs or genitalia of the male reproductive system consist of the following:

- The *scrotum,* a pouch or sac suspended from the pubis (front section of the pelvis), encloses and supports the two testes. The scrotum is divided into left and right halves by a midline septum, so there is a compartment for each of the two testes. The production of viable sperm in the testes is highly dependent on the maintenance of an appropriate temperature. In cold temperatures, the scrotum pulls the testes up close to the body and its heat. In warm temperatures, the scrotum is more flaccid and loose, allowing the testes to hang lower and away from body heat.
- The *penis*, composed of erectile tissue, functions both as the organ of copulation and excretion.

The internal organs of the male reproductive system include the following:

- The *testes* are significant in the production of spermatozoa (sperm cells) and the secretion of male sex hormones (primarily testosterone). These hormones influence the appearance of secondary sex characteristics, development of the body, and behavior.
- A *system of ducts* transports the spermatozoa from the testes to the outside of the body. These ducts include the epididymis, which is continuous with the vas deferens and merges with the seminal vesicle ducts, ending in the ejaculatory duct. The latter duct ejects semen containing spermatozoa and various fluids into the urethra, which carries both semen and urine.
- The *prostate gland* surrounds the upper section of the urethra just below the bladder. It secretes an alkaline fluid containing enzymes (fibrinolysin and acid phosphatase) and is significant in activating sperm. Because of its location near the rectum, the prostate gland can be palpated by digital (finger) examination of the rectum, and this should always be included in physical examinations to detect evidence of prostate enlargement or cancer.
- Two *Cowper's glands,* each about the size of a pea, are found on either side of the urethra and below the prostate. These secrete a clear, mucous-like substance into the urethra that becomes part of the semen and acts as a lubricant.
- Two *seminal vesicles* are situated near the lower surface of the bladder and secrete a thick fluid that mixes with the sperm from the testes (see Figure 12.2).

AGE-RELATED CHANGES IN THE MALE REPRODUCTIVE SYSTEM

Aging changes in males involve the following:

- Fewer viable sperm are produced and motility of sperm dssecreases. Nevertheless, most healthy men continue to produce enough viable sperm to fertilize ova well into older age.
- The amount and consistency of the seminal fluid remains about the same, while the ejaculatory force takes more time to achieve.
- Decreases in testosterone levels probably occur with age, and the testes become less firm and are smaller.
- Increased time is needed to attain an erection. They are less firm and often need direct stimulation to retain rigidity.
- An increase in the size of the prostate gland often accompanies aging. Enlargement of the prostate may compress the urethra and inhibit or prevent the flow of urine (Eliopoulos, 2018).

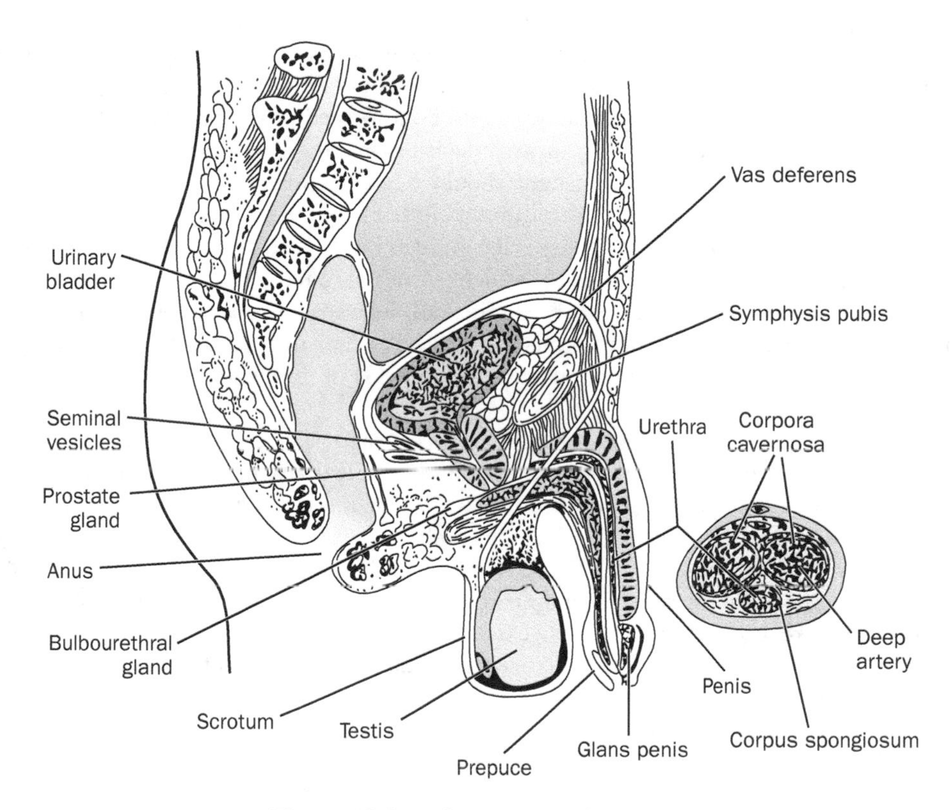

Figure 12.2 Male urogenital system.

AGE-RELATED DISORDERS OF THE MALE REPRODUCTIVE SYSTEM

Prostate Enlargement (Benign Prostatic Hyperplasia)

Enlargement of the prostate gland, or benign prostatic hyperplasia, is a significant age-related change in males. This topic is discussed in the chapter on the urinary system.

Cancer of the Prostate

Prostate cancer, the second most common cancer in men (after skin cancer), increases with age and occurs most often in men in their 60s, 70s, and 80s, but it can also occur in those who are younger. Risk factors include age (older than 50), family history of prostate cancer, ethnicity (African American men have the highest rate of prostate cancer), and diet (ingesting large amounts of fat, especially from well-done red meat, increases testosterone and other hormone production and speeds up growth of prostate cancer). A diet low in fiber also seems to increase the risk, as does exposure to environmental toxins (Ignatavicius, 2013b). In the early stages there are usually no symptoms. Later, there may be obstruction of the urethra, urinary frequency, blood in the urine, some discomfort in the perineal and rectal areas, nausea, weakness, and pain in the lower back, pelvis, and upper thighs. Still later, if the disease has spread, pathological fractures and lower extremity edema may be present.

Guidelines for prostate screening are available and should be monitored as they are updated. For men with average risk, no organization endorses routine prostate cancer screening due to concerns of potential overdiagnosis (identifying disease that may have never caused symptoms or harm) and the potential serious side effects of treatment (ACS, 2020). Men aged 50+, with at least a 10-year life expectancy, should be informed of the benefits, risks, and uncertainties of screening and have an "informed decision-making" discussion with their health provider. However, African American men should begin having this discussion with their health provider at age 45. Diagnostic procedures include palpating the prostate gland by digital examination and the use of the prostate-specific antigen (PSA) test, although this test remains controversial. PSA is a blood serum test that determines elevated PSA levels. In pathological conditions of the prostate, PSA levels are significantly elevated. Elevated levels, however, do not necessarily indicate malignancy because such levels occur in both benign and malignant disorders of the prostate. A transrectal ultrasound (TRUS) of the prostate and a biopsy often confirm a diagnosis.

Various treatment procedures are available, and because prostate cancer is usually slow growing, some men opt for monitoring the prostate regularly rather than immediate treatment. Treatment choices ideally are a decision between the individual and healthcare providers depending on the stage of the disease. Options may include combinations of external beam radiation, cryosurgery (freezing the cancer cells with liquid nitrogen), minimally invasive surgeries, prostatectomy (surgical removal of the prostate and nearby lymph nodes), and radioactive seed implants (brachytherapy). Treatment varies to a large extent depending on the stage of the cancer (Linton, 2007; Tabloski, 2014). Nonsurgical management techniques include radiation therapy, hormonal therapy, and sometimes chemotherapy.

SEXUALITY

According to the now classic research by Masters and Johnson (1966, 1970), the human sexual response can be categorized into four phases: excitement, plateau, orgasm, and resolution. For both males and females, sexual response in each of these phases changes somewhat with age (Atkinson, 2006).

In older men:

1. The excitement phase is longer than for younger men; it may take several minutes to achieve an erection, and the erection is also somewhat weaker.
2. The plateau phase, in which sexual tension is enjoyable, can be maintained longer than in younger men.
3. Less seminal fluid may be ejaculated. Older men may not experience an orgasm each time they engage in sexual intercourse. Changes in the orgasmic process sometimes cause anxiety in older men, who view their responses as less sexually adequate than in their younger years.
4. In the resolution phase, when sexual organs resume their unstimulated state after ejaculation, a longer time is required to again be responsive to sexual stimulation. Also, in most older men an erection is lost more rapidly after ejaculation.

In older women:

1. In the excitement phase, lubrication of the vagina is a slower process, requiring a longer period of sexual stimulation before adequate lubrication occurs. Atrophic vaginal changes may make intercourse uncomfortable for some older women, but, as indicated earlier, topical application of appropriate creams or hormone therapy usually relieves the discomfort.
2. In the plateau phase, there is less elevation of the uterus, but this is apparently of little significance in obtaining sexual satisfaction. This phase is a bit longer for older women than for younger women.

3. In orgasm, uterine contractions are fewer and perhaps spastic, causing possible pain in some women. This phase is also somewhat shorter for older women.
4. Resolution, or return to the unstimulated state, is quicker for older women than for younger women.

Sexuality is a basic human need and is defined as a state of emotional, social, mental, and physiological well-being as it relates to sexuality. It offers one the chance to show affection, passion, and admiration for another person while enhancing personal growth and communication. Sexual expression reduces stress and increases feelings of belonging and being loved. Furthermore, sexuality affirms life and engenders new growth and development of the person (Touhy, 2017).

None of the age-related physiological changes in sexual responsiveness in either sex should impose significant limitations on sexual activity and enjoyment in older age. By understanding the altered response patterns, and learning ways to adapt to age-related changes most healthy older adults can continue to have active and satisfying sexual relationships well into advanced old age. Older adults, though, do need accurate information about normal aging changes in the reproductive system, and some may need professional assistance to help them learn appropriate adaptations. Research studies have confirmed that many older men and women can and do remain sexually active into very old age.

Many factors are involved in continuing to have sexual relationships in older age. One issue of significance is that of erectile dysfunction. The term *impotence* has been replaced by *erectile dysfunction,* or *ED,* which is the lack of ability to achieve or maintain an erection adequate for satisfactory sexual functioning. ED increases with age, and estimates suggest that 50% of men ages 40 to 70 and nearly 70% of those 70 and older experience ED (Agronin, 2004). It is now recognized that physiological factors are primarily the most common cause, not psychological factors (although they are not unimportant), and certainly it is not due to aging per se.

Physiological factors likely to cause ED are effects of medications and various pathological conditions (especially diabetes and alcoholism). Psychological factors that may contribute to ED are depression, anxiety, relationship difficulties, fear of failure, and self-esteem issues. Many men with ED still do not seek professional healthcare, although with the availability of various medications to deal with ED and increased media attention to this difficulty, more men are likely to seek help than ever before (Miller, 2009). Side effects of these medications must always be considered, however. Other treatment options are constriction device therapy, implants, and surgery and counseling. Female sexual dysfunction is also receiving more attention from healthcare providers as women become less inhibited about discussing their sexual issues.

Variables found throughout the literature affecting sexual responsiveness in both men and women include the following:

1. A variety of different medications have side effects that influence sexual activity.
2. Many health problems, especially chronic diseases, affect sexual relations. Examples are diabetes, myocardial infarction, stroke, chronic obstructive pulmonary disease, arthritis, Parkinson's disease, and renal disease. Surgeries that are disfiguring or even perceived as disfiguring may impair the expression of sexual behavior. Examples are mastectomy, colostomy, prostate surgery, and hysterectomy.
3. A very important point is that sexual activity in younger years is a significant factor because many professionals suggest the best indicator of sexual activity in older age is positive, satisfying sexual relationships in younger years. For those who have never enjoyed sex, older age may be a convenient excuse to stop participating in sexual behaviors.
4. For many women in our society, sexual relations cease because they lack a partner. Statistically, there are more older women than older men, thus reducing the availability of a partner for the remainder of their lives.

5. Those in nursing homes often lack privacy, which is a major deterrent to continuing sexual relationships. In fact, interest in sex or overt sexual behavior is sometimes interpreted as abnormal and in some long-term care settings may result in ridicule or even subtle punishment. Sexual behavior in older adults who have dementia poses other challenges for caregivers and family. In our society, some people still have very ageist views on this topic and believe that sexual interest and activity are not appropriate for anyone who is older. However, more long-term care settings now reserve rooms for conjugal visits or make some provision for privacy. Continuing in-service programs on life-span sexuality are necessary to improve caregivers' attitudes in this important area of life.

Culturally derived sexual fears and myths common among older adults include fear of having a heart attack or stroke during sex, fear that sexual activity shortens life, fear of social criticism for continued sexual interest and activity, fear of engaging in masturbation or homosexual behaviors because of cultural or religious beliefs, and fear of defying the cultural bias against older women associating with younger men.

For many years, sexual activity had been generally considered to be reserved for procreation and, for women, as something to be endured but not necessarily enjoyed. For those who subscribe to this view, sexual interest and expression in the later years are not appropriate behaviors. In addition, children sometimes view their parents, especially older parents, as sexless and become disturbed when a parent expresses interest in dating or remarrying. Although these are antiquated ideas, remnants of such beliefs and stereotypes still linger to some extent. Culturally, there has been substantial societal and legislative change in the acceptance of homosexuality and homosexual relationships. However, for some older homosexual adults, there can still be anxiety and isolation due to perceived stigma from others.

Older adults who do experience difficulties in their sexual lives should seek sexual health counseling or sex therapy. Cultural stereotypes about aging and sex have been major deterrents to the availability of treatment for sexual dysfunction and dissatisfaction in older adults. Another issue of great importance for older adults is HIV/AIDS infections, which are much more prevalent than most people think. This topic is addressed in the chapter on the immune system. As information about life-span sexuality becomes more widely accessible to all ages in our society, hopefully more sensitive attention will be devoted to helping older adults meet this basic human need that continues throughout life (Eliopoulos, 2018).

REFERENCES

Agronin, M. (2004, Summer). *Sexuality and aging: An introduction* (pp. 12–13). CNS Long-Term Care.

American Cancer Society. (2020). *Cancer facts & figures 2020*. https://www.cancer.org/content/dam/cancer-org/research/cancer-facts-and-statistics/annual-cancer-facts-and-figures/2020/cancer-facts-and-figures-2020.pdf

Atkinson, P. J. (2006). Intimacy and sexuality. In S. E. Meiner & A. G. Lueckenotte (Eds.), *Gerontologic nursing* (3rd ed., pp. 268–280). Mosby Elsevier.

Choma, K. K. (2017). Assessment of the reproductive system. In S. L. Lewis, L. Bucher, M. M. Heitkemper, & M. M. Harding (Eds.), *Medical–surgical nursing* (10th ed., pp. 1183–1222). Elsevier.

Crowley, L. (2011). *Essentials of human disease*. Jones & Bartlett.

Eliopoulos, C. (2018). *Gerontological nursing* (9th ed.). Wolters Kluwer.

Ignatavicius, D. (2013b). Care of male patients with reproductive problems. In D. Ignatavicius & L. Workman (Eds.), *Medical–surgical nursing* (7th ed., pp. 1629–1651). Saunders Elsevier.

Ignatavicius, D. (2013a). Care of patients with gynecologic problems. In D. Ignatavicius & L. Workman (Eds.), *Medical–surgical nursing* (7th ed., pp. 1611–1628). Saunders Elsevier.

Linton, A. D. (2007). Genitourinary system. In A. D. Linton & H. W. Lach (Eds.), *Matteson & McConnell's gerontological nursing* (3rd ed., pp. 484–524). Saunders Elsevier.

Marieb, E. N., & Hoehn, K. (2019). *Human anatomy and physiology* (11th ed.). Pearson.
Masters, W. H., & Johnson, V. E. (1966). *Human sexual response.* Little, Brown.
Masters, W. H., & Johnson, V. E. (1970). *Human sexual inadequacy.* Little, Brown.
Mauk, K., & Silva-Smith, A. (2018) Gerontological nursing: Competencies for care. In K. Mauk (Ed.), *Gerontological nursing* (4th ed., pp. 305–385). Jones and Bartlett.
Miller, C. A. (2009). *Nursing for wellness in older adults* (5th ed.). Wolters Kluwer/Lippincott Williams & Wilkins.
Tabloski, P. (2014). *Gerontological nursing* (3rd ed.). Pearson Education.
Touhy, T. A. (2017). Intimacy and sexuality. In T. A. Touhy & K. Jett (Eds.), *Ebersole & Hess' toward healthy aging* (9th ed., pp. 445–462). Elsevier.
Winton, M. B. (2019). Endocrine function. In S. Meiner & J. J. Yeager (Eds.), *Gerontologic nursing* (6th ed., pp. 521–542). Elsevier.

13

The Endocrine System

STRUCTURE OF THE ENDOCRINE SYSTEM

The regulation and integration of body activities depend on the nervous system and the endocrine system. In contrast to the exocrine glands, which secrete substances through tubes or ducts that then empty into a body cavity or onto a specific surface, the endocrine system functions through ductless glands that secrete chemicals called hormones directly into the blood or lymph. Whereas nervous system activity depends on electrochemical nerve impulses taking only milliseconds, it may take seconds, hours, or even days for the endocrine system to transport hormones via the blood or lymph to organ sites. Furthermore, responses to hormones tend to be more prolonged than those initiated by the nervous system (Marieb & Hoehn, 2019).

The pituitary gland and the brain are especially closely related and influence each other. The complex neuroendocrine system consists of the hypothalamus in the brain; the pituitary gland; other glands such as the thyroid, adrenals, and gonads (the ovaries and testes); tissues controlled specifically by the pituitary and the aforementioned glands; and other tissues that produce hormones.

A specific hormone influences activity of only certain cells known as target cells. Hormones alter target cell activity by increasing or decreasing rates of normal cellular activity. Certain chemicals in the blood play a primary role in controlling hormone secretion. Such substances may be hormones from other glands or, in some cases, no hormonal chemicals. For example, secretions of the thyroid gland are triggered by hormones produced in the pituitary gland. On the other hand, blood sugar, a nonhormonal substance, triggers hormone production in the pancreas. Sometimes hormones secreted by gland B will, when stimulated by hormones from gland A, in turn exert some control on gland A's secretions. Endocrine activity is a highly complicated feedback loop in which one action triggers others so that the necessary internal equilibrium may be maintained in spite of constantly varying demands, both internal and external. In these ways the endocrine system contributes significantly to the maintenance of equilibrium (homeostasis) in the body. Significant body processes that are integrated and controlled by hormones are reproduction; processes of growth and development; activation of body defenses against stressors; electrolyte, water, and nutrient balance of the blood; and control of cell metabolism and energy balance.

The major glands of the endocrine system include the pituitary, thyroid, parathyroid, adrenal, pineal, and thymus glands. Other organs that contain endocrine tissue and are considered major endocrine organs are the pancreas, ovaries, and testes (see Figure 13.1). Organs that produce hormones in addition to these major organs include the gastrointestinal tract, heart, skeleton, and adipose tissue, among others (Marieb & Hoehn, 2019).

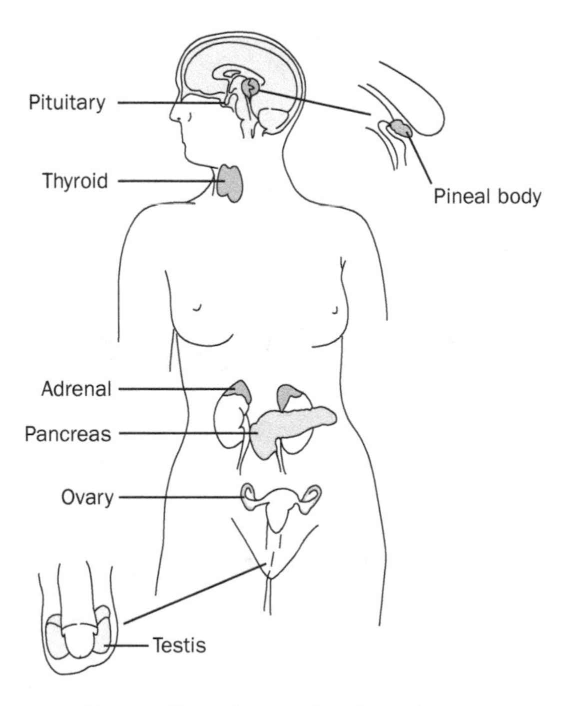

Figure 1.1 Major endocrine glands.

Pituitary Gland

In collaboration with the hypothalamus, the pituitary exerts control over a wide array of body functions. The pituitary gland is attached to the base of the brain and has two distinct lobes, the anterior and the posterior. Historically referred to as the master gland, the pituitary secretes a greater number of hormones than any other gland in the endocrine system. However, the hypothalamus of the brain produces chemicals that stimulate the pituitary gland to produce hormones, so it also exerts major controlling functions.

The *anterior lobe* of the pituitary secretes a variety of hormones, including the following:

- Growth hormone (GH), which primarily regulates growth of the body skeleton and muscles
- Thyroid-stimulating hormone (TSH), which regulates the development and activity of the thyroid gland
- Adrenocorticotropic hormone (ACTH), which is necessary for normal development and functioning of the adrenal gland and is often helpful in dealing with stressors
- Follicle-stimulating hormone (FSH) and luteinizing hormone (LH), which are produced by both sexes and are necessary for the normal development and functioning of the gonads
- Prolactin (PRL), a hormone that stimulates lactation (milk production) in nursing mothers

The *posterior lobe* of the pituitary releases two hormones:

- Oxytocin, which stimulates the breasts to release milk after childbirth
- Vasopressin, or antidiuretic hormone (ADH), which causes increased reabsorption of water from the urine back into the blood, resulting in less urine produced and an increase in blood volume

ADH is very significant in regulating water balance in the body. These two hormones are actually produced in the hypothalamus but are stored in the pituitary and released as needed.

Thyroid Gland

The thyroid gland is a two-lobed structure situated on either side of the upper part of the trachea. It assists in the metabolic functions of virtually every cell in the body. Energy is produced by oxidative reactions in the cells, with the oxidation rate primarily controlled by the thyroid. An increase in thyroid hormone secretion raises the basal metabolic rate, whereas a decrease in hormone secretion lowers the basal metabolic rate. The thyroid gland produces two major hormones:

- Thyroid hormone (TH), the major hormone concerned primarily with body metabolic rate and body heat production. An iodine deficiency results in a deficiency of thyroid hormone.
- Calcitonin, a thyroid hormone that lowers blood calcium levels. It is most important in childhood when skeletal growth occurs rapidly.

Parathyroid Glands

The parathyroid glands are located close to the thyroid gland. Parathyroid hormone (PTH) is the most important hormone regulating calcium balance of the blood. Calcium levels are especially significant in muscle contraction, nerve impulse transmission, and blood clotting.

Adrenal (Suprarenal) Glands

The two adrenal glands are located on the upper part of each kidney and are composed of two distinct parts: an adrenal cortex (the outer layer) and an adrenal medulla (the inner part), which is part of the sympathetic nervous system.

The hormones produced by the adrenal cortex are called corticosteroids and are extremely important in the regulation of electrolytes, especially sodium and potassium. Sodium regulation is crucial for overall body homeostasis. They are also important in regulating levels of fats and proteins in the body and in controlling body reactions to stress. Such stresses to the body include extremes of heat or cold, excessive muscular activity, burns, infections, and other types of trauma. Specifically, the adrenal cortex steroid hormones are classified as (a) glucocorticoids (cortisol), which affect glucose metabolism; (b) mineralocorticoids (aldosterone), essential for fluid and electrolyte balance; and (c) androgens, substances stimulating development of male characteristics, produced in small but important amounts.

The adrenal medulla plays a significant role in activating responses to emergencies (the fight or flight reactions). The catecholamines, epinephrine and norepinephrine, produce the same arousal effects as the sympathetic nervous system, although they usually cause briefer or short-term responses.

Pancreas

The pancreas, a gland with both exocrine and endocrine functions, is located in front of the first and second lumbar vertebrae and partially behind the stomach. The exocrine function of the pancreas is to produce an enzyme-rich juice that is sent through a duct or tube into the duodenum to help with the digestion of food. The endocrine portion of the pancreas is called the islets of Langerhans. The two hormones produced here are insulin and glucagon (a form of carbohydrate), both of which are important in regulation of blood glucose levels.

Insulin lowers blood glucose levels and also influences protein and fat metabolism. It lowers blood sugar levels by increasing the supply of glucose and other simple sugars to muscles, connective tissue, and white blood cells, where they can be metabolized. It also regulates the breakdown of glycogen (the form in which carbohydrates are stored in the body) to glucose and the conversion of amino acids or fatty acids to glucose.

Insufficient insulin leads to diabetes, characterized by above-normal amounts of sugar (glucose) in the blood and urine. Too much insulin in the body, however, produces low blood sugar (hypoglycemia). If the level of blood sugar becomes too high or too low, death occurs. When low blood glucose occurs, glucagon is released from the pancreas targeted to the liver, causing the liver to change glucagon to glucose, which then enters the bloodstream and raises blood sugar levels. This also increases the formation of glucose in the liver. Glucagon action is the opposite of insulin action.

Pineal Gland, Thymus Gland, and Gonads

The *pineal gland* is a very small structure in the diencephalon of the brain, and its endocrine function is not completely understood. The major secretion of the pineal gland is melatonin, a hormone derived from serotonin. Changing melatonin levels seem to affect day/night cycles (circadian rhythm), which may then affect body temperature, sleep, and appetite. It may also control the timing of puberty (Marieb & Hoehn, 2019). The *thymus gland*, located in the chest, seems to be significant in the development of the immune system. The thymus is large in children but shrinks in size with age. In older age it is composed mainly of fibrous connective tissue and adipose tissue (Marieb & Hoehn, 2019). The *gonads* (ovaries and testes) have a dual role. They produce ova and sperm (the reproductive cells of the body) and also hormones. Female ovaries produce ova; estrogens and progesterone, the hormones responsible for maturation of the female reproductive organs; development of secondary sex characteristics; and the cyclic changes in the uterus necessary for menstrual cycles. The testes of the male produce sperm and the hormone testosterone. Testosterone is responsible for the maturation of male reproductive organs, development of secondary sex characteristics, and production of sperm, as well as sex drive.

AGE-RELATED CHANGES AND DISORDERS OF THE ENDOCRINE SYSTEM

Because the endocrine system is so complex and interrelated, it is difficult to ascertain the effects of aging on specific glands. In most glands there is some atrophy of tissues and a decreased rate of secretion with age, but the implications of these changes have not been clearly identified. The majority of endocrine organ systems function well into old age except those that have been exposed to environmental pollutants or hypersecretory disorders. The process of aging may modify the rates of hormonal secretion or target cell receptor sensitivity (Marieb & Hoehn, 2019).

Homeostatic failure is the foundation for age-related declines in physiological performance in the endocrine system. Age-related changes in glands can result in critical health problems associated with metabolism of electrolytes, glucose, water, and minerals, but this is highly individualized and the concentrations of major hormones necessary for homeostatic equilibrium are not universally changed with aging. It does appear that aging in the endocrine system contributes to increased difficulties in regulating various mechanisms that ultimately result in lessened vitality in older adults (Hill, 2006).

Older adults' symptoms of endocrine disorders are often atypical, nonspecific, or muted, or they may mimic other disorders such as depression or fatigue. Sometimes in this age group there are no presenting symptoms even when disease is present. In addition, signs or symptoms of endocrine disease may be changed or masked by concomitant illnesses or medications. Endocrine pathologies are classified as those based on hyposecretion, hypersecretion, and hyporesponsiveness of various glands.

Aging and the Pituitary Gland (Hypophysis)

The pituitary gland appears to maintain its overall functioning reasonably well into old age, even though its hormonal secretions generally decline with advancing age. Both vascularization and connective tissue decrease which can or cannot affect hormonal production (Marieb & Hoehn, 2019). There are no highly significant age-related diseases of the pituitary gland. GH secretion decreases with age and GH deficiency may result in decreased lean body mass, decreased bone density, and increased fat mass. Generally, though, the decrease in GH does not have important health consequences. Oversecretion of GH can result in acromegaly (increased growth in bones of the hands, feet, face, and increased size of viscera), but this is quite rare.

Aging and the Thyroid Gland

Thyroid hormone synthesis and release activity decreases somewhat with age as does the basic metabolic rate. This decrease is caused by mild hypothyroidism or increased body fat relative to muscle since the muscle is more metabolically active than fat (Marieb & Hoehn, 2019). Two pathologies, however, hyperthyroidism and hypothyroidism, are not uncommon in older adults. The signs and symptoms of both in older adults are often atypical and may be overlooked, mistaken for other pathological conditions, or attributed to normal aging.

Hyperthyroidism

Hyperthyroidism is caused by overproduction of thyroid hormone. Thyrotoxicosis, a toxic condition, results when there is hyperactivity of the thyroid gland. In older people hyperthyroidism often is due to Grave's disease or toxic goiter. It may also be caused by excessive intake of pharmaceutical thyroid hormone. The symptoms of hyperthyroidism in older adults are usually subtle and not the classic symptoms seen in younger people (Linton et al., 2007; Tabloski, 2014). Older adults often have fewer symptoms and a different complex of symptoms than younger people. Symptoms may also be masked by coexisting diseases. Cardiac complications and congestive heart failure are often the most common symptoms of hyperthyroidism in older adults, but other symptoms such as weight loss, fatigue, heat intolerance, heart palpitations, nervousness, depression, and tremors can also occur. Goiter, or enlarged thyroid gland, may or may not be present.

Diagnosis involves measuring serum levels of thyroid hormones in the blood, other specific laboratory tests, physical examination, and usually a thyroid scan. There are three usual treatment options for hyperthyroidism: medication to reduce the overactive thyroid, surgery to remove the hyperfunctioning tissue, or radioactive sodium iodine to destroy the gland. Currently the preferred treatment strategy is radioactive sodium iodine, although it takes more time to accomplish results than the other two choices. Drug therapy is effective and quite fast, but side effects must be carefully considered. Because of other coexisting medical conditions surgery usually is not a treatment of first choice in older adults (Tabloski, 2014).

Hypothyroidism

Decreased thyroid hormone results in hypothyroidism, and a deficiency in iodine is the most common cause of hypothyroidism. Primary hypothyroidism is due to dysfunction of the thyroid gland itself; secondary hypothyroidism is caused by a deficit in the pituitary gland that causes thyroid malfunctioning; and transient hypothyroidism is caused by thyroiditis, or discontinuing thyroid hormone therapy. Symptoms of hypothyroidism are easily attributed to some other disorder, and the true problem is not always diagnosed. Fatigue, depression, weight gain, cold intolerance, muscle cramps, constipation, dry skin, mental confusion, and loss of appetite may be attributed to aging rather than to thyroid disease. Hypothyroidism is quite common in older adults and affects women much more commonly than men, often between 30 and 60 years of age (Linton et al., 2007).

Diagnosis is achieved through a detailed health history, physical examination, and laboratory tests to evaluate thyroid function. The TSH test is the best screening procedure available and allows for earlier diagnosis of thyroid disorders than in the past. Treatment involves replacement of the deficient hormone. Levothyroxine is often prescribed, starting with low doses and increasing gradually to prevent cardiovascular and neurologic side effects. Continued dosage monitoring is necessary and treatment is usually lifelong.

Aging and the Parathyroid Glands

Aging seems to produce no dramatic change in the parathyroids other than a gradual reduction in glandular activity with age. Although not necessarily age related, both hyperparathyroidism and hypoparathyroidism sometimes occur in older adults. Changes in the PTH may contribute to osteoporosis.

Hyperparathyroidism

Hyperparathyroidism is caused by oversecretion of PTH, which regulates calcium and phosphate levels by increasing bone resorption of calcium, renal reabsorption of calcium, and vitamin D activation. This disease is seen more commonly in women than in men and increases after age 50 in both men and women. Symptoms typically include apathy, fatigue, muscular weakness, constipation, and psychiatric manifestations. However, many older adults are asymptomatic or have only mild symptoms of depression, muscle weakness, or malaise. Diagnosis depends on assessing serum calcium levels and levels of PTH. Ultrasound, CT scanning, and MRI are also useful to evaluate the parathyroid glands. Treatment of choice usually is partial or complete removal of parathyroid tissue, or medical management if the individual is a poor surgical risk. Continued ambulation is crucial, accompanied by high fluid and moderate calcium intake. Various drugs are used to lower PTH levels.

Hypoparathyroidism

Hypoparathyroidism results from lessened PTH secretion, often after excessive removal of parathyroid tissues surgically. Symptoms arise from elevated blood phosphate and decreased concentration of blood calcium. Tetany (generalized muscle contractions) is the most common manifestation of this disease. Maintaining appropriate calcium levels is crucial. Diagnosis is by physical examination, EEG, CT scans, and laboratory studies to assess hormones and phosphate levels. Treatment may involve drug therapy to raise serum calcium levels and a diet high in calcium but low in phosphorus. Monitoring calcium levels is important because excessively high calcium levels can be hazardous to kidney function (Konick-McMahan, 2007; Workman, 2013).

Aging and the Adrenal Glands

A gradual reduction in adrenal secretions occurs with age as the adrenal glands experience structural changes. The usual control of cortisol remains if the individual is not stressed and is healthy (Marieb & Hoehn, 2019). Disorders of the adrenal glands may occur at any age, but none are clearly related to the aging process. Two diseases of the adrenal gland encountered in older persons, as well as younger adults, are Addison's disease and Cushing's syndrome.

Addison's Disease

Addison's disease is caused by a deficiency of hormones produced by the adrenal cortex and is characterized by muscle weakness, lack of appetite, weight loss, hypotension, fatigue, dark pigmentation of the skin, gastrointestinal symptoms, low blood glucose, low serum sodium, and high serum potassium. All three groups of adrenal corticosteroids are reduced in Addison's disease. The most common cause is an autoimmune response. Early diagnosis is difficult because symptoms typically do not appear until the disease is advanced. Diagnosis is confirmed by laboratory tests, especially by low levels of adrenocortical hormones in the blood or urine. CT scan or MRI is useful in localizing tumors, calcification, or enlargement. Treatment involves daily replacement of deficient adrenal hormones to prevent a recurrence of adrenal insufficiency, and individuals must be taught about long-term management. Options include oral doses, injections of corticosteroids, or androgen replacement therapy. Those with Addison's disease have difficulty adapting to physical or emotional stress and must learn to effectively manage their hormone replacement therapy on a daily basis.

Cushing's Syndrome

Cushing's syndrome, the opposite of Addison's disease, reflects excessive activity of the adrenal cortex with increased levels of corticosteroids. In older adults, the most common cause of Cushing's syndrome is a pituitary tumor or the side effects of corticosteroid medications. If these drugs are medically necessary, they should be prescribed in the minimum dosage and reviewed often as to dosage and need (Konick-McMahan, 2007; Workman, 2013).

Characteristic symptoms are fatty deposits around the face (moon face) and trunk with thin arms and legs, weakness, sleep disturbances, weight gain, and retention of sodium and water. Various laboratory tests such as urine, blood work, saliva tests, and possibly a CT scan or MRI (if a tumor is suspected) are used for diagnosis. Treatment is surgery and/or radiation therapy if a tumor (of the pituitary or adrenal cortex) is found. Drug therapy to control cortisol production is used if surgery is contraindicated or as an adjunct to surgery.

Aging and the Pancreas

Pancreatic secretions decrease somewhat with age, but generally not enough to significantly disrupt body functioning. However, glucose tolerance starts to diminish around age 40, and blood glucose levels elevate to higher levels and return to resting levels more gradually in older adults (Marieb & Hoehn 2019). Two significant changes in blood sugar levels that do affect older adults are hyperglycemia and hypoglycemia.

Hyperglycemia

Common symptoms of hyperglycemia (high blood sugar levels) are increased thirst, increased frequency of urination, fatigue and muscle weakness, a tendency to develop lesions or infections that do not heal, and decreased ability to sense heat and cold. In older adults, lethargy, fatigue, and subtle changes in mental functioning may be the only indications of hyperglycemia. Because some degree of glucose intolerance tends to occur with aging, a clinical diagnosis must differentiate between this age-related change and the clinical disease diabetes. Diagnosis usually involves fasting plasma glucose concentration or an oral glucose tolerance test.

Hypoglycemia

Symptoms of hypoglycemia (abnormally low blood sugar levels) can occur suddenly and unexpectedly and vary considerably from one person to another. Some possible causes are intense exercise, alcohol intake, or medication mismanagement. Symptoms of mild hypoglycemia are sweating, tremor, nervousness, hunger, and heart palpitations. Moderate hypoglycemia involves central nervous system impairments such as confusion, inability to concentrate, headache, slurred speech, irrational behavior, drowsiness, and uncoordinated movements. In severe hypoglycemia, there is disoriented behavior, seizures, difficulty arousing from sleep, and even loss of consciousness.

Because pancreatic secretions decrease with age and body tissues often become decreasingly sensitive to insulin, these two changes together undoubtedly contribute to *diabetes mellitus,* a disease extremely common in older age and a growing public health concern in the United States (Jett, 2016). The National Diabetic Statistic Report (Centers for Disease Control and Prevention, 2020) indicates that in 2018, 10.5% of the U.S. population had diabetes. Of those age 65 and over 26.8% or 14.3 million people had diagnosed or undiagnosed diabetes. Of those individuals 18 years of age or older it is believed that 34.5% had prediabetes and were unaware of it. A glucose level between 100 and 125 mg/dL is consistent with prediabetes. Individuals with prediabetes are more likely to develop type 2 diabetes. Type 2 diabetes is defined as impaired fasting glucose or impaired glucose tolerance or both. Delaying or preventing the development of type 2 diabetes may be accomplished by exercise, healthy eating, and not being overweight. There are also ethnic and minority group differences, with diabetes more common in Native Americans, African Americans, and Hispanics than in Caucasians.

Diabetes is defined as a group of chronic metabolic diseases in which there is hyperglycemia (increase in blood sugar) resulting from deficits in insulin production (specifically by beta cells of the pancreas), or insulin resistance (when body tissues do not respond to insulin), or both. There are several classifications of diabetes, including prediabetes, type 1, and type 2. There are other causes of hyperglycemia (pancreatitis, carcinoma, drug-induced diabetes), gestational diabetes, and conditions in which glucose levels are high but not high enough for a diagnosis of clinical diabetes. Specific criteria for a diagnosis of diabetes are as follows:

- A1C 6.5 or higher
- Symptoms of diabetes (weight loss, frequency of urination, and excessive hunger and thirst)
- Random plasma glucose equal to or greater than 200 mg/dL
- Fasting plasma glucose equal to or greater than 126 mg/dL
- Two-hour plasma glucose level greater than or equal to 200 mg/dL (Dickinson, 2017)

Type 1 Diabetes (Formerly Called Insulin-Dependent or Juvenile Diabetes)

This type of diabetes is a multisystem disease involving beta cell destruction, causing absolute insulin deficiency or pancreatic destruction from a viral infection such as mumps, measles, encephalitis, or influenza. It usually occurs in people younger than age 30, but it can also occur at any age. Considered an autoimmune disorder, the insulin-producing beta cells of the pancreas are destroyed, so external sources of insulin are necessary or the individual will die. The process of cell destruction usually occurs for years before symptoms become obvious. Typical symptoms are weight loss, excessive thirst, frequent urination, and excessive hunger. If insulin is not taken, the individual will develop diabetic ketoacidosis (DKA), a life-threatening condition. A genetic susceptibility to type 1 diabetes is well established, and immunologic as well as environmental factors may be contributors. With better education and more effective methods of control, however, more people with type 1 diabetes are now living longer.

Type 2 Diabetes (Formerly Called Non-Insulin-Dependent or Adult-Onset Diabetes)

Type 2, the more common type of diabetes, involves hyperglycemia caused by insulin resistance, in which cells do not respond to insulin due to unresponsive insulin receptors that are lacking in number or both. Second, there is a diminished ability for the pancreas to produce insulin. Third, the liver is not appropriately producing glucose. Fourth, there is modified production of hormones cytokines by adipose tissue (Dickenson, 2017).

In addition to genetic predisposition, lifestyle (especially obesity and lack of exercise), ethnicity, and age-related changes in metabolism increase the risk of older adults developing diabetes. Although some people with type 2 diabetes do take insulin, they do not necessarily depend on insulin intake to sustain life, as do those with type 1 diabetes. The pancreas produces some insulin, but either it is not enough for body needs or it is not effectively used in body tissues. People with type 2 diabetes can often control the disease (at least for many years) by lifestyle changes, diet, exercise, and medications. However, it must be stressed that diabetes requires constant attention to food intake (both quantity and type), blood glucose monitoring, evaluation of all medications ingested, and regular exercise. It can be a rather challenging disease to manage.

Chronic complications of types 1 and 2 diabetes often occur from 10 to 15 years after onset and include coronary, peripheral vascular, and cerebrovascular disease; diabetic retinopathy affecting small blood vessels in the eye; neuropathy of the kidney; and other neuropathies affecting sensorimotor and autonomic nerves. It is the leading cause of adult blindness, lower limb amputation, and end-stage kidney disease, and is a major factor in strokes and heart disease (McLeod, 2013).

Acute complications associated with diabetes include hypoglycemia and hyperglycemia, both of which are life threatening if not treated appropriately. Acute complications usually result from an unbalanced treatment regimen.

Diagnosis and Treatment for Diabetes

The presenting signs of diabetes may be quite different in older adults compared with younger adults, creating complex issues for accurate diagnosis and treatment. For example, excessive thirst, frequent urination, and weight loss are not always evident, and symptoms that do exist may be attributed to aging. The existence of other chronic illnesses further complicates diagnosis and treatment. Some medications taken by older adults may contribute to hyperglycemia, whereas others potentiate hypoglycemia. Diagnosis depends on results of fasting plasma glucose as well as a 2-hour oral glucose tolerance test, a test for elevated glycosylated hemoglobin levels and AIC levels. A complete physical examination, health history, and comprehensive geriatric assessment are also necessary.

The overall goal of treatment for diabetes is to stabilize insulin activity and blood glucose levels within normal ranges to reduce the risk of acute and chronic complications associated with diabetes. Treatment continues over a lifetime, and treatment regimens need to be varied as research advances occur. The American Diabetes Association recommends blood pressure be less than 130/80 and low-density lipoprotein cholesterol (LDL-C) be less than 100 mg/dL. Lifestyle modifications should concentrate on reducing saturated fat, trans fat, and cholesterol intake and increasing omega-3 fatty acid intake and fiber. Weight loss, if necessary, and increasing physical activity are also highly recommended. Generally, there are five important components of treatment or management of diabetes: diet and meal planning, exercise, monitoring, medications, and education.

Diet

The diet must be balanced, nutritious, and one that aids in maintaining appropriate body weight. General recommendations for nutrient balance are the following, realizing that each person's meal plan should be individualized to take lifestyle and health goals into consideration:

The individual should be taught how to manage their dietary intake of carbohydrate, fats, and proteins. Guidance from a dietitian can be very helpful. Carbohydrates need to be maintained in a healthy range. This range is dependent upon blood glucose levels, weight, level of activity, age, preference of the individual, and medications they are taking. Some alcohol may be ingested if the person does not have a problem with excessive alcohol ingestion and monitors their blood glucose levels. Moderate consumption of alcohol is considered to be 1 drink per day for women and 2 drinks per day for men (Dickinson, 2017). A diet close to the individual's usual diet encourages compliance. Dietary counseling and extensive education are necessary if the patient is to understand the complexities of sound dietary choices. Ideally, diabetic diets for older adults need to take into consideration ethnic and cultural food preferences, economic factors, physical problems such as ill-fitting dentures, and resistance to modifying lifetime eating habits.

Exercise

Exercise is essential to the health of people with diabetes because it lowers blood glucose, reduces cardiovascular risk factors, improves circulation, strengthens muscles, helps in weight control, and is useful in stress management. Weight-bearing exercises are especially necessary to help promote bone density because people with diabetes may be at higher risk for osteoporosis. An appropriate exercise program needs to be devised under the supervision of a qualified professional, and older adults need to understand how to use it safely and be motivated to continue it.

Monitoring

Health factors to monitor regularly in people with diabetes include the following:

- Foot care, because of lessened sensory acuity, slower wound healing in those with diabetes, and the danger of serious infection. Toenails should be cared for, preferably by a podiatrist, and proper shoes worn.
- Eye care, because of an increased likelihood of glaucoma, cataracts, and diabetic retinopathy.
- Urinary system functioning, because renal function is easily impaired and urinary tract infections are common.
- Periodic cardiovascular evaluation, because arteriosclerosis is often associated with diabetes; heart function may be impaired if vascular disease goes undetected and untreated.
- Blood sugar levels and diet, because these require careful attention in people with diabetes.

Medications

Medications for type 2 diabetes are divided into insulin and oral antidiabetic drugs. Insulin is used by some older people with diabetes and may be injected into the body one or more times a day. Special instruction is necessary to teach individuals these procedures, especially those who have visual or manipulative difficulties. A number of special devices available to assist with injections such as prefilled syringes, insulin syringe magnifiers, and blood glucose meters that talk or display results in large print are all helpful and easy to use. Insulin pumps and nasal sprays may also be used. A medic alert bracelet or necklace should always be worn by the individual with diabetes.

Oral antidiabetic drugs may be necessary for people with type 2 diabetes who cannot control their diabetes by diet alone. Several oral medications are now available to treat type 2 diabetes. Choice of drug depends on the primary care practitioner's assessment of the individual's need. It is still necessary to monitor blood sugar and follow a prescribed diet. Newer medications are constantly being developed to facilitate optimal management of this complex disease, but all must be used carefully with constant monitoring.

Education

The variety of complications associated with diabetes make effective management of this disease difficult. These include the following:

- Leg ulcers or pressure ulcers that make mobility difficult
- Blindness, deafness, swallowing, and communication problems that influence compliance
- Chronic health conditions such as cardiac or renal diseases that increase the risk of adverse drug reactions
- Increased risk of hypoglycemia caused by inadequate food intake, poor glucose monitoring, or acute illness
- Recurrent infections such as lung, bladder, or others that increase the risk of hyperglycemia
- Inadequate teaching about all aspects of the disease

Ongoing education is a must for the appropriate management of diabetes. Diabetes educators and patient support groups are helpful and often necessary in increasing compliance and understanding monitoring methods and medication adjustment. Because diabetes is a major health management problem, treatment should always be under the supervision of a primary care practitioner who works closely with the older diabetic adult. Several smart phone apps are available including Glucose Buddy, Diabetes Companion, and Calorie King, among others, to assist the individual in managing their diabetes.

Six areas are of special importance to those working with older adults who have diabetes to improve quality of care:

- Polypharmacy
- Depression
- Cognitive impairment
- Urinary incontinence
- Excessive falls
- Pain

Each of these should be assessed and monitored regularly to promote quality of life for those dealing with this difficult disease (Tabloski, 2014).

SUMMARY

Compared with other organ systems of the body, the endocrine glands do not show consistent and predictable age-related changes other than gradual slowing of functions and perhaps somewhat less efficient functioning in very old age. Because of the complex interrelationships between the various endocrine glands and between the endocrine and nervous system, specific age-related changes and their behavioral significance cannot always be clearly identified. Diseases associated with endocrine functioning may occur at any age; thus, age-related specificity is not a characteristic of endocrine system functions as a whole.

Overall, the various endocrine glands undergo gradual decline in function with age, causing some additional stress on the organism's ability to maintain homeostatic equilibrium and to adapt. However, barring disease, the endocrine system seems to remain remarkably stable over time.

REFERENCES

Centers for Disease Control and Prevention. (2020). National diabetes statistic report. https://www.cdc.gov/diabetes/library/features/diabetes-stat-report.html

Dickinson, J. K. (2017). Diabetes mellitus. In S. L. Lewis, L. Bucher, M. M. Heitkemper, & M. M. Harding (Eds.), *Medical-surgical nursing* (10th ed., pp. 1120–1155). Elsevier.

Hill, C. (2006). Endocrine function. In S. E. Meiner & A. G. Lueckenotte (Eds.), *Gerontological nursing* (3rd ed., pp. 535–560). Mosby Elsevier.

Jett, K. (2016). Physiological changes with aging. In T. A. Touhy & K. Jett (Eds.), *Toward healthy aging* (9th ed., pp. 308–318). Elsevier.

Konick-McMahan, J. (2007). Endocrine system. In S. L. Lewis, M. M. Heitkemper, S. R. Dirksen, P. G. O'Brien, & L. Bucher (Eds.), *Medical-surgical nursing* (7th ed., pp. 1234–1252). Mosby Elsevier.

Linton, A. D., Hooter, L. J., & Elmers, C. R. (2007). Endocrine system. In A. D. Linton & H. W. Lach (Eds.), *Matteson & McConnell's gerontological nursing* (3rd ed., pp. 525–571). Saunders Elsevier.

Marieb, E. N., & Hoehn, K. (2019). *Human anatomy and physiology* (11th ed.). Pearson Education Limited.

McLeod, M. (2013). Care of patients with diabetes mellitus. In D. Ignatavicius & L. Workman (Eds.), *Medical-surgical nursing* (7th ed., pp. 1410–1462). Saunders Elsevier.

Tabloski, P. (2014). *Gerontological nursing* (3rd ed.). Pearson Education.

Workman, L. (2013). Care of patients with problems of the thyroid and parathyroid glands. In D. Ignatavicius & L. Workman (Eds.), *Medical-surgical nursing* (7th ed., pp. 1393–1409). Saunders Elsevier.

14

The Immune System

STRUCTURE OF THE IMMUNE SYSTEM

The Lymphatic System

Integral to the effective functioning of the immune system is the lymphatic system, composed of (a) numerous lymphatic vessels that transport back to the blood any fluids that have escaped from the blood vascular system; and (b) lymph organs and tissues, essential in body defenses and resistance to disease.

Lymph organs and tissues include the following (Marieb & Hoehn, 2019):

- *Lymph nodes*, small nodes that cluster around lymph vessels, contain macrophage cells, which destroy invading cells or debris in the lymphatic system, and lymphocytes, which are white blood cells also crucial in an immune response of the body.
- *The spleen*, the largest lymph organ, is located in the left side of the abdominal cavity just below the diaphragm. The basic functions of the spleen are to provide for the immune response; to cleanse the blood of toxins, debris, bacteria, viruses, and defective blood cells; to store some red blood cell breakdown products for later reuse; and to release others into blood.
- *The thymus gland*, most important in the early years of life, atrophies throughout life and is almost nonexistent in older age. Its function is to produce hormones causing T lymphocytes to become effective against specific pathogens (substances capable of producing a disease).
- *The tonsils* form a ring of lymphatic tissues around the entrance to the pharynx (throat). They remove many pathogens entering the body through food or inhaled air.
- *Peyer's patches*, similar to tonsils, are clusters of lymph nodes in the ileum of the small intestine positioned to destroy bacteria before they enter the intestinal tract.
- *The appendix*, a part of the large intestine, contains many lymphoid follicles. It can destroy bacteria before they reach the actual intestinal wall and can also produce lymphocytes for long-term immunity.

For an infection to occur there must first be a source of infection, which may be from inside the person's body (endogenous) or from the environment (exogenous). Second, there must be a way for pathogens to enter the body and have access to susceptible tissues in order to produce disease. Entry may be through penetration of the skin barrier or mucous membranes, direct contact of pathogens with exposed tissues or mucous membranes (as happens with sexually transmitted diseases), or by ingestion through the mouth and gastrointestinal tract (e.g., food poisoning, hepatitis A). Inhalation is another route of entry by pathogens (e.g., the common cold, influenza, bacterial pneumonia). Third, for an infection to occur there must be a susceptible host. These factors are unique to each individual and determine whether a person remains well or becomes ill.

The general functions of the immune system are to recognize that which is *self* (the body's own cells) from that which is *nonself* (foreign substances) and to remove or destroy those substances not identified as belonging to the body. Both the nervous system and the endocrine system collaborate with the immune system in body regulation and in defending the body against foreign substances (Aldwin & Gilmer, 2004; Thames, 2011). An autoimmune response occurs when the immune system does not correctly identify the body's own cells and begins to attack and destroy them. Rheumatoid arthritis is an example of an autoimmune disease.

Specific functions of the immune system are (a) *defense*, or protection against antigens (substances recognized as foreign to the body that activate immune system responses); (b) *homeostasis*, involving the digestion and removal of damaged cells; and (c) *surveillance*, continuous monitoring for the presence of foreign substances or antigens in the body.

Another way to look at the complex immune system involves its three general lines of defense against invading substances. The first and most basic line of defense is anatomical surface barriers to invaders, such as the skin and mucous membranes. In addition to the purely physical barriers there are also biochemical barriers, certain secretions that can destroy foreign substances if the physical barriers are not sufficient. The body may also use mechanical means such as fever, vomiting, or diarrhea to get rid of foreign substances.

The second line of defense is an inflammatory response, innate internal defense if the actions of the first line of defense are not effective. Inflammation produces cells and fluids that can attack and destroy invaders. These responses of the first and second line of defense are all identified as *nonspecific* mechanisms in that they respond to all invading substances in essentially the same way; they are not able to discriminate among different kinds of invaders.

The third line of defense is an adaptive defense, the actual immune response, which targets and responds to *specific* invading substances. For example, B cells and T cells are two types of lymphocytes that are specific in that they respond to only a particular type of invader (Marieb & Hoehn, 2019; Tabloski, 2014).

Yet another classification of immune system activity involves innate (nonspecific) immunity and acquired (specific or adaptive) immunity. *Innate*, or nonspecific, immunity is the immunity one has to a particular antigen without ever having been previously exposed to it. Mechanisms of innate immunity include the previously noted first line of defense, the physical barriers, skin and mucous membranes, biochemical secretions, and mechanical means such as vomiting or diarrhea. If these are not successful, the second line of defense comes into play, involving inflammation as well as the actions of certain types of cells called monocytes, neutrophils, and natural killer (NK) cells. Actually, monocytes turn into macrophages, especially important in innate immunity. Macrophages act through phagocytosis, a process in which they surround and consume the invading antigen. They also stimulate NK cells to help in destroying invaders. Another action of innate immunity involves the *complement* system, a collection of proteins that can kill antigens as well as initiate inflammatory responses. These actions are all part of innate or nonspecific immunity.

Acquired, or specific, immunity involves both B cells and T cells, types of lymphocytes. Two types of acquired immunity are humoral and cell-mediated immunity. *Humoral immunity* defends the body against extracellular antigens found in the blood or other body fluids through the production of antibodies generated by B-cell lymphocytes. Antibodies are antigen-attacking proteins, which when released into the blood bind with their *specific* targeted antigen to help destroy it. *Cell-mediated immunity* depends primarily on the activated T-cell lymphocytes to assist in attacking and destroying invading antigens. Both B cells and T cells have the ability to become "memory" cells and to "learn" to identify a specific antigen and attack it if it recurs in the body at a later time (Heineman et al., 2010).

Thus, innate immunity is a first line of defense against invading or foreign substances, and its action is general or nonspecific, whereas acquired immunity is highly specific in its actions to remove specific invaders whenever they occur. The processes necessary for full-body protection by the immune system are innate defenses and adaptive defenses and surface barrier and internal defenses (Marieb & Hoehn, 2019).

AGE-RELATED CHANGES IN THE IMMUNE SYSTEM

Immunosenescence refers to age-related changes that occur in the immune system and result in a decrease in immune functioning, but there is considerable variation in the effectiveness of the immune system in older adults.

Generally, however, the following age-related changes affect the immune system of older adults:

- Aging changes in skin and mucous membranes reduce the effectiveness of the immune system's first line of defense.
- The thymus gland decreases in size and activity and is eventually unable to supply enough new T cells to deal with antigens invading the body, so that older adults may then be more vulnerable to infection and disease.
- Age-related changes result in a reduction of immune responsiveness in both innate and acquired immunity.
- In humoral immunity the effects of aging on B cells is not clear.
- In cell-mediated immunity there is some decline in T-cell function.
- There is apparently an increase in autoimmunity but little agreement on the specific reasons or to what extent it occurs.

Immune system changes in older adults result in an increased incidence of infections, possibly increased tumors, and an increased likelihood of autoimmune disorders. Common infections are often more severe, with slower recovery and reduced chances of developing adequate immunity after an infection (Tabloski, 2014).

AGE-RELATED DISORDERS OF THE IMMUNE SYSTEM

Common infections in older adults with reduced immune system efficiency include tuberculosis, pneumonia, influenza, herpes zoster (shingles), and urinary tract infections, particularly in women. Cancer occurs more commonly in older adults, especially lung cancer, breast cancer, and prostate cancer, and seems to be related to reduced immune system surveillance for abnormal cell growth.

It is estimated that about 5% of North Americans have an autoimmune disease and two-thirds of them are women. Some of the more frequent autoimmune disorders that may affect older adults as well as younger people are multiple sclerosis; rheumatoid arthritis; systemic lupus erythematosus (SLE), a systemic disease especially affecting the kidneys, lungs, and skin; pernicious anemia, a deficiency that interferes with absorption of vitamin B_{12}; glomerulonephritis, a severe impairment of kidney functions; Graves' disease, a thyroid dysfunction; myasthenia gravis, a neuromuscular disease; and multiple sclerosis, a disease destroying the white matter (myelinated nerve fibers) of the central nervous system (Marieb & Hoehn, 2019).

HIV/AIDS in Older Adults

HIV is a retrovirus that results in immunosuppression causing immune deficiency syndrome (AIDS), making individuals more susceptible to infections. Weaker immune systems in older age make them more susceptible to HIV/AIDS than younger people. Most HIV/AIDS in older adults is not the result of blood transfusions, as commonly believed, but results from unprotected sex. Older adults tend to be less likely to use safe sex practices since many believe that sexually transmitted diseases do not happen to those in older age, making it not necessary to practice safe sex, as for example in using condoms for protection. There has not been as much educational information about HIV/AIDS targeted at older adults as there has been for younger people (Haber, 2016).

Older adults with HIV/AIDS may not be diagnosed early because HIV/AIDS symptoms such as fatigue, weakness, night sweats, weight loss, increasing infections, and anorexia tend to be similar to those associated with aging. There are increasing numbers of older adults living with HIV disease due to HIV treatment, which has reduced the number of deaths resulting from opportunistic infections. In addition this age group is being infected at a higher rate, which is expected to grow (Kwong, 2017). Further complicating the situation, older adults generally report fewer symptoms, have chronic illnesses, and take more medications that may conflict with those prescribed for HIV. Many lack a psychosocial support network, or sometimes just choose not to use it. The reticence of both healthcare providers and older adults to discuss sexual issues makes both education and treatment more difficult.

Transmission of HIV/AIDS necessitates contact with infected body fluids such as blood, semen, vaginal secretions, and breast milk. The most common mode of transmission is unprotected sexual contact with an HIV-infected partner (Bradley-Springer et al., 2007). It may be sexually transmitted through oral, anal, or vaginal intercourse. The Centers for Disease Control and Prevention reports that 50% of individuals diagnosed with HIV/AIDS in the United States in 2018 were at least 50 years of age. There are several reasons for this age group to be infected, including less likelihood for condom use, decrease in the immune function, and the belief that they are not at risk of contracting the disease (Reuben et al., 2018). Many older adults are never tested for HIV believing they are not at risk for contracting the disease.

Diagnosis of HIV requires laboratory tests to detect HIV-specific antibodies and antigens in the blood. It is estimated that 14% of persons who have HIV have not been tested and are unaware they have the disease, which contributes to transmitting it to others (Kwong, 2017).

Treatment typically involves the use of antiretroviral drugs. The goal of treatment is to reduce the viral load to the lowest possible level. Special attention to interactions and untoward effects of all medications being used is important. Goals for treating HIV infection with drugs are (a) to decrease the viral load, (b) delay the progression of the disease, (c) maintain or increase CD4 T-cell counts, (d) prevent the transmission of the disease, and (e) prevent opportunistic diseases and symptoms of HIV-related symptoms. The disease cannot be cured but the viral load can be decreased and the progression of the disease delayed (Kwong, 2017). Newer drugs allow most people to live longer than previously, so there are now many more older adults aging with HIV/AIDS whose needs must be recognized and met.

Both older adults and healthcare providers need to be better educated to recognize the high risk of HIV/AIDS in older age and to understand the modes of transmission of HIV infections as well as the importance of HIV testing (Touhy, 2016). The Centers for Disease Control and Prevention now recommends HIV testing as a routine component of medical care if any risk factors for HIV are identified, and this must include older adults as well as other ages.

SUMMARY

Various issues related to aging affect the competency of the immune system and should be specifically addressed when working with older adults. They include nutritional and dietary status as well as nutritional balance, iron and trace element deficiencies, vitamin deficiencies, fat intake (because excessive amounts of unsaturated fat may adversely affect immune functions), stress, depression, bereavement and grief, and the presence or absence of a social support system. Other lifestyle variables important to assess include the person's activity level, smoking habits, medications, and use of herbs, because certain herbs may negatively affect the immune system (Thames, 2011).

REFERENCES

Aldwin, C. M., & Gilmer, D. F. (2004). *Health, illness, and optimal aging*. Sage.

Bradley-Springer, L., Shaw, C. A., & Lewis, S. L. (2007). Infection in human immunodeficiency virus infection. In S. L. Lewis, M. M. Heitkemper, S. R. Dirksen, P. G. O'Brien, & L. Bucher (Eds.), *Medical-surgical nursing* (7th ed., pp. 243–270). Mosby Elsevier.

Haber, D. (2016). *Health promotion and aging* (9th ed.). Springer Publishing Company.

Heineman, J. M., Hamrick-King, J., & Sewell, B. S. (2010). Review of the aging of physiological systems. In K. L. Mauk (Ed.), *Gerontological nursing* (2nd ed., pp. 128–231). Jones & Bartlett.

Kwong, J. (2017). Infection and human immunodeficiency virus infection. In S. Lewis, L. Bucher, M. McLean Heitkemper & M. M. Harding (Eds.), *Medical-surgical nursing* (10th ed., pp. 213–233). Elsevier.

Marieb, E. N., & Hoehn, K. (2019). *Human anatomy and physiology* (11th ed.). Pearson.

Reuben, D. B., Hess, K. A., Pacala, J. T., Pollack, B. J., Potter, J. F., & Semla, T. P. (2018). *Geriatrics at your fingertips* (20th ed.). American Geriatrics Society.

Tabloski, P. (2014). *Gerontological nursing* (3rd ed.). Pearson Education.

Thames, D. (2011). Infection. In S. Meiner (Ed.), *Gerontologic nursing* (4th ed., pp. 283–292). Mosby Elsevier.

Touhy, T. A. (2016). Intimacy and sexuality. In T. A. Touhy & K. Jett (Eds.), *Ebersole & Hess' toward healthy aging* (9th ed., pp. 445–462). Mosby Elsevier.

15

Aging With Lifelong Disabilities

INTRODUCTION

The Americans With Disabilities Act of 1990 (ADA, 1990) defines disability as "a physical or mental impairment that substantially limits one or more major life activities." Therefore, the scope of what is considered a disability is large and varied. For instance, people may have difficulties associated with communication, such as impairment in hearing, vision, and speech. Others may have issues with physical functioning and mobility, such as walking or grasping objects. Aside from those with sensory and physical disabilities, there are also those with intellectual and developmental disabilities.

The experience of having a disability is widespread throughout the population. According to the U.S. Centers for Disease Control and Prevention (CDC), 61 million Americans (26%) have a disability (2019a). Disability becomes even more commonplace with increasing age; 40% of adults age 65 years and older have a disability (CDC, 2019a). Therefore, the population of those with disabilities is substantial both in terms of numbers and in the resources and assistance they need to ensure that their needs are supported so that they may fully participate in the community and optimize their aging process.

Throughout this book, there are descriptions of changes in function that can accompany both normative aging and pathological aging caused by diseases. Losses in function in the respiratory system and the cardiovascular system, for example, can lead to significant reduction in mobility because of shortness of breath and fatigue. Osteoarthritis of the knees and spine may also affect mobility, whereas osteoarthritis in the hands may reduce manual dexterity, making it difficult to perform tasks like opening pill bottles, writing, and holding and manipulating objects such as cutlery and cups. Complications of diabetes can result in an older adult becoming an amputee. Aging-related eye disorders, such as cataracts, macular degeneration, and glaucoma, impair vision and can sometimes result in blindness. There are numerous other examples of chronic conditions that may lead to an older adult adjusting to a physical or sensory disability in later life. All these can be described as acquired disabilities from aging-related diseases. In addition, one cannot overlook the growing population of people who have had lifelong disabilities and who are aging. This special population can have particular aging issues that require increased attention and vigilance from health professionals (Perkins & Moran, 2010).

CONGENITAL VERSUS ACQUIRED DISABILITY

Many people with lifelong physical or sensory disabilities, although considered disabled by others, do not consider themselves as such. Those who were born (congenital) with a disability or manifested a disability at a very young age have spent their lifetimes successfully overcoming the challenges and limitations their disabilities have presented. Indeed, societal attitudes, low expectations, and lack of accessibility of physical environments are more limiting than the disability itself. Because people with lifelong disability have never experienced life without the presence of their particular disability, or the disability arose early enough in childhood that they have little memory of prior functioning, it is actually difficult for many of them to imagine what it would be like if they did not have their particular disability. Some may describe it as "what you have never had, you never miss." Indeed, for many people with disabilities, a more pressing concern is for others not to see them as disabled, but to look beyond their disability and focus on their actual abilities. The majority of people with a congenital disability are well-adjusted to the limitations their disability presents. Many are fully aware of their range of ability and are extremely creative in finding alternative methods of accomplishing particular tasks.

For those who have acquired disabilities in adulthood or later life, adapting can present more of a challenge. This group of people have had to adjust to a significant and permanent reduction in their functional abilities. There is not only the physical adjustment to this change, which results in having to learn new ways of doing one's daily activities, but there is also the psychological adjustment to that loss of function (and in some cases the actual change in the physical body that has occurred, e.g., from amputation). The psychological challenges presented by a new disability may need time, encouragement, and support to successfully overcome. Whether the newly acquired disability is due to accident, trauma, or the complication of a disease, it is essential that a positive attitude be encouraged. Many people mourn the perceived loss of their independence, but in many cases individuals find alternative ways of accomplishing tasks, often with the aid of assistive technologies. There is a natural period of grief for the loss of those previous abilities, but whatever a person's age, physical and occupational therapy can still help to maximize their independence. Specialized support groups can be particularly beneficial in providing emotional support from other people who have been through similar experiences. Furthermore, the resources and information that support groups share about special assistive technologies and adaptations they have made to their homes, cars, and work environments can be invaluable to the person with a newly acquired disability. These groups provide much needed role models and a shared sense of community.

THE COMPLEXITY OF AGING WITH DISABILITIES

How the aging process is experienced by those with disabilities is not only a question of how the individual disability itself affects the aging process but also a matter of the greater ramifications the disability may have for the aging person. Does it result in an increased risk for an aging-related change or aging-related disease? The most accurate answer is that it can actually be a very complex matter that often depends on the type of disability. For example, someone already visually impaired may develop age-related eye diseases that further reduce visual functioning; however, in the practical sense, this person is already readily adapted to life with visual impairment. In this case, the effect of aging-related decline may be less significant than it would be for someone in the general population who is adjusting to new and late-life vision loss. The same can be said of those who have lifelong hearing impairment versus someone coping with aging-related hearing loss. Think again about those with sensory disabilities. They often have highly developed acuity in the other senses; therefore, the effect of aging-related change in the sense already impaired may be minimal. But what about the aging-related changes in the other senses that those individuals rely on more? Aging-related hearing loss in a person with lifelong visual impairment is, in fact, substantially more detrimental than it would be in the non–visually impaired older adult. Similarly, aging-related vision loss in a person with lifelong hearing impairment is again more significant than it would be in their non-hearing-impaired counterpart. A person who is blind also relies greatly on the sense of touch (e.g., for reading Braille), and sensitivity decreases with age. In addition, the person could develop a condition, such as neuropathy or osteoarthritis, that affected their dexterity, and consequently their ability to manipulate and process information from the environment through the sense of touch would be drastically reduced. This is an outcome not even considered by someone without visual impairment.

Those with physical disabilities may develop additional issues from changes in their musculoskeletal system. As is the case with sensory disabilities, compensatory use, in this case of their limbs, may lead to increased risk for developing osteoarthritis. For example, an amputee who is missing a limb uses the remaining limbs to a much greater extent than if all limbs were present. Therefore, amputees may be at greater risk of developing osteoarthritis in their present limbs, because there has been significantly greater usage ("wear and tear") over the life span of those limbs. The effect of osteoarthritis may be much more significant in someone who has only one hand. For those who do not have arms, and instead use their feet to perform the activities of daily living, consider the significance of developing osteoarthritis in the ankles, knees, and hips. Not only is their ability to ambulate at risk, but also the performance of all those activities they are very well adapted to perform with their feet, such as eating, drinking, and dressing, becomes much more difficult with loss of dexterity and flexibility from osteoarthritis. Therefore, each individual who is aging with a disability faces considerable challenges from the aging process. Healthcare professionals need to be aware of and sensitive to the cascade effect that any aging-related change can have on the independence of an aging person with a disability.

DEVELOPMENTAL AND INTELLECTUAL DISABILITIES

The previous section has highlighted aging issues of those with sensory or physical disabilities. Depending on their particular disabilities, some of these people are considered to have developmental disabilities. People with a developmental disability may have congenital disabilities or they may have disabilities that manifested during childhood or adolescent development. The federal definition of developmental disabilities is stated in the Developmental Disabilities Assistance and Bill of Rights Act of 2000; it refers to "developmental disability" as a severe, chronic disability of an individual that

- Is attributable to a mental or physical impairment or combination of mental and physical impairments
- Is manifested before the person attains age 22
- Is likely to continue indefinitely
- Results in substantial functional limitations in three or more of the following areas of major life activity: (a) self-care, (b) receptive and expressive language, (c) learning, (d) mobility, (e) self-direction, (f) capacity for independent living, and (g) economic self-sufficiency
- Reflects the person's need for a combination and sequence of special, interdisciplinary, or generic services, individualized supports, or other forms of assistance that are lifelong or of extended duration and are individually planned and coordinated

Generally, some common conditions that are considered developmental disabilities include cerebral palsy (CP), epilepsy, autism, and intellectual disabilities (IDs), as well as blindness and deafness. Historically, nationwide prevalence of developmental disabilities has been difficult to obtain, but more sophisticated monitoring of developmental disabilities by the CDC in children estimates that about one in six, or about 17%, of children aged 3 through 17 years have one or more developmental disabilities (CDC, 2019b). Because the previous section has already covered issues of vision and hearing loss, the remainder of this section will focus on those aging with IDs, Down syndrome (DS), and CP.

According to the American Association on Intellectual and Developmental Disabilities, *intellectual disability* is defined as significant limitations both in intellectual functioning and in adaptive behavior, as expressed in conceptual, social, and practical adaptive skills, that originates before age 18 (Schalock et al., 2010). Traditionally the level of ID present has been described as mild, moderate, severe, or profound (Jacobson & Mulick, 1996). These levels are associated with increasingly lower intelligence quotient (IQ) scores. The threshold for onset of ID is considered to begin when intellectual functioning is assessed at 2 or more standard deviations (SDs) below the population mean on standardized IQ tests (an IQ of 70 or less). Also, the number of concurrent limitations in domains of adaptive functioning (AF) is also considered. AF has three domains—conceptual, social, and practical—and generally refers to the skills that allow a person to live and function independently in daily life (e.g., dressing, grooming, reading, managing finances, appropriate behavior in social interactions, forming friendships). The true prevalence of IDs has been difficult to determine but is usually estimated to be approximately 1% of the population.

In many cases the specific cause (etiology) of ID is unknown. In some cases, specific syndromes are associated with ID (e.g., DS). In other cases, developmental disabilities may be associated with ID in some but not all cases. An example is CP; some individuals with this condition have physical or sensory disabilities but not ID. Another example is autism; some people with autism may have ID but many do not. Therefore a person who has an ID always has a developmental disability; however, a person with a developmental disability does not necessarily have an ID. It is crucially important for healthcare professionals to be mindful of this distinction and not to make automatic assumptions of ID when communicating with a person who has an apparent physical disability. This is especially true when talking to a person who may have speech difficulties (as may be the case for a person with CP) but actually has no ID.

The Aging Population With Intellectual Disabilities

As with the general population, the life expectancy of those with IDs has increased substantially during the last century. In fact, the increase in life expectancy has been greater in the ID population in recent years. The development and implementation of more proactive, holistic, and individualized care has undoubtedly been a major factor in this striking increase in longevity (Haley & Perkins, 2004). Other factors include improvements in the medical treatment of commonly associated comorbidities, such as epilepsy, as well as successful treatment of recurrent respiratory infections that are especially problematic for many children with IDs. In fact, it has been suggested that the life expectancy of those with less severe disabilities may match that of the general population in the not too distant future (Janicki et al., 1999).

At present, though, there are still differences between the life expectancy of the general population and the life expectancy of persons with ID in general, and persons with DS in particular (Janicki et al., 1999). Nevertheless, the gain in the latter two populations over the last few decades has been substantial. For example, between the 1930s and 1990s, the mean age at death for people with ID increased 47 years, from 18.5 to 66.2 years (Braddock, 1999). People with DS are noted to have a reduced life expectancy compared with others in the ID population, but they too have experienced increased longevity. In the early 1900s the majority of children with DS did not even live to be teenagers, let alone survive into adulthood. Their life expectancy was a grim 9 years (Selikowitz, 1990). Between 1984 and 1993, mean age of death for those already 40 and older had reached 55.8

years (Janicki et al., 1999). As a very general guide, one can think of the ID population as having a life expectancy of around 10 years less than the general population, and those with DS having a life expectancy around 20 years less than the general population. Nevertheless, because both the ID and DS populations' life expectancies are increasing faster than those of the general population, the disparity between these populations is likely to be greatly reduced in the future.

General Aging Issues

Physical changes as a result of aging or aging-related health conditions can present some significant additional challenges to persons with intellectual or developmental disabilities. Greater longevity can also bring additional functional impairment, morbidity, and mortality from early age-onset conditions associated with their intellectual disability, from both the progression over the life span and also their interactions with older age-onset issues (Evenhuis et al., 2001).

Sensory impairments are particularly problematic for the aging ID population (Schrojenstein Lantman-de Valk et al., 1997). Generally, aging-related changes in vision (presbyopia) and hearing (presbycusis) may be present in rates similar to those in the general population, as are age-related pathologies of vision (i.e., cataracts, macular degeneration, glaucoma, and diabetic retinopathy), but the effect is often more severe because of much higher rates of preexisting, childhood-onset visual and auditory pathological conditions (Evenhuis, 1995a, 1995b; Schrojenstein Lantman-de Valk et al., 1994).

Musculoskeletal issues also tend to arise more commonly in this population. By the age of 60, around 30% of people with ID will also have significant mobility and gait issues, and by 75 years, 60% will have such difficulties (Evenhuis, 1999). Issues such as hypotonia (reduced muscle tone), musculoskeletal deformities, and lack of weight-bearing exercise put the ID population at an increased risk of developing osteoporosis (Center et al., 1998).

Because a person with ID often has comorbid conditions, the use of medications over the life span may also have long-term consequences (Evenhuis et al., 2001). Bone mineralization problems can develop from chronic use of some anticonvulsants (Phillips, 1998), and particularly troublesome tardive dyskinesias can arise from long-term neuroleptic use (Wojcieszek, 1998).

Research focusing on the aging population of people with autism has received far less attention within the disability field, though this is likely to improve due to better recognition and diagnosis in the future (Perkins & Berkman, 2012). A more substantial body of literature exists and has noted that aging people with DS or CP have an accelerated aging process because of complications and interactions with many secondary medical aspects accompanying their conditions. They warrant special attention to highlight their unique aging issues.

Specific Aging Issues With Down Syndrome

As a result of genetic and endocrinological factors, people with DS are likely to retain a reduced life expectancy compared with both the ID population and the general population. Despite this fact, gains in their life expectancy over the last 30 years are nothing short of dramatic. Accordingly, the number of individuals with DS who live into old age is going to increase considerably. The aging of an adult with DS is a complex issue, but with due vigilance to their special needs many more will age successfully.

DS is one of the most well-known genetic conditions associated with intellectual disability. It is estimated to occur in approximately 1 in every 733 live births across all racial and socioeconomic groups (CDC, 2006). Although increasing maternal age is associated with increased risk of having a child with DS, the majority of children with DS are born to mothers age 35 or younger. Most people with DS have mild to moderate levels of ID.

DS is often known as trisomy 21, indicative of the third copy of chromosome 21 that causes DS in 95% of cases. Another cause of DS is translocation, in which chromosome 21 material is incorrectly rearranged so that it attaches to another chromosome. Even rarer is the type of DS called mosaicism, which occurs when there is a mixture of cell lines (i.e., some have a normal set of chromosomes and others have trisomy 21). Depending on the mix of cell lines that are trisomy 21 compared with unaffected cells, a milder expression of DS characteristics may result. In fact, one of the longest-living persons with DS, a woman who died at the age of 83, had mosaic-type DS (Chicoine & McGuire, 1997).

General characteristics of a person with DS include hypotonia (decreased muscle tone), hyperextensive joints, small stature (average height is around 5 feet for men, 4 feet 9 in. for women), flattened occiput, poorly developed bridge of nose, small mouth, high arched palate, recurrent conjunctivitis, and blepharitis (caused by lack of lysozyme, an enzyme produced by the immune system that is present in tears). Many are born with congenital heart defects, in particular septal defects, or holes in the septum (the tissue separating the atria and ventricles of the heart). People with this defect are often referred to as having a "hole in the heart," and depending on the exact location and size, some will require surgical repair in infancy, whereas others will heal themselves over time.

People with DS are very prone to developing sensory problems. They can have numerous vision issues, including myopia, hyperopia, nystagmus (involuntary flickering of the eyes, which makes fixation on specific objects and tasks like reading more difficult), strabismus, and amblyopia. There are two types of hearing loss, conductive and sensorineural. People with DS are at greater risk of either or both types of hearing loss. Children with DS often develop conductive hearing loss caused by frequent and undiagnosed middle ear infections (otitis media) in childhood. The hair cells of the cochlea (cilia) may be congenitally absent, causing deafness. They are also more prone to early degeneration of cilia, leading to sensorineural hearing loss. Persons with DS may actually experience a more progressive, earlier hearing loss, starting as young as their early 20s. Therefore, great vigilance is required to monitor aging-related hearing and vision loss, as well as aging-related diseases (e.g., cataracts, glaucoma, macular degeneration) in adults with DS. They already have considerably more vision and hearing impairments, which are then further compounded by normal aging-related loss and onset of aging-related diseases. The importance of regular vision and screening is therefore crucial at all ages but becomes vitally important as an adult with DS ages.

Compared with the general population, the population with DS is at greater risk to develop a variety of health conditions. They are more likely to develop hypothyroidism, diabetes, epilepsy, sleep apnea, and leukemia than is the general population. Issues such as endocrinological dysregulation and hypotonia in persons with DS also result in the increased prevalence of osteoporosis. Increased risk of developing the diseases and conditions already mentioned can result in formal and informal caregivers coping with a variety of medical comorbidities in an older adult with DS. Significantly, however, one of the major factors associated with the reduced life expectancy of older adults with DS is that they are genetically more susceptible to developing Alzheimer's disease (AD).

Down Syndrome and Alzheimer's Disease

Adults with DS are at a significantly greater risk of developing earlier-onset AD (Wisniewski et al., 1985; Zigman et al., 1996). One notable statewide study reported a prevalence of 56% of adults with DS older than age 60 with dementia (Janicki & Dalton, 2000). This prevalence rate in the general population is not reached until the 85+ age group. It should be noted that, generally, those with ID (from other etiologies) also have similar prevalence rates as seen in the general population.

For adults with DS, various studies have reported a mean age of onset between 51 and 57 years, which is still considerably younger than both the non-ID population and persons with ID from other etiologies (Janicki & Dalton, 2000; Prasher & Krishnan, 1993; Zigman et al., 2002). Not only is there a greater risk of developing AD at a younger age, but the duration of the disease can be much shorter because the decline in cognitive functioning is oftentimes more rapid (Hammond & Beneditti, 1999; Tyler & Shank, 1996). The increased risk for AD has been attributed to the fact that the gene coding for amyloid precursor protein (APP) is located on chromosome 21. Because of the additional chromosome 21 material, this gene is overexpressed, resulting in an increased formation of amyloid plaques—one of the major characteristics of AD (Hof et al., 1995). It has long been established that by the age of 40 nearly all individuals with DS will develop the characteristic neuropathological brain lesions that are associated with AD. These include the amyloid plaques, Hirano bodies, and neurofibrillary tangles (Wisniewski et al., 1985). Nevertheless, it must be stressed that in many cases the presence of these lesions does not occur in sufficient concentration for the clinical manifestation of AD (Perkins & Small, 2006). There is also a common misconception that all people with DS will automatically develop AD; in fact, many people with DS will age without ever developing Alzheimer's. For example, a case study of the successful aging of a 70-year-old man with trisomy 21 DS had tracked his cognitive abilities over 16 years, and he had not displayed any sign of AD-related cognitive decline (Krinsky-McHale et al., 2008).

Treatment of AD in adults with DS is no different than it is for the general population; the same medications have been shown to be effective for both populations (Prasher, 2004). Obtaining a timely diagnosis and detecting the cognitive decline in adults with DS is a significant challenge. Because vision and hearing loss can result in comprehension and communication difficulties that could be mistaken for dementia symptoms, it is essential that the sensory abilities of an aging person with DS are regularly tested. Hypothyroidism, which is more prevalent in those with DS, can also be the cause of mental confusion and apathy rather than the onset of dementia. As with the general population, all other possible diagnoses must be ruled out; the fact that persons with DS may have more of these medical issues makes it imperative that a thorough physical examination be undertaken. Nevertheless, assessment of dementia in an adult with DS presents difficulties. Many of the standard screening instruments used in memory disorder clinics are not valid for the DS population; Screening instruments that are available specifically for persons with ID and DS include the Down Syndrome Dementia Scale (Gedye, 1995) and the Dementia Screening Questionnaire for Individuals with Intellectual Disabilities (DSQIID; Deb et al., 2007). The National Task Group on Intellectual Disabilities and Dementia Practices (NTG) released their NTG-Early Detection Screen for Dementia (NTG, 2013). This is an adaptation from the DSQIID, and the NTG designed it to be a user-friendly screen for staff and family caregivers to note functional decline and health problems that can be invaluable for further formal assessment. Some agencies providing residential services to adults with DS have found it very useful to compile video recordings of a person every year or so from the mid-40s onward because this helps to form a useful baseline for later comparison. The person with DS could be filmed performing basic daily tasks such as walking, dressing, eating, writing, or drawing specific objects. Videotaping is particularly helpful in recording more subtle changes in behavior.

The major difference between AD in people with DS and the general population is the lower age of onset and the more rapid progression of the disease. In addition, sometimes people with DS who have AD will also develop epilepsy, often myoclonic epilepsy in which the person will have the "twitches" or sudden jerks. Often antiepileptic medications are effective, but the development of late-onset epilepsy with AD in persons with DS commonly indicates that there will also be acceleration in cognitive decline.

Solid Tumor Cancers and Down Syndrome

One area in which those with DS do seem to have an advantage over the general population is their reduced risk of developing solid tumor cancers (e.g., breast cancer and colon cancer). This has been somewhat of an enigma for many years. Previously it was thought that the lower life expectancy of persons with DS meant that they simply did not live long enough to develop age-related cancers. This explanation has not been borne out with the increase in life expectancy; older adults with DS still are not developing solid tumor cancers in the numbers that would be expected. Compared with the general population, they seem to be protected from developing cancers that form solid masses, though one should remember that their rates of developing leukemia are significantly increased (Hasle et al., 2000). One of the explanations for this may rest again with the additional chromosome 21, in this case overexpressing the gene Ets2 (Sussan et al., 2008). The Ets2 gene is usually linked to increased tumor production, but it appears that having overexpression of this gene, as occurs with persons with DS, actually counters the risk of developing cancer.

Overall, persons with DS have a variety of medical issues throughout their lives, and their aging process is challenging. Nevertheless, great strides have been made in their life expectancy, and many of their comorbid conditions are manageable chronic conditions. As their medical and social care has improved, common assumptions regarding the abilities of people with DS increasingly are being challenged. It may surprise some people to learn that some people with DS have driver's licenses and have completed marathons, and a teenager with DS has climbed Mount Everest! In the last century, people with DS were lucky to survive their first decade, but now they are outliving their parents and siblings, many of whom have been their primary caregivers. This underscores the importance of the need for healthcare professionals to be aware of and sensitive to aging in persons with DS in order to better support the endeavors of both formal and informal caregivers.

Aging With Cerebral Palsy

CP is a nonprogressive neurological disorder that affects movement and posture. *Cerebral* refers to the brain, and *palsy* refers to disorder of movement. In the vast majority of cases, the cause of CP is unknown, but it occurs as a result of damage in areas of the brain that control motor function such as the motor cortex, basal ganglia, and cerebellum. In some cases possible causes include infection, malnutrition, or trauma to the brain in early childhood. Most often this damage has occurred during pregnancy; in a small number of cases the injury may occur during labor and delivery or between birth and age 3. Around 60% of people with CP also have additional developmental disabilities, including ID, vision impairment, hearing loss, and epilepsy. ID can be present in up to 50% of all people with CP (Liptak & Accardo, 2004). Severe visual impairment and blindness is present in 11% (Johnson, 2002), and many more contend with mild to moderate visual impairment. Visual impairments can be oculomotor problems (i.e., strabismus and amblyopia), acuity problems (i.e., myopia), visual field loss, and cortical visual impairment resulting in incorrect processing of images by the brain. Hearing loss is also problematic and can affect between 4% and 13% of people with CP (Reid et al., 2011). In the general population, epilepsy has a prevalence of 5% to 7%, whereas the prevalence of epilepsy in children and adults with CP ranges between 15% and 55%, and up to 71% for those with both CP and IDs (Wallace, 2001). In addition, up to 80% have some impairment in speech (Odding et al., 2006).

It is critical to note that the brain damage that has caused CP does not progress with age, but because the musculoskeletal system is neurologically affected, orthopedic deformities commonly develop. People with CP can have difficulties with mobility, posture, balance, eating, controlling body movements, and purposeful movement. They may have increased or decreased muscle tone (hypertonia or hypotonia) or muscle tone that fluctuates.

Types of Cerebral Palsy

There are four types of CP: spastic, athetoid, ataxic, and mixed. Each type has a specific movement issue. Spasticity is the most common type of CP, affecting up to 80% of people with CP. These people have hypertonicity; the excessive muscle tone causes stiffness and difficulty in movement, and permanent shortening of muscles and tendons can result. These can lead to contractures resulting in deformity and reduction in the range of motion in the affected limbs. There are three types of spasticity: hemiplegia (one side of the body is affected), diplegia (lower limbs are affected more than the arms), and quadriplegia (all four limbs are affected).

Athetoid CP is characterized by constant and uncontrollable movements of the head, eyes, mouth, and limbs and can also cause difficulty with speech. People with athetoid CP have difficulty maintaining a fixed position. This presents a challenge in keeping a firm grasp on anything; therefore tasks such as writing, eating and drinking, and getting dressed can be difficult. It can also cause difficulties with reading if the ability to visually track text across the page is affected. In ataxic CP, the major issue is poor balance and coordination, which can make walking difficult. Sometimes there may be tremors and difficulty with depth perception. A fourth type, mixed CP, occurs when there is a combination of more than one of the previous three types—spastic, athetoid, or ataxic CP. It is common for individuals to display both spastic and athetoid types of CP.

CP is incurable, so care is focused on the treatment and prevention of possible complications. CP is one of the most expensive medical issues to treat across the life span. Physical therapy is essential, especially in children, to promote maximum joint motion during development. Physical therapy is important across all age groups to help maintain muscle tone and bone structure and prevent dislocation of joints. Occupational therapy helps those with CP to maximize their function in activities that promote independence. To assist with gait issues, orthotic devices such as ankle–foot orthoses are often worn to improve stability and speed of walking. Other aids, such as rolling walkers and individually custom-molded wheelchairs, are sometimes needed. Surgical procedures may also be performed to improve muscle spasticity and contractures. The most common surgical procedures are transfers of tendons, lengthening or release of tendons, and osteotomies (when bone is cut to shorten, lengthen, or change alignment). Another type of surgery, selective dorsal rhizotomy, works by severing nerves in the spine to allow muscle relaxation.

Muscle relaxant drugs, such as baclofen, are often prescribed to ease hypertonia and can now be conveniently administered intrathecally (directly into the spine) in conjunction with a surgically implanted pump in the abdomen. Botox is also used as an effective muscle relaxer, and injections can last 4 to 6 months. Antiepilepsy medication is prescribed to treat epilepsy.

Because CP can bring many challenges to musculoskeletal system functioning, it is the main body system prone to precocious aging in CP. This is due to lifelong abnormal postures, movements, and secondary conditions that can worsen with increasing age (Svien et al., 2008). These include impairment in motion and increasingly problematic contractures (Andersson & Mattsson, 2001). Reduction in mobility is very common with increasing age (Strauss et al., 2004). Spinal alignment issues can get progressively worse, and this may affect respiratory, bowel, and bladder function in an aging adult with CP. Respiratory diseases are much more common in older adults with CP, including chronic obstructive pulmonary disease, respiratory influenza, and pneumonia (Strauss et al., 1999). Changes in posture with increasing age can also lead to greater difficulty in eating, especially chewing and swallowing; around 50% of adults with CP will be affected by this (Ferrang et al., 1992). Sometimes this condition results in choking episodes in which a food particle incorrectly enters the lungs, decays, and causes infection. This is known as aspiration pneumonia. In terms of respiratory diseases, aspiration pneumonia is actually one of the most common causes of death across all ages in those who have CP (Strauss et al., 1999). Changes in muscle tonicity, reduced mobility, and postural changes can also lead to slower digestion, gastric reflux, and increased risk of chronic constipation; these problems are routinely encountered by many with CP over their life spans.

Declining mobility, excessive wear and tear on joints, reduction in muscle and bone mass, restricted patterns of movement, and abnormal contact and compression between joint surfaces all contribute to the very high rates of osteoporosis and osteoarthritis found in the CP population at a much younger age than would be seen in the general population (King et al., 2003; Murphy et al., 1995). Vigilance in maintaining the integrity of skin and prevention of pressure sores, a concern especially for the nonambulatory across their life spans, becomes even more important with increasing age.

Unfortunately, one of the most problematic issues is that the numerous stresses on the musculoskeletal system may result in chronic pain for many persons with CP, and this becomes even more evident with increasing age. The experience of pain is widespread, and it is of great concern that many individuals with CP who are suffering chronic pain do not seek formal help from their physicians to better manage their symptoms (Engel et al., 2002). The other major effect of musculoskeletal stress is an increasing level of physical fatigue with age compared with that experienced by the general population (Jahnsen et al., 2003). This can be very detrimental to maintaining mobility and can reduce physical vitality to the point that quality of life is much diminished.

Other aging-related diseases have also been found to be more prevalent in the CP population compared with the general population. These include cardiovascular disease, cerebrovascular disease, and cancer (Strauss et al., 1999). A major concern with these high prevalence rates is that lack of regular screening and communication difficulties may result in delayed diagnosis, thereby reducing the possibility of early detection and treatment (Strauss et al., 1999).

Obesity and the Importance of Health Promotion in Adults With Intellectual Disability

In all adults, irrespective of whether they have a disability or not, there is the need to endorse the message of good well-balanced nutrition, regular weight-bearing and cardiovascular exercise, and avoidance of poor lifestyle habits (e.g., smoking and drinking alcohol). However, there is justifiable concern that promoting a healthy lifestyle is not being encouraged among those with IDs with as much vigor and coverage as it is in the general population. In fact, the troubling trend of increasing rates of obesity in the general population appears to be much worse in the ID population. Even more concerning is that this trend is already evident; children with ID have a prevalence rate of 28.9% compared with 15.5% of children without ID (Segal et al., 2016). There has been a philosophical shift in the guiding philosophy of care and support for persons with ID. Promoting self-direction, autonomy, inclusion, and community living is of the utmost importance, but attention needs to be directed at education about healthy living practices for those living independently or with family members.

Because obesity is a major issue, any aging-related diseases and chronic conditions associated with obesity, such as cardiovascular disease, diabetes, and osteoarthritis, risk becoming even more commonplace if wellness programs to improve general fitness are not effectively implemented. Indeed, there is the distinct possibility that some of the dramatic gains in life expectancy may be slowed or halted by the lack of widespread health promotion in the ID population. Therefore, healthcare professionals should promote health and wellness programs to adults aging with IDs and should note that this segment of the population is expected to derive great benefit from such programs.

SUMMARY

People aging with lifelong disabilities have to face the same aging issues as those that arise for the general population; however, their disabilities, or associated medical conditions, may result in more adverse outcomes from the aging process, and they undoubtedly have more complex challenges. Irrespective of their disabilities, it is important that they receive the same health promotion and prevention initiatives as the general population. Better diet and appropriate exercise regimens appear to be areas requiring considerable improvement. It is important to recognize that people aging with disabilities are actively encouraged to participate in wellness and screening programs such as getting mammograms and Pap smears, screening for colon cancer and bone mass loss, and monitoring of cholesterol, blood pressure, and blood sugar levels. Regular dental, vision, and hearing assessments are also extremely important, yet can often be overlooked.

Overall, there is justifiable reason to be more concerned about the aging process of those who have disabilities. A major risk factor is the lack of awareness by healthcare professionals and caregivers of all the unique problems that can and do arise. Nevertheless, with appropriate support and attention to their particular needs, many people with disabilities will also enjoy increasing longevity and age successfully. Healthcare professionals perform a vital role in assisting aging persons with disabilities to maximize their abilities, thus increasing the likelihood of maintaining independence into older adulthood.

REFERENCES

Americans With Disabilities Act of 1990, Pub. L. No. 101–336, 104 Stat. 328 (1990).

Andersson, C., & Mattsson, E. (2001). Adults with cerebral palsy: A survey describing problems, needs, and resources with special emphasis on locomotion. *Developmental Medicine and Child Neurology, 43*, 76–82. https://doi.org/10.1111/j.1469-8749.2001.tb00719.x

Braddock, D. (1999). Aging and developmental disabilities: Demographic and policy issues affecting American families. *Mental Retardation, 37*, 155–161. https://doi.org/10.1352/0047-6765(1999)037%3C0155:AADDDA%3E2.0.CO;2

Center, J., Beange, H., & McElduff, A. (1998). People with mental retardation have an increased risk for osteoporosis: A population study. *American Journal on Mental Retardation, 103*, 19–28. https://doi.org/10.1352/0895-8017(1998)103%3C0019:pwmrha%3E2.0.co;2

Chicoine, B., & McGuire, D. (1997). Longevity in a woman with Down case study. *Mental Retardation, 35*, 477–479. https://doi.org/10.1352/0047-6765(1997)035%3C0477:LOAWWD%3E2.0.CO;2

Deb, S., Hare, M., Prior, L., & Bhaumik, S. (2007). Dementia screening questionnaire for individuals with intellectual disabilities. *British Journal of Psychiatry, 190*, 440–444. https://doi.org/10.1192/bjp.bp.106.024984

Developmental Disabilities Assistance and Bill of Rights Act of 2000, H.R. 4920 (106th) (2000).

Engel, J. M., Kartin, D., & Jensen, M. P. (2002). Pain treatment in persons with cerebral palsy. *American Journal of Physical Medicine & Rehabilitation, 81*, 291–296. http://doi.org/10.1097/00002060-200204000-00009

Evenhuis, H. M. (1995a). Medical aspects of ageing in a population with intellectual disability: I. Visual impairment. *Journal of Intellectual Disability Research, 39*, 19–26. https://doi.org/10.1111/j.1365-2788.1995.tb00910.x

Evenhuis, H. M. (1995b). Medical aspects of ageing in a population with intellectual disability: II. Hearing impairment. *Journal of Intellectual Disability Research, 39*, 27–33. https://doi.org/10.1111/j.1365-2788.1995.tb00910.x

Evenhuis, H. M. (1999). Associated medical aspects. In M. P. Janicki & A. J. Dalton (Eds.), *Dementia, aging and intellectual disabilities: A handbook* (pp. 103–118). Brunner/Mazel.

Evenhuis, H., Henderson, C. M., Beange, H., Lennox, N., & Chicoine, B. (2001). Healthy ageing—Adults with intellectual disabilities: Physical health issues. *Journal of Applied Research in Intellectual Disabilities, 14*, 175–194. https://doi.org/10.1046/j.1468-3148.2001.00068.x

Ferrang, T. M., Johnson, R. K., & Ferrara, M. S. (1992). Dietary and anthropometric assessment of adults with cerebral palsy. *Journal of the American Dietetic Association, 92*(9), 1083–1086.

Gedye, A. (1995). *Dementia scale for Down's syndrome: Manual.* Gedye Research and Consulting.

Haley, W. E., & Perkins, E. A. (2004). Current status and future directions in family caregiving and aging people with intellectual disability. *Journal of Policy and Practice in Intellectual Disabilities, 1*, 24–30. https://doi.org/10.1111/j.1741-1130.2004.04004.x

Hammond, B., & Beneditti, P. (1999). Perspectives of a care provider. In M. P. Janicki & A. J. Dalton (Eds.), *Dementia, aging, and intellectual disabilities* (pp. 32–41). Brunner-Mazel.

Hasle, H., Clemmensen, I. H., & Mikkelsen, M. (2000). Risks of leukemia and solid tumours in individuals with Down's syndrome. *Lancet, 355*, 165–169. https://doi.org/10.1016/S0140-6736(99)05264-2

Hof, P. R., Bouras, C., Perl, D. P., Sparks, I., Mehta, N., & Morrison, J. H. (1995). Age-related distribution of neuropathologic changes in the cerebral cortex of patients with Down's syndrome. *Archives of Neurology, 52*(4), 379–391. https://doi.org/10.1001/archneur.1995.00540280065020

Jacobson, J., & Mulick, J. A. (Eds.). (1996). *Manual of diagnosis and professional practice in mental retardation.* American Psychological Association.

Jahnsen, R., Villien, L., Stanghelle, J. K., & Holm, I. (2003). Fatigue in adults with cerebral palsy in Norway compared with the general population. *Developmental Medicine and Child Neurology, 45*, 296–303. https://doi.org/10.1111/j.1469-8749.2003.tb00399.x

Janicki, M. P., & Dalton, A. J. (2000). Prevalence of dementia and impact on intellectual disability services. *Mental Retardation, 38*, 276–288. https://doi.org/10.1352/0047-6765(2000)038%3C0276:PODAIO%3E2.0.CO;2

Janicki, M. P., Dalton, A. J., Henderson, C. M., & Davidson, P. W. (1999). Mortality and morbidity among older adults with intellectual disability: Health services considerations. *Disability and Rehabilitation, 21*, 284–294. https://doi.org/10.1080/096382899297710

Johnson, A. (2002). Prevalence and characteristics of children with cerebral palsy in Europe. *Developmental Medicine & Child Neurology, 43*, 713–717. https://doi.org/10.1017/S0012162201002675

King, W., Levin, R., Schmidt, R., Oestreich, A., & Heubi, J. E. (2003). Prevalence of reduced bone mass in children and adults with spastic quadriplegia. *Developmental Medicine and Child Neurology, 45*, 12–16. https://doi.org/10.1111/j.1469-8749.2003.tb00853.x

Krinsky-McHale, S. J., Devenny, D. A., Gu, H., Jenkins, E. C., Kittler, P., Murty, V. V., Schupf, N., Scotto, L., Tycko, B., Urv, T. K., Ye, L., Zigman, W. B., Silverman, W., & Taylor, S. J. (2008). Successful aging in a 70-year-old-man with Down syndrome: A case study. *Intellectual and Developmental Disabilities, 46*, 215–228. https://doi.org/10.1352/2008.46:215-228

Liptak, G. S., & Accardo, P. J. (2004). Health and social outcomes of children with cerebral palsy. *Journal of Pediatrics, 145*, S36–S41. https://doi.org/10.1016/j.jpeds.2004.05.021

Murphy, K. P., Molnar, G. E., & Lankasky, K. (1995). Medical and functional status of adults with cerebral palsy. *Developmental Medicine and Child Neurology, 37*, 1075–1084. https://doi.org/10.1111/j.1469-8749.1995.tb11968.x

National Task Group on Intellectual Disabilities and Dementia Practice. (2013). *NTG-Early detection screen for dementia.* https://www.the-ntg.org/ntg-edsd

Odding, E., Roebroeck, M. E., & Stam, H. J. (2006). The epidemiology of cerebral palsy: Incidence, impairments and risk factors. *Disability Rehabilitation, 28*, 183–191. https://doi.org/10.1080/09638280500158422

Perkins, E. A., & Berkman, K. A. (2012). Into the unknown: Aging with autism spectrum disorders. *American Journal on Intellectual and Developmental Disabilities, 117*(6), 478–496. https://doi.org/10.1352/1944-7558-117.6.478

Perkins, E. A., & Moran, J. A. (2010). Aging adults with intellectual disabilities. *Journal of the American Medical Association, 304*(1), 91–92. https://doi.org/10.1001/jama.2010.906

Perkins, E. A., & Small, B. J. (2006). Aspects of cognitive functioning in adults with intellectual disabilities. *Journal of Policy and Practice in Intellectual Disabilities, 3*, 181–194. https://doi.org/10.1111/j.1741-1130.2006.00078.x

Phillips, J. (1998). Complications of anticonvulsant drugs and ketogenic diet. In J. Biller (Ed.), *Iatrogenic neurology* (pp. 397–414). Butterworth-Heinemann.

Prasher, V. P. (2004). Review of donepezil, rivastigmine, galantamine and memantine for the treatment of dementia in Alzheimer's disease in adults with Down syndrome: Implications for the intellectual disability population. *International Journal of Geriatric Psychiatry, 19*, 509–515. https://doi.org/10.1002/gps.1077

Prasher, V. P., & Krishnan, V. H. (1993). Age of onset and duration of dementia in people with Down's syndrome: Integration of 98 reported cases in the literature. *International Journal of Geriatric Psychiatry, 8*, 915–922. https://doi.org/10.1002/gps.930081105

Reid, S. M., Modak, M. B., Berkowitz, R. G., & Reddihough, D. S. (2011). A population-based study and systematic review of hearing loss in children with cerebral palsy. *Developmental Medicine & Child Neurology*, *53*, 1038–1045. https://doi.org/10.1111/j.1469-8749.2011.04069.x

Schalock, R. L., Borthwick-Duffy, S. A., Bradley, V. J., Buntinx, W. H. E., Coulter, D. L., Craig, E. M., Gomez, S. C., Lachapelle, Y., Luckasson, R., Reeve, A., Shogren, K., Snell, M. E., Spreat, S., Tasse, M. J., Thompson, J. R., Verdugo-Alonso, M. A., Wehmeyer, M. L., & Yeager, M. H. (2010). *Intellectual disability: Definition, classification, and systems of supports* (11th ed.). American Association on Intellectual and Developmental Disabilities.

Schrojenstein Lantman-de Valk, H. M., van den Akker, M., Maaskant, M. A., Haveman, M. J., Urlings, H. F., Kessels, A. G., & Crebolder, H. F. (1997). Prevalence and incidence of health problems in people with intellectual disability. *Journal of Intellectual Disability Research*, *41*, 42–51. https://doi.org/10.1111/j.1365-2788.1997.tb00675.x

Schrojenstein Lantman-de Valk, H. M. J., Haveman, M. J., Maaskant, M. A., & Kessells, A. G. (1994). The need for assessment of sensory functioning in ageing people with mental handicap. *Journal of Intellectual Disability Research*, *38*, 289–298. https://doi.org/10.1111/j.1365-2788.1994.tb00396.x

Segal, M., Eliasziw, M., Phillips, S., Bandini, L., Curtin, C., Kral, T. V., Sherwood, N. E., Sikich, L., Stanish, H., & Must, A. (2016). Intellectual disability is associated with increased risk for obesity in a nationally representative sample of U.S. children. *Disability and Health Journal*, *9*(3), 392–398. https://doi.org/10.1016/j.dhjo.2015.12.003

Selikowitz, M. (1990). *Down syndrome: The facts*. Oxford University Press.

Strauss, D., Cable, W., & Shavelle, R. (1999). Causes of excess mortality in cerebral palsy. *Developmental Medicine and Child Neurology*, *41*, 580–585. https://doi.org/10.1017/s001216229900122x

Strauss, D., Ojdana, K., Shavelle, R., & Rosenbloom, L. (2004). Decline in function and life expectancy of older persons with cerebral palsy. *NeuroRehabilitation*, *19*, 69–78.

Sussan, T. E., Yang, A., Li, F., Ostrowski, M. C., & Reeves, R. H. (2008). Trisomy represses ApcMin-mediated tumours in mouse models of Down's syndrome. *Nature*, *451*, 73–75. https://doi.org/10.1038/nature06446

Svien, L. R., Berg, P., & Stephenson, C. (2008). Issues in aging with cerebral palsy. *Topics in Geriatric Rehabilitation*, *24*, 26–40. http://doi.org/10.1097/01.TGR.0000311404.24426.45

Tyler, C. V., & Shank, J. C. (1996). Dementia and Down syndrome. *Journal of Family Practice*, *42*, 619–621.

U.S. Centers for Disease Control and Prevention. (2006). Improved national prevalence estimates for 18 selected major birth defects—United States, 1999–2001. *Morbidity and Mortality Weekly Report*, *54*(51, 52), 1301–1305. https://www.jstor.org/stable/23316574

U.S. Centers for Disease Control and Prevention. (2019a). *Disability impacts all of us*. https://www.cdc.gov/ncbddd/disabilityandhealth/infographic-disability-impacts-all.html

U.S. Centers for Disease Control and Prevention. (2019b). *Facts about developmental disabilities*. https://www.cdc.gov/ncbddd/developmentaldisabilities/facts.html

Wallace, S. J. (2001). Epilepsy in cerebral palsy. *Developmental Medicine & Child Neurology*, *43*, 713–717. https://doi.org/10.1017/s0012162201001281

Wisniewski, K., Wisniewski, H. M., & Wen, G. Y. (1985). Occurrence of neuropathological changes and dementia of Alzheimer's disease in Down's syndrome. *Annals of Neurology*, *17*, 278–282. https://doi.org/10.1002/ana.410170310

Wojcieszek, J. (1998). Drug-induced movement disorders. In J. Biller (Ed.), *Iatrogenic neurology* (pp. 215–230). Butterworth-Heinemann.

Zigman, W. B., Schupf, N., Sersen, E., & Silverman, W. (1996). Prevalence of dementia in adults with and without Down syndrome. *American Journal on Mental Retardation*, *100*, 403–412.

Zigman, W. B., Schupf, N., Urv, T., Zigman, A., & Silverman, W. (2002). Incidence and temporal patterns of adaptive behavior change in adults with mental retardation. *American Journal on Mental Retardation*, *107*, 161–174. https://doi.org/10.1352/0895-8017(2002)107<0161:IATPOA>2.0.CO;2

16

Special Topics (Alcoholism, Falls, Foot Care, Pain, Abuse and Neglect)

ALCOHOLISM

Statistics regarding alcohol abuse in older adults are contradictory because this is a challenging area to research. Determining what constitutes excessive alcohol intake or alcohol abuse is also difficult. For example, if older adults say they drink a glass of wine at lunch and before dinner, it is important to know whether a "glass" means a 5-ounce wine glass or a 16-ounce tumbler. Many older adults are somewhat vague in reporting alcohol consumption and may not present the true picture. Also, denial is extremely common in many who drink excessively. Binge drinking is also found among this population. Binge drinking is described as four or more alcoholic drinks at one time for women and five for men (Haber, 2016). In addition, older adults are usually retired and not as visible to society at large as younger, employed people; thus, they are able to hide the effects of drinking from others, even from family, more successfully.

In older adults, alcohol abuse can be found in two general groups: those who began to drink earlier in life (30s or 40s or younger) and continue into older age (usually more likely to be men than women) and those who began drinking in older age. Approximately one-third of older alcoholics are late-onset drinkers. Numerous factors contribute to late-onset alcohol abuse in both men and women. Risk factors for alcohol-related disorders are genetic predisposition; male gender; history of psychological/psychiatric problems, particularly depression; limited education; and poverty (Tabloski, 2014). Particular risk factors for women are widowhood, loss of friends or family, and loss of home and relocation. Other age-related factors that may contribute to alcohol abuse in older persons are boredom in retirement, loss of health, financial difficulties, loss of self-esteem or competency, and any other life transition that overwhelms the individual's ability to cope and adapt. Idiosyncratic or highly individualized responses to difficulties in life must also be considered. In addition, cumulative losses are especially difficult for older adults to manage, and these usually occur more often in older age. Perhaps the most significant aspect of alcohol abuse is the effect it has on a person's life. Alcoholism is a progressive, chronic illness that causes physical, psychological, social, and financial problems, all of which may be intensified in older age.

Physical Effects of Alcohol Abuse in Older Adults

Physical effects of alcohol abuse include the following:

- *Central nervous system:* slurred speech, visual impairments, body instability with the likelihood of falls, confusion, memory impairment, increased depression, and impaired thinking and problem-solving.
- *Cardiovascular:* increased blood pressure, irregular heartbeat, enlarged heart; alcohol also contributes to coronary artery disease and increases the risk of heart attack.
- *Gastrointestinal:* contributes to malnutrition and gastritis (may lead to serious anemia); nausea and vomiting may lead to dangerous electrolyte imbalances.
- *Metabolic effects:* may lead to hypoglycemia (low blood sugar) or to acidosis (excessive acidity of the body fluids); chronic alcohol intake may result in osteomalacia if vitamin D is not metabolized normally, and fractures are then likely. Liver toxicity is a well-known complication of chronic alcohol abuse and often leads to cirrhosis, a serious liver disease.

These are common physical effects of alcohol abuse in older adults; other responses to alcohol are more individualized. Chronic alcohol intake, then, can have serious health consequences, especially for older adults whose health status may already be precarious. There should be special concern in this age group who often consume large numbers of medications both prescribed and over the counter. About 100 of the most often prescribed medications ingested by them interact negatively when taken with alcohol (Haber, 2016). For instance, alcohol and pain medications can cause stomach and intestinal bleeding; antihistamines and alcohol result in extreme drowsiness and lack of concentration; antibiotics and alcohol can lead to nausea, vomiting, headaches, and possibly convulsions; and antidiabetic medications and alcohol often lead to unpredictable severe reactions.

Aging Changes and Alcohol

Certain aspects of the aging process increase the effects of alcohol in the body. With age, tolerance for alcohol usually decreases. Two significant changes are (a) lessened reserve capacity in all organ systems of the body, which lessens the ability to deal appropriately with alcohol effects; and (b) the body's composition changes, so there is less lean body mass and an increase in fatty tissue accompanied by less total water. Alcohol is distributed in total body water; therefore, one drink produces higher blood alcohol concentrations than it would in a younger person, and the effects will last longer in an older adult. These changes may have even more serious effects in women. Alcohol use can be challenging to assess because most of its symptoms mimic other disease conditions of older age (Yeager, 2019).

Identification and Treatment

Some signs of possible alcohol abuse (although there are many others) are drinking secretly; gaps in memory; unwillingness to discuss drinking; making excuses for drinking; hiding alcohol; neglecting appearance and self-care; disregarding home, bills, or pets; becoming aggressive or abusive; appearing depressed; having frequent accidents; a poor diet; and withdrawing from social activities and contacts with people. Screening for alcohol use is desirable as a way to detect excessive alcohol use and as a possible opportunity for intervention. One screening device commonly used is the Short Michigan Alcoholism Screening Test—Geriatric version (SMAST-G; Tabloski, 2014); another possible screening tool is the Alcohol Use Disorders Identification Test (AUDIT). Both of these have been quite effective in the identification of alcohol abuse in older people (Eliopoulos, 2018). There are still other screening possibilities but not all are considered effective with older adults. If alcohol abuse is suspected, encourage the individual to seek treatment, although this may be difficult as denial is common in older drinkers. However, treatment of older alcoholics is generally quite successful, often more successful than for younger persons, and especially so for those who began to drink later in life. Psychological support from family and friends is important, as is the older person's belief that treatment will actually help.

Treatment regimens usually include detoxification, education, counseling, medications as necessary, and aftercare or continued treatment and support (Yeager, 2019). Groups such as Alcoholics Anonymous (AA) for alcoholics and Al-Anon for families of alcoholics may also be helpful in the recovery process. For those who misuse alcohol but are not truly alcohol dependent or addicted, brief therapeutic interventions such as one or several short therapeutic sessions have been found to be effective. Rather than confrontation, brief interventions use motivation, education, goal setting, and direct feedback to encourage positive behavior changes (Frandzel, 2008). Such interventions need to be made more readily available to older people.

FALLS

Falls are a leading cause of both morbidity and mortality in those older than age 65 and constitute a major public health problem in the United States. One out of three adults 65 and older fall each year, and among older adults, falls are the leading cause of both fatal and nonfatal injuries. Older people who fall and require hospitalization have a 50% chance of dying within 1 year. In long-term care facilities, older residents fall and some fall repeatedly. Injury rates are much higher in these settings (Touhy, 2016a). Death may not result from the fall itself but from its consequences, such as immobility, embolism, or infections.

Fractures are the most serious consequence of falls, and at least 95% of all fractures in older people are from falling (Touhy, 2016a). The most common fractures are hip, leg, ankle, forearm, upper arm, hand, and pelvis, with hip fracture the most feared (Diebold et al., 2010). Women experience approximately 80% of all hip fractures, although fracture rate increases in both women and men with age. Osteoporosis is often a major contributing factor in fractures after a fall because weakened bones both cause falls and are responsible for fractures that are the result of a fall. Still other injuries attributed to falls are soft tissue injuries such as joint dislocations, sprains, and hemarthrosis (blood in a joint cavity). In older age many lose functional independence after falls and injury, which has numerous extremely serious implications for quality of life.

Falls are a symptom of underlying problems but become a major focus of attention when they do occur. Most falls result from a number of contributing factors rather than one clearly identified reason; therefore, a comprehensive fall assessment is crucial and needs to consider possible interactions among the multiple risk factors involved. Risk factors for falls are generally classified as intrinsic, or resulting from body changes predisposing an individual to falling, and extrinsic, or environmental factors and precipitating causes that may contribute to falling. Young and healthy older adults tend to fall more often from extrinsic risk factor while those who are older with one or more health conditions tend to fall due to intrinsic risk factors and ambulatory issues. The more risk factors an individual has, the greater likelihood for falling (Touhy, 2016a).

Intrinsic factors include the following:

1. *Sensory impairments.* Deficits in vision, hearing, vestibular functions, and proprioception (awareness of posture, movement, and equilibrium changes) all increase risks for falling.
2. *Central nervous system disorders.* Stroke, Parkinson's disease, and normal pressure hydrocephalus (a disorder involving cerebrospinal fluid in the brain) are examples of major

central nervous system pathological conditions that predispose older adults to falling. Cardiovascular problems also increase the possibility of falls. Additionally, any disturbance in the body's balance or any changes in gait patterns may be expected to contribute to instability and falls.

3. *Cognitive changes.* Those with dementia have an increased risk for falls. Depression may also be a factor in falls in older adults. Any cognitive change that distracts the individual from attention to the environment increases their accident rate.
4. *Musculoskeletal changes.* Those with arthritis or muscle weakness difficulties are obviously more predisposed to falls. Thus, any difficulty involving muscles, bones, and joints affects stability and increases the likelihood of falling. Foot problems such as calluses, bunions, even deformed toenails and poorly fitting shoes may reduce the accuracy of proprioceptive information and increase the possibility of falling.
5. *Drugs.* Medications are a major contributor to falls in older adults. Drugs with a central depressant effect (minor tranquilizers, sedatives, hypnotics) affect postural stability, as do antihypertensive drugs that cause postural hypotension. Some drugs such as major tranquilizers and tricyclic antidepressants have a central depressant effect and may also cause postural hypotension. Other medications such as anti-inflammatory drugs cause dizziness. Numerous other medications have a selective effect on the stability of a given individual; consequently, all medications should always be evaluated in older adults who are subject to falling. Using four or more medications has been found to be a highly significant risk factor for falls (Zoorab & Sidani, 2008).

Extrinsic factors include the following:

1. *Activity.* Most falls occur while the individual is engaged in some type of activity, usually a relatively nonhazardous activity such as walking, changing body position, or while engaged in other activities of daily life. In actuality, relatively few falls occur during potentially hazardous movements such as climbing on ladders or chairs.
2. *Specific environmental factors.* The majority of falls happen in homes, although not all falls result in injury. Examples of typical hazards in the home are stairs (more accidents occur going down than going up), furniture, small objects on the floor, slippery floors, poor lighting, and certain patterns on floors or carpeting that may cause visual perception distortions. (See Appendix A for further information about safety in the home.) Animals too are often the cause of falls when they are tripped over or the individual is pulled down by them. In a long-term care setting, falls often result from furniture, such as beds or chairs and commodes, that are too high or too low.

Precipitating causes include dizziness, drop attack, slipping, tripping, low blood pressure, and others (Sanders, 2019).

Not only is the physical toll of falls and accidents expressed in pain, restricted mobility, and reduced independence, but there are psychological ramifications as well. Many older adults who have fallen (and even some who have not actually had a fall) worry excessively about falling (fallophobia), which may unduly curtail their activities. They then experience disuse effects and become further limited and even more prone to falls. Falls can cause anxiety and depression as the individual sees themselves becoming more dependent on others and losing personal autonomy. Some older adults, however, persist in denying any limitations associated with growing older and may engage in risky behaviors to prove they are not restricted in any way. These actions may well increase the possibility of falls and accidents. Falls can also affect caregivers who have to deal with the restrictions imposed by a fall.

Evaluation and Prevention of Falls

Evaluation involves a thorough and comprehensive assessment of both intrinsic and extrinsic factors contributing to fall risk. Recognize that not all older people define falls in the same way and it is important to also assess slips, tripping, and near-falls, which older adults may not report when asked about falling. Many older people and families consider falling to be a part of normal aging and do not attend to it unless it precipitates a medical crisis. The older adult, family, and caregivers need to be educated about falls to provide important information necessary for an accurate assessment of fall risk and management should one occur. In order to accurately assess extrinsic factors, a visit to the home is necessary to observe the older adult functioning in a familiar environment because verbal reports are often not accurate or comprehensive enough for a thorough assessment (Miller, 2009).

Numerous fall assessment tools are available that can be useful either in long-term care facilities or in the home to identify those at risk and to assist in devising effective interventions to prevent future falls. Two instruments often used are the Morse Fall Scale (1989) and the Hendrich II Scale (Hendrich et al., 2003). A fall diary, completed by the person or caregiver, can be very helpful. Recorded are the date and time of the fall, what the person was doing at the time, the symptoms they had, and whether or not they had an injury. This should be completed each time a fall occurs and shared with the primary caregiver (Sanders, 2019). Another simple but useful tool to determine risk of falls is the "Timed Up and Go" (TUG) test. It assesses the time it takes for a person to stand up from a standard armchair, walk about 10 feet, then turn and walk back to the chair, and sit down. A practice trial is allowed, then two timed trials, and the results of the timed tests averaged and compared with norms available to determine fall risk (Jacobs & Fox, 2008).

In prevention it is desirable to modify the environment to be safer, deal with physical problems that may contribute to falling, and educate the individual and the family about falls and accident prevention to help reduce their fears yet be reasonable in activities they may undertake. Interventions to prevent falls must include the following:

- Exercise, especially balance, gait training, and muscle-strengthening exercises. Exercises should be devised for the individual by a professional trained to work with older adults. A consistent exercise routine is probably one of the best intervention/prevention behaviors one can do to prevent falls.
- Medication management, which assesses all medications used, not just prescription drugs, as well as how the individual actually uses the medications. Certain classes of medications, such as the selective serotonin reuptake inhibitors, tricyclic antidepressants, neuroleptic drugs, benzodiazepines, anticonvulsants, some antiarrhythmic medications, and psychoactive drugs, have been related to increased risk of falling. In addition, because older adults tend to have idiosyncratic responses to any medication, all medications must be assessed periodically to determine the actual effects on behavior that may increase the risk of falling.
- Environmental modifications are based on a thorough home safety assessment to remove or modify the environment to be safer and more supportive for the older individual. This may also include education and training in recognizing potentially dangerous areas in the home as well as training in the proper use of necessary assistive devices such as canes, walkers, and wheelchairs.
- Multidimensional programs for intervention are necessary because so many factors, both intrinsic and extrinsic, contribute to falls in older adults. Goals are to reduce fall risk, maintain quality of life, and preserve the individual's personal autonomy as much as possible. Additional assistive products to help reduce falls and injury are being developed such as hip protectors, low beds, improved walkers, and other devices that will become more widely available as research in this area continues (MacCulloch et al., 2007; Touhy, 2016a).

Bedridden individuals or those in a wheelchair are also at risk of fall and may benefit from the use of an alarm. A sensor is attached to the bed or chair and to the person. When attempting to get up the sensor falls off and causes an alarm to sound, signaling those caring for the person. Several other devices are available that may be helpful to ensure the safety of the person (Sanders, 2019). Some clinicians and research do not support the use of alarms, believing they may promote immobility, which in turn may result in muscle weakness, imbalance, decubiti, and constipation. Other issues include being disturbed by the alarm, agitation, depression, and even isolation. They recommend specific exercises to strengthen the core muscles, mobility, and movement (Kaldy, 2018).

FOOT CARE

With each step, our feet orchestrate the balance and motion of the skeleton. More than 1,000 tons per day are absorbed by the 26 bones, 19 muscles, 33 joints, and 107 ligaments in each of our feet. Because most of us are on our feet about 80% of our waking time, the wear and tear on our feet over the years is immense. It is understandable, then, that feet require special attention and care. Estimates suggest that approximately 90% of those older than age 65 suffer from some type of foot problem (Touhy, 2016b). Disorders involving feet can result in falls, impair an older adult's ability to be independent, limit ability to carry out activities of daily living, and even restrict social contact with others. However, oftentimes many foot problems are preventable or at least readily treatable. Foot disorders evolve over many years and range from common conditions such as corns or calluses to involvement caused by systemic diseases such as peripheral vascular disease, diabetes, or arthritis. Unfortunately, many people, including some healthcare professionals, do not attend to proper foot care in older adults.

Foot care is often difficult for older adults who are obese or have limited vision, arthritis, coordination, or manipulation problems. Older adults with excessive body weight place greater stress on their feet, or they may become inactive, which then encourages degeneration of muscles and further limitation of movement. Heavier persons usually have wide feet with low metatarsal arches, fat deposits on the dorsal part of the foot, and thickening of the sole of the foot. Wide shoes are needed to accommodate these changes.

Age-related changes in the foot include the following (Luggen & Jett, 2008):

- Fat pads shrink and degenerate on the sole of the foot, reducing the cushioning effect.
- Skin may become more dry, inelastic, and cool.
- Subcutaneous tissue on the bottom and sides of the foot thins.
- Toenails are more brittle; they tend to thicken and are more likely to develop fungal infections.
- Foot range of motion may diminish because of degenerative joint diseases.

Foot Disorders and Treatment

Disorders of the foot are categorized into skin (dermatological), nail, structural deformity, arthritic conditions including gout, and conditions that are complications of systemic disease (Touhy, 2016b). Some of the most common foot disorders among older persons are dry skin, corns and calluses, toenail disorders, structural deformities, arthritis, neuropathy associated with diabetes, and circulatory and inflammation disorders.

Skin

The skin of older persons' feet becomes dry, and pruritus, dermatitis, or eczema can cause itching with subsequent scratching, skin breakdown, and eventual infection. The daily use of creams containing lanolin will lubricate the skin. If itching becomes a problem, an oral antihistamine or topical corticosteroid may be helpful.

Corns and Calluses

Corns and calluses are the most common podiatric complaints of older adults and result from excessive friction and pressure over a bony prominence such as a hammer toe (a joint that is contracted) or a bunion (a swelling of the bursa usually caused by shoes that do not fit properly). Excessive tissue should be removed, preferably by a podiatrist, and padding applied to reduce friction over the area. The use of medicated pads is not recommended because they may result in burns or infections (Meiner, 2011). Wider and softer shoes are highly recommended. Calluses are similar to corns but occur on soles and heels of feet as a result of irritation and friction from improper shoes. Treatment is the same as for corns.

Toenail Disorders

Toenail disorders include fungal infections, ingrown toenails, and inflammation around the nail or nail bed. Poorly fitting shoes or a foot injury predisposes the individual to inflammation, reduced blood supply to the area, or an infection. Fungal disease causes the nails to become brittle and thickened, and the disease may spread from one nail to the next. Antifungal medications placed around the nail bed are often used but may not be effective; a podiatrist should be consulted for proper treatment. Oral antifungal medications such as Terbinafine may be prescribed. Onychogryposis (ram's horn–shaped nails) form long spirals that, if not treated, invade the soft tissues, causing pain and inflammation. These conditions need to be treated by a podiatrist.

Structural Deformities

Structural deformities such as hammertoes and bunions result from foot changes throughout a lifetime and from wearing improper shoes. Women who have worn high-heeled shoes are at particular risk for structural deformities. Such conditions usually require special foot padding, special shoes, or even surgery. They can be extremely painful and severely limit ambulation.

Arthritic Conditions

Rheumatoid arthritis, osteoarthritis, and gout of the foot are arthritic conditions resulting in swollen, deformed joints and an altered gait. Bearing weight on these joints is difficult, and walking is often impaired and painful. Special shoes, antiarthritic medications, and surgery are generally effective treatments.

Diabetes

Diabetes mellitus is a major cause of foot or leg amputation. Diabetic neuropathy (lack of sensation in the foot), ulcerations, infections, and gangrene are common foot problems found in people with diabetes. With adequate foot care and appropriate preventive measures, most of these complications can be avoided. Major preventive considerations include meticulous skin care, avoiding injury to the skin, keeping the feet clean and dry, and maintaining blood sugar within normal limits.

Circulatory and Inflammation Disorders

Arterial diseases cause diminished blood supply to the foot, resulting in numbness, tingling, pain, infection, and ulcerations. Treatment involves the use of vasodilator medications, excellent foot care by both the individual and a podiatrist, and the immediate treatment of infections. Bursitis (inflammation of the bursae), a systemic disease, is usually found in the large toe joint and is a very painful condition. It may require long rest periods, steroid injections, anti-inflammatory drugs, and specialized shoes (Jessett & Helfand, 1991).

Many of these foot problems require the regular care of a podiatrist or an RN prepared in this care. Others require the periodic care of a pedicurist.

Proper Foot Care

Recommendations for foot care should include the following:

1. Maintain a well-balanced diet.
2. Wash the feet daily in warm water using mild soap. Dry well, especially between the toes, and apply an emollient such as Vaseline. Care should be taken, however, to avoid falling.
3. Inspect the feet daily for dry skin, toenail problems, sores, blisters, cracks, changes in skin temperature indicating possible circulatory impairment, changes in color or sensation, thickened toenails, reddened areas, swelling, pain, ulcers, drainage (indicative of infection), and odor. A family member may need to do these inspections if an older person has visual or manipulative difficulties.
4. Cut nails straight across to avoid skin punctures and ingrown toenails.
5. Wear shoes that fit. Those with diabetes or circulatory problems should be especially careful to wear enclosed leather shoes that still allow the feet to breathe.
6. Never go barefoot because of the danger of injury and consequent infection.
7. Have corns and calluses cut and treated by a podiatrist.
8. Use range-of-motion exercises for the feet to preserve mobility.
9. Maintain adequate circulation of blood to the feet by walking, stretching, or doing specific exercises. A foot massage or a warm foot bath also promotes circulation.

Foot care is extremely important in older adults for health, comfort, and to maintain mobility and independence, although it does not always receive proper attention and treatment. Those in the helping professions need to be aware of the ramifications of inadequate attention to foot care and encourage intervention before difficulties arise.

PAIN

There are various definitions of pain but all agree that it is an unpleasant experience and involves both mind and body. Because it is subjective and multidimensional, including psychological, sensory, emotional, and social factors, it is difficult to objectively assess and treat. The most practical definition of pain is that it is whatever the individual says it is, existing whenever the person says it does (McCaffery, 1968). Among the various myths about aging in our society is one implying that pain is a normal and usual part of older age. Older persons who have pain are commonly told they have to "learn to live with it," suggesting that little can be done to control or manage it. Older adults are more likely than younger people to develop a variety of health problems involving pain, ranging from pain caused simply by lack of activity (disuse), a common but often unrecognized source of pain in older age, to that caused by malignant diseases such as cancer. Specifically, many older people suffer from arthritis, osteoporosis, trigeminal neuralgia, postherpetic neuralgia after an attack of shingles, angina, injuries from falls, and various types of muscle injuries. Musculoskeletal problems account for the most common causes of pain in older age. Thus, the presence of multiple medical problems and multiple sources of pain makes pain management in older adults a definite challenge.

The prevailing perspective in gerontology that aging is not synonymous with illness or disease encourages older adults to take a greater proactive role in modifying their lifestyles and being informed consumers of their own healthcare in order to avoid or postpone disease and frailty for as long as possible. These issues are also relevant in pain control and management. The study of pain (algology) has increased dramatically over the years, but many of the principles of pain management currently available are still not used extensively with older adults. Older persons are greatly underrepresented in both pain research and in pain clinics and are often not offered options for pain control other than medications. However, research and clinical experiences of those who work in this area indicate that most older persons are quite able to learn and use a variety of pain management techniques (Luggen, 2000).

Classification of Pain

Pain may be classified as acute or persistent.

Acute Pain

Acute pain is associated with physiological responses to actual tissue damage and is of rapid onset. When the damaged tissues heal, acute pain decreases or disappears, typically in days or weeks. Acute pain can usually be treated quite well with medications or surgery, and these are generally effective with older adults, although practitioners need to be alert to the greater likelihood of idiosyncratic reactions to medications in this age group because of various individual age-related physical changes.

Persistent Pain

Pain that persists beyond the time expected for an injury to heal or pain associated with a chronic pathological condition causing continuing pain for months or years is called *persistent,* formally known as *chronic pain*. Pain specialists define persistent pain as pain persisting for at least 3 to 6 months in the absence of identifiable physical reasons to account for it. Initially when pain is first diagnosed, a cause initiated by the mechanism in generating the pain is identified. Gradually as the pain becomes persistent several mechanisms develop with central sensitization. Treatment then needs to address both the initial reason for pain as well as the central sensitization (Schneiderman & Orizondo, 2017). Persistent pain may lead to a persistent pain syndrome in which the individual's life is disrupted in a number of ways: Persons with pain become preoccupied with the pain; they are irritable and lose interest in activities previously enjoyed; their sleep patterns are disturbed, which often leads to sleep deprivation; they have high levels of anxiety and depression, use pain medications excessively, have disturbed interpersonal relationships, and become pessimistic because efforts to control their pain have been ineffective.

Persistent pain is sometimes further classified as follows:

- Nociceptive pain arising from peripheral or visceral pain receptors (nociceptors) such as soft tissue injuries or osteoarthritis. Nociceptive pain usually responds well to analgesic medications and nonpharmacological interventions.
- Neuropathic pain arises from pathology of peripheral nerves or within the central nervous system such as trigeminal neuralgia, neuropathies, central poststroke pain, and postherpetic neuralgia after an attack of shingles. Neuropathic pain does not respond as well as nociceptive pain to the usual analgesics, although antidepressants and anticonvulsant medications are sometimes effective (Ignatavicius, 2013; Tabloski, 2014).

- Mixed or unspecified pain typically has unknown causes and may be a mixture of nociceptive and neuropathic pain. Examples are recurrent headaches and compression fractures with nerve root irritation, common with osteoporosis.
- Another source of persistent pain may stem from psychological/psychiatric disorders. These pain situations are best treated with psychiatric approaches rather than analgesic medications.

Although pain is not an inevitable part of normal aging, many older adults do suffer from persisting pain. One estimate is that 50% of older people in the community have significant pain problems, and perhaps as many as 70% to 80% of those institutionalized have pain that is undertreated (Jett, 2016). The incidence of pain, however, may be underestimated as older people do not always report pain because they expect it in older age or they do not want to complain. In addition, pain may be demonstrated as confusion, restlessness, fatigue, or aggression in older adults, leading to a delay in treatment or to misdiagnosis (Miller, 2009).

Older adults with cognitive impairment may pose a unique challenge since they may be unable to convey the presence of pain. These individuals are often undermedicated (Alderman, 2019). Crying, restlessness, pacing, aggressive behaviors, inability to sleep, elevation in vital signs, decreased socialization, and altered food intake may present as signs of pain. Great care must be taken not to minimalize the presence of pain among these individuals. Seeking help from the family or other caregivers may be of value. Cultural issues too should be considered. In some cultures the pain is expected to be endured, while in others verbal expression and crying out may be the norm. Men especially may not express the presence and intensity of pain (Eliopoulos, 2018).

As for assessment, the first step in comprehensive pain assessment is a thorough history and physical examination. Appropriate pain management depends on understanding the underlying disease or illness. A variety of pain assessment tools are readily available, including the commonly used numeric pain scale, in which the individual rates pain from 0 (no pain) to 10 (worst possible pain), and the Faces Pain Scale. Others commonly used are verbal descriptive pain intensity scales using categories such as "no pain," "little pain," "lot of pain," and "too much pain," or the visual analog scales in which pain is rated on a scale from "no pain" to "pain as bad as it could get." The McGill Pain Questionnaire is also a commonly used assessment tool (Eliopoulos, 2018). Pain logs and pain diaries are also used to obtain information about frequency and severity of pain (Tabloski, 2014). Other aspects of pain assessment necessary in older adults are assessments of degree of functional impairment, cognitive impairment, depression, and quality of life. A comprehensive pain assessment needs to consider all possible factors that could contribute to the reported pain.

Medications are most often used for treatment of pain. Analgesics are categorized into three groups: nonopioids, opioids, and adjuvant medications. The term *narcotic* is no longer used in clinical practice and has been replaced by opioid and nonopioid medications. Some individuals in this age group have taken opioids for pain control over the years. Others are or have used marijuana, which is now legalized in some states for medicinal or recreational use. These situations may present special safety challenges to healthcare professionals. The nonopioid drugs used for mild pain include acetaminophen, tramadol, and nonsteroidal anti-inflammatory drugs (NSAIDs). However, NSAIDs must be used judiciously in older adults because of potentially dangerous side effects (Jett, 2016). Opioids are more powerful medicines used for moderate to severe pain and include codeine, morphine, methadone, and fentanyl. Adjuvant medications are those not primarily indicated for treatment of pain but that have been shown to be effective in many pain situations. Examples are antidepressants or anticonvulsants that can be used with nonopioid or opioid drugs, corticosteroids, local anesthetics, and muscle relaxants (Adams et al., 2019).

Recognizing that older adults tend to experience medication side effects more often than younger persons, all pain medications must be carefully prescribed and medications adjusted periodically once used. General advice for drugs given to older adults is to "start low and go slow." The World Health Organization developed a pain relief ladder for use with older adults in persistent pain. Three steps refer to different levels of pain intensity and allows for individualizing therapeutic regimens for pain control. For example, step 1 addresses mild pain and recommends nonopioid analgesics with an adjuvant medication if deemed appropriate. If pain persists in a mild to moderate range, step 2 suggests adding a weak opioid or opioid combination with an adjuvant medicine if appropriate. For pain that persists or becomes worse, step 3 advocates stronger opioids and adjuvants. In this system one could also start at step 3 if the pain is excruciating and not follow through step 1 and step 2, but then reduce the medications as pain is relieved (World Health Organization [WHO], 1996).

A great variety of nonpharmacological treatments for pain are available, and often the most effective pain treatment involves both pharmacological and nonpharmacological approaches. Examples of nonpharmacological pain treatments are transcutaneous electrical nerve stimulation (TENS), acupuncture and acupressure, biofeedback, relaxation, guided imagery, hypnosis, meditation, exercise, massage, music, cognitive behavioral therapy, and various other complementary therapies (Tabloski, 2014).

Specific pain programs designed to help those with persistent pain are generally of three types and may include both inpatient and outpatient treatment.

1. *Syndrome-oriented pain centers.* These centers focus on treating a specific persistent pain problem such as headaches or back pain. They may use only one treatment method or they may use a variety of treatment techniques.
2. *Modality-oriented pain centers.* This type of pain center uses only one specific treatment technique such as, for example, behavior therapy or acupuncture.
3. *Comprehensive pain centers.* Services in these centers include initial assessment, treatment, and follow-up. The staff is multidisciplinary and usually includes physicians, nurses, physical therapists, occupational therapists, massage therapists, rehabilitation specialists, psychologists, and social workers, all working together as a coordinated team. A variety of pain management techniques are used and the treatment program is individualized for each client.

Comprehensive pain centers offer the following:

1. *Medical treatment.* This can include the judicious use of specific medications; injection of muscle trigger points (any place on the body that when stimulated causes sudden pain in a specific area); anesthetizing muscles in spasm; peripheral nerve blocks (anesthetizing a nerve or nerves to prevent transmission of pain impulses); continuous pain medications delivered by pumps via intravenous, subcutaneous, epidural, or intrathecal routes; and perhaps surgery if indicated. None of these procedures is used alone for treatment of persistent pain, but the specific medical treatment deemed most appropriate is used in combination with other pain management strategies.
2. *Medication adjustment.* Many people with persistent pain use a high number of drugs leading to possible overdose or abuse (Eliopoulos, 2018). An important aspect of pain treatment is to reduce usage of such medications over time and eliminate their habitual use. Only medications absolutely necessary for pain control are retained, and individuals are taught careful management of these drugs.
3. *Physical modalities.* Examples of physical adjuncts used for effective pain management include massage, hot or cold packs, acupuncture, TENS, and exercise. These and other physical methods are used to relieve pain and rehabilitate the individual while they are learning various other pain management techniques to be used during and after completion of the pain center program.

4. *Psychological techniques.* Various psychological intervention techniques are taught to the person with pain to increase their sense of personal control and independence. These often involve relaxation training and biofeedback; behavior modification principles and cognitive behavior therapy to teach more appropriate behaviors for effective pain management; psychological strategies to deal more effectively with emotional factors involved in pain such as anxiety, depression, frustration, and irritability; teaching individuals to properly pace life activities; and teaching more effective coping strategies such as visualization/imagery and self-hypnosis (Jett, 2016; Tabloski, 2014).

Many older adults are excellent candidates for pain treatment programs and are able to benefit from a multimodal approach just as well as younger persons. Individuals need to understand, though, that they must be active participants in their pain treatment programs, that education for both the pain patient and family members is extremely important, and that it is of utmost importance to choose an accredited pain center or program with an experienced and qualified staff. Although persistent pain may not be curable, learning effective techniques to manage it and thus exert some personal control over it dramatically helps to improve the quality of life at any age.

ABUSE AND NEGLECT

Older adult abuse is becoming more prevalent in the United States and is a serious health problem. Research indicates that 1 in every 10 older adults experiences some form of elder abuse. It can occur in any setting and socioeconomic group (Miller, 2017).

Elder abuse "is a single, or repeated act or lack of appropriate action, occurring within any relationship where there is an expectation of trust or dependence which causes harm or distress to an older person" (WHO, 2020, para. 1). Identifying existing numbers of older adults who are abused is difficult due to underreporting of abuse and neglect by health professionals and the use of inappropriate screening instruments. Since there is no national abuse reporting system, statistics on the occurrence of abuse are often not accurate. Older adults may not report abuse for fear of reprisal. Assessing abuse is difficult with individuals who may be cognitively impaired and not able to report an abusive situation (Mauk, 2018).

Seven types of elder abuse are identified by the National Center on Elder Abuse (n.d.).

1. *Neglect:* The failure or refusal of caregivers or other responsible persons to provide protection and shelter, food or adequate healthcare to a vulnerable older adult.
2. *Self-Neglect:* The behavior of an older adult that threatens their own safety and health such as lack of adequate nutrition or not taking prescribed medications.
3. *Emotional or Psychological:* Abuse causing distress, anguish, or mental pain on the older adult verbally or nonverbally.
4. *Sexual Abuse*: Abuse encouraging an older adult to engage in non-consensual sexual behavior or to observe unwanted sexual behavior.
5. *Financial or Material Exploitation:* Taking, concealing, or misusing assets or property of a vulnerable older adult.
6. *Abandonment:* The intentional placing of a vulnerable older adult at risk for harm by anyone who may be responsible for the custody or care of the person.
7. *Physical Abuse:* The use of physical force that may result in bodily injury, physical pain, or impairment.

Individuals who have a mental or physical illness are at greater risk for abuse or neglect. Other issues relate to prejudice and stereotyping of individuals with disabilities. Social isolation and little support are related to several forms of abuse. At greater risk for abuse are those cared for by individuals who are mentally ill, alcoholic, or drug abusers who live with the older person or who are dependent financially (Miller, 2017).

Abuse and neglect also occur in long-term care facilities. Resident-to-resident aggression as well as abuse and neglect by staff, family, or other individuals exists. Being responsible for the care of large numbers of ill older adults, inadequate staffing ratios and a lack of formal required educational programs all potentiate the risk for mistreatment of older adults (Pickens et al., 2011).

REFERENCES

Adams, P. M., Urban, C. Q., & Sutter, R. E. (2019). *Pharmacology: Connections to nursing practice* (4th ed.). Pearson.

Alderman, J. (2019). Pain. In S. E. Meiner & J. J. Yeager (Eds.), *Gerontologic nursing* (6th ed., pp. 213–229). Elsevier.

Diebold, C., Fanning-Harding, F., & Hanson, P. (2010). Management of common problems. In K. Mauk (Ed.), *Gerontological nursing* (2nd ed., pp. 461–469). Jones & Bartlett.

Eliopoulos, C. (2018). *Gerontological nursing* (8th ed.). Wolters Kluwer.

Frandzel, S. (2008). Substance abuse invisible but growing issue among elderly. *CNS Seniorcare, 7*, 14–15.

Haber, D. (2016). *Health promotion and aging* (7th ed). Springer Publishing Company.

Hendrich, A. L., Bender, P. S., & Nyhuis, A. (2003). Validation of the Hendrich II Fall Risk Model: A larger concurrent case/control study of hospitalized patients. *Applied Nursing Research, l6*, 19–24. https://doi.org/10.1053/apnr.2003.yapnr2

Ignatavicius, D. (2013). Pain: The fifth vital sign. In D. Ignatavicius & L. Workman (Eds.), *Medical–surgical nursing* (7th ed., pp. 39–63). Saunders Elsevier.

Jacobs, M., & Fox, T. (2008, March/April). *Using the "Timed Up and Go/TUG" test to predict risk of falls.* Assisted Living Consult, p. 16.

Jessett, D. F., & Helfand, A. D. (1991). Foot problems in the elderly. In M. S. J. Pathy (Ed.), *Principles and practice of geriatric medicine* (2nd ed., pp. 1301–1307). John Wiley & Sons.

Jett, K. (2016). Pain and comfort. In T. A. Touhy & K. Jett (Eds.), *Ebersole & Hess' toward healthy aging* (8th ed., pp. 323–337). Elsevier.

Kaldy, J. (2018). The Buzz: Facilities are going alarm-free. *Caring for the Ages, 19*(12), 16–17. https://doi.org/10.1016/j.carage.2018.12.008

Luggen, A. (2000). Pain. In A. G. Lueckenotte (Ed.), *Gerontologic nursing* (2nd ed., pp. 281–301). Mosby.

Luggen, A. S., & Jett, K. (2008). Biological maintenance needs. In P. Ebersole, P. Hess, T. A. Touhy, K. Jett, & A. S. Luggen (Eds.), *Toward healthy aging* (7th ed., pp. 157–193). Mosby Elsevier.

MacCulloch, P. A., Gardner, T., & Bonner, A. (2007). Comprehensive fall prevention programs across settings: A review of the literature. *Geriatric Nursing, 28*, 307–311. https://doi.org/10.1016/j.gerinurse.2007.03.001

Mauk, K. (2018). *Gerontological nursing: Competencies for care* (4th ed.). Jones and Bartlett Learning.

McCaffery, M. (1968). *Nursing practice theories related to cognition, bodily pain and man-environment interactions.* UCLA Students Store.

Meiner, S. (2011). *Gerontological nursing* (4th ed.). Elsevier.

Miller, C. A. (2009). *Nursing for wellness in older adults* (5th ed.). Wolters Kluwer/Lippincott Williams & Wilkins.

Miller, C. A. (2017). *Elder abuse and nursing: What we need to know and can do*. Springer Publishing Company.

Morse, J., Morse, R., & Tylko, S. (1989). Development of a scale to identify the fall-prone patient. *Canadian Journal on Aging, 8*, 366–377. https://doi.org/10.1017/S0714980800008576

National Center on Elder Abuse. (n.d.). *Types of abuse*. https://ncea.acl.gov/Suspect-Abuse/Abuse-Types.aspx

Pickens, S., Halphen, J. M., & Dyer, C. B. (2011). Elder mistreatment in long-term care settings. *Annals of Long Term Care, 19*(8), 30–36.

Sanders, D. (2019). Safety. In S. E. Meiner & J. J. Yeager (Eds.), *Gerontologic nursing* (6th ed., pp. 174–199). Elsevier.

Schneiderman, J., & Orizondo, C. (2017). Chronic pain: How to approach 3 common conditions. *Clinical Reviews, 27*(10), 38–49.

Tabloski, P. (2014). *Gerontological nursing* (3rd ed.). Pearson Education.

Touhy, T. A. (2016a). Mobility. In T. A. Touhy & K. Jett (Eds.), *Ebersole & Hess' toward healthy aging* (8th ed., pp. 200–222). Elsevier.

Touhy, T. A. (2016b). Elimination, sleep, skin and foot care. In T. A. Touhy & K. Jett (Eds.), *Ebersole & Hess' toward healthy aging* (8th ed., pp. 164–199). Elsevier.

World Health Organization. (1996). *Cancer pain relief* (2nd ed.). World Health Organization.

World Health Organization. (2020). *Elder abuse – Key facts.* https://www.who.int/news-room/fact-sheets/detail/elder-abuse

Yeager, J. J. (2019). Drugs and aging. In S. E. Meiner & J. J. Yeager (Eds.), *Gerontological nursing* (6th ed., pp. 257–277). Elsevier.

Zoorab, R., & Sidani, M. (2008, April). *Minimizing the risk of falls in elderly patients* (pp. 75–81). Clinical Advisor.

17

Health Promotion and Exercise

INTRODUCTION

Health, as defined by the World Health Organization (1946), "is a state of complete physical, mental, and social well-being not merely the absence of disease or infirmity" (p. 100). Health and wellness for older adults can be best described as a potential balance between a person's internal and external environment plus an individual's physical, social, emotional, and cultural functioning (Touhy, 2012).

Older adults often view themselves as healthy when able to carry out the activities of daily living and live full productive lives despite the presence of disability and disease. In this light, an appropriate definition of health for older persons is the ability to live and function in society and to exercise self-reliance and autonomy to the maximum extent feasible but not necessarily as total freedom from disease (Filner & Williams, 2000). Hansen-Kyle (2005) describes health in the latter years as "a process of slowing down, physically and cognitively, while resiliently adapting and compensating in order to optimally function and participate in all areas of one's life physical, cognitive, social, and spiritual" (p. 52). Rather than continuing the more traditional healthcare modalities of managing disease and illness, the focus must be on (a) promoting biological wellness such as healthy eating, sleeping, controlling health problems, avoiding smoking and seeking the best quality healthcare, (b) promoting social wellness, facilitating interaction with others and pets, (c) functional wellness support staying engaged and active to the highest level possible, (d) environmental dimension involving creating physical spaces that support healthy living, and (e) psychological dimension assisting openness and respect for individuals spiritual beliefs (Jett & Touhy, 2016).

Older age is often considered to be a time of illness, disability, and pain, yet based on current evidence it is reasonable to suggest that many of the health problems commonly associated with older age need not exist or at least may be delayed to old-old age. Fries (1980) proposed and, along with colleagues (Vita et al., 1998), later proved the theory of the "compression of morbidity" in which life-threatening chronic problems can be delayed to the very end of the life span by judicious use of preventive health measures. Thus, the population of older adults will be larger and healthier than previously, but there will also be a larger number coping with age-related and age-determined disability, frailty, and comorbidity (Nakasato & Carnes, 2006). Among the leading causes of illness and death in people older than age 65 are strokes, cancer, heart disease, lung disease, falls and fractures, depression, dementia, diabetes, influenza, and pneumonia. Many of these could be prevented or their progress slowed, resulting in substantial improvement in quality of life. Most people, no matter how long they live, will experience a period of time being disabled (Haber, 2016). Preventable risk factors include hypertension, elevated blood sugar, smoking, and obesity.

HEALTH PROMOTION

Health promotion that focuses on assisting individuals to achieve maximum health status is closely linked to successful aging and involves individual responsibility (Resnick, 2011). Additionally, it includes activities individuals can engage in proactively to enhance health (Nelson, 2018). Rowe and Kahn (1998) define successful aging as the ability to maintain three important states: (a) low risk for disease or disabilities, (b) high level of physical and mental functioning, and (c) engaging actively in life. Healthy aging not only involves the older adult stage of life but should begin in the prenatal period and extend throughout life with a focus on health promotion activities and adequate healthcare (Touhy, 2016). Resnick (2011) identifies significant areas of health promotion for older adults:

- Control of smoking
- Mind–body health
- Maximal nutritional status
- Weight control
- Social health
- Self-care, medically
- Spiritual healthcare
- Health, environmental

Addressing health in a holistic manner should involve incorporating all these aspects into a person's life. Maintaining a positive attitude while realistically facing concomitant age-related changes and diseases can result in a happy, fulfilled, transcendent life throughout the later years.

DISEASE PREVENTION

The prevalence of chronic disease escalates as we age. More than 80% of individuals older than age 70 have one chronic health problem and 50% experience multiple health problems (Touhy, 2012). Chronic and even communicable diseases in older age pose a tremendous financial burden on society, and they constitute a large percentage of the national health budget. As increasing numbers of people live to older age, this percentage will escalate unless health promotion and disease prevention measures and programs are adopted by individuals of all ages.

Hickey and Stilwell (1991) discuss the differences between health promotion for the young and those who are old. For youth, the focus is primarily on exercising regularly, eating a good diet, and not smoking, which are goals for any age. However, for older adults, the focus is on early treatment and modification of lifestyle to slow the progression of chronic diseases, minimizing disability and maintaining functional independence for as long as possible.

Ideally, those who arrive at old age in a healthy state will be able to maintain their health for a long time. Preventive health strategies are important, though, and are beneficial for individuals in all stages of life. Fried (1990) discusses various health issues specifically affecting older adults as (a) the presence of new chronic or acute diseases; (b) diseases already present that will result in death; and (c) the dependence, disability, functional losses, hospitalizations, and eventual death that are the outcomes of chronic illnesses. All these issues may be influenced by the knowledge of and adherence to the following preventive measures.

- *Primary prevention.* Identifying and targeting risk factors in individuals and preventing disease before it starts, such as, for example, eating a diet with lower levels of fat and salt to reduce the incidence of cardiovascular disease or getting recommended immunizations. Regular exercise, smoking cessation, stress management, moderate alcohol use, being socially active, and having cognitive stimulation all involve primary prevention.
- *Secondary prevention.* Preventing disease while still asymptomatic or unreported and identifying unrecognized health problems in the early stages by screening and assessment; for example, identifying high cholesterol levels or blood pressure; performing screening tests for cancer; and assessing mobility, cognition, nutritional status, pain levels, and skin integrity are necessary during a visit to the primary care practitioner.
- *Tertiary prevention.* Minimizing the overall effects of an established disease by accurate diagnosis, appropriate treatment, and rehabilitation that aims to prevent the progression of symptoms and complications. Locating and securing appropriate services and equipment such as assistive devices are also very important.
- *Quaternary prevention.* Reducing the disability resulting from chronic symptoms while at the same time attempting to maintain the attained level of functioning (Resnick, 2011).

Health promotion and disease prevention measures are valuable no matter when in life they are initiated.

Primary and Secondary Disease Prevention Guidelines

Primary and secondary disease prevention should include the following:

1. Dental checkups are recommended annually or biannually.
2. Important immunizations for older adults are the yearly influenza vaccine, the pneumonia vaccines Prevnar and Pneumovax, the shingles vaccine Shingrix, and the tetanus-diptheria and tetanus-diptheria-pertussis vaccines.
3. The Affordable Care Act initiated a program that prevents illness and promotes health. It includes an initial physical examination and counseling followed by a yearly wellness visit with a healthcare professional. Various periodic screenings for the endocrine, neurological, gastrointestinal, sensory, skin, cardiovascular, urinary, musculoskeletal systems may be included when appropriate and needed. Also included are psychological counseling and assistance with obesity, alcoholism, and smoking. More specific information is available at www.medicare.gov.

Many of the policies related to The Affordable Act result from research reviewed by the U.S. Preventative Services Task Force (USPSTF) a volunteer, independent, group of national experts on prevention and evidenced-based medicine. The UPSTF, along with the CDC Advisory Committee on Immunization Practices (ACIP), and the Health Resources and Services Administration (HRSA) have produced a very useful free tool called *My Health Finder* (https://health.gov/myhealthfinder) that provides a comprehensive list of screening and vaccines recommended for a user's age, sex, smoking status, and whether they are sexually active.

4. Periodic assessments are advised regarding the health habits and behaviors of older adults, including tobacco use, drug or alcohol abuse, sexual behavior, food intake, exercise

regimen, sleep patterns, mental health status, and quality of their living environment, particularly regarding injury and violence. Some advocate the use of a health contract/calendar to self-manage the assuming of healthy behaviors. With the assistance of a clinician or health educator, a behavior change goal is identified along with a plan to reach the goal. A formal contract is then drawn up including the goal and the specific plan of action signed and dated by the person and the health professional and reviewed periodically by them. Several forms of health contracts exist, allowing the individual to choose one they are comfortable with (Haber, 2018; Resnick, 2011).

Tertiary Disease Prevention Guidelines

Secondary and tertiary prevention often overlap. Tertiary preventive measures are directed toward maintaining functional autonomy and enhancing quality of life when disease is already present. The following are recommended:

1. Evaluation of the individual's physical and functional impairments. This might include assessing drug use and side effects, incontinence, falls, foot problems, depression, immobilization, disorientation, visual or hearing impairments, and how dependent the person is on others.
2. Assessment of factors in the person's environment, such as room temperature; accessibility of bathroom, kitchen, and bedroom; presence of safety hazards (such as scatter rugs); ability to contact others in emergencies; ability to secure food and maintain an adequate diet; availability of adequate clothing; and knowledge of how to access support services.
3. Continual monitoring of rehabilitative therapies such as physical, occupational, or speech therapy to restore and maintain functional capabilities for as long as possible.
4. Monitoring of health status and personal hygiene by healthcare providers.
5. Assessment of family and social services available in the community to assist the elder in compensating for losses.

Fried (1990) suggests that tertiary prevention is the most important focus of care for older adults because it deals with preventing dependency and disability, maintaining functioning, and improving quality of life in the later years. All four types of prevention—primary, secondary, tertiary, and quaternary—can have a major effect on the lives of older adults with an approach based on the use of risk profiles and individual health assessments to determine the best possible outcomes.

STRATEGIES FOR CHANGE

Healthy People 2030 is the fifth edition of an initiative begun in 1979, when the Surgeon General Julius Richmond issued a landmark report entitled *Healthy People: The Surgeon General's Report on Health Promotion and Disease Prevention* (U.S. Department of Health and Human Services [USDHHS], 2020a). It is composed of goals and objectives with 10-year targets, 2020 to 2030, to guide the promotion of health and prevent disease to better improve the health of all people of the United States. Its vision statement is "A society in which all people can achieve their full potential for health and well-being across the lifespan" (p. 1). There are five major areas and 62 topics: health conditions (e.g., cancer diabetes, dementias); health behaviors (e.g., physical activity, tobacco use); populations (e.g., older adults, women, people with disabilities); settings and systems (e.g., public health, health insurance, housing and homes), and social determinants of health (e.g., economic stability, healthcare access and quality).

There are 20 Healthy People 2030 objectives for older adults (USDHHS, 2020b).

1. Increase the proportion of older adults with physical or cognitive health problems who get physical activity.
2. Reduce the rate of pressure ulcer-related hospital admissions among older adults.
3. Reduce the rate of hospital admissions for diabetes among older adults.
4. Increase the proportion of older adults with diagnosed Alzheimer's disease and other dementias, or their caregiver, who are aware of the diagnosis.
5. Reduce the proportion of preventable hospitalizations in older adults with dementia.
6. Increase the proportion of adults with subjective cognitive decline who have discussed their symptoms with a provider.
7. Reduce infections caused by *Listeria*.
8. Reduce the rate of hospital admissions for urinary tract infections among older adults.
9. Reduce fall-related deaths among older adults.
10. Reduce the proportion of older adults who use inappropriate medications.
11. Reduce the rate of emergency department visits due to falls among older adults.
12. Reduce the proportion of older adults with untreated root surface decay.
13. Reduce the proportion of adults aged 45 years and over who have lost all their teeth.
14. Reduce the proportion of adults aged 45 years and over with moderate and severe periodontitis.
15. Reduce hip fractures among older adults.
16. Increase the proportion of older adults who get screened for osteoporosis.
17. Increase the proportion of older adults who get treated for osteoporosis after a fracture.
18. Reduce the rate of hospital admissions for pneumonia among older adults.
19. Reduce hospitalizations for asthma in adults aged 65 years and over.
20. Reduce vision loss from age-related macular degeneration.

Healthy People initiatives are oriented toward health and not disease. Credence is also given to lifestyle, socioeconomic state, and other issues that complicate one's ability to be healthy (Haber, 2016). The Healthy People 2030 website provides much detailed information for the consumer and for healthcare providers (https://health.gov/healthypeople).

MEDICARE, MEDICAID, AND THE AFFORDABLE CARE ACT

Medicare offers health insurance for individuals age 65 or older, for those with certain disabilities under age 65 plus, and for people of any age with end-stage renal disease. Part A, Hospital Insurance, covers inpatient hospital care, skilled nursing care, home healthcare and hospice care. Part B, Medical Insurance, covers doctors and other healthcare services, home healthcare, outpatient care, durable medical equipment, and many preventive services. Part C, Medicare Advantage plan, covers all services and benefits under parts A and B, and usually includes Medicare prescription drug coverage Part D. Plans have a yearly limit on out-of-pocket expenses. This plan is managed by private insurance companies that must follow Medicare rules. They may offer extra benefits such as a fitness membership or a certain dollar amount of over-the-counter supplies from designated stores. Medicare Part D is a Medicare prescription drug coverage program which helps to cover prescription drug costs. This program is administered by Medicare; approved drug plans must comply with Medicare rules.

The Patient Protection and Affordable Care Act of 2010 (USDHHS, 2019) expanded preventive services that are covered by Medicare. Within the first year of enrolling in Medicare Part B, a preventive physical examination (IPPE) also called the Welcome to Medicare Examination is covered by Medicare. This includes a medical and social history review related to health status plus education and counseling to discuss the available preventative services such as shots, certain health screenings, and referrals to other care if needed. There is also a yearly wellness visit that can be scheduled to develop and or update a personalized plan to prevent disease or disability based on risk factors and current health status. The physician requests that the individual complete a "Health Risk Assessment" during the visit to develop a personalized disease prevention plan to assist them in staying healthy. When scheduling this yearly visit, it is advised that the person inform the doctor's office that they would like to schedule their yearly "Wellness" visit. With the implementation of the Affordable Care Act, various types of screening, medical supplies, various shots, health counseling rehabilitation, and other services are available and may be covered by Medicare. Each year, the U.S. government sends each Medicare enrollee a book *Medicare and You*. Referral to this book provides detailed information about the Medicare program and those services that are covered or not. For further information, access the Medicare website www.medicare.gov or call 800-MEDICARE, that is, 800-633-4227.

Medicaid offers healthcare coverage for eligible low-income adults, children, pregnant women, older adults and people with disabilities. Since it is administered by each state, the policies for payment and eligibility to receive services vary considerably from state to state. Medicaid also provides coverage for people with disabilities and frail older adults who require nursing home care or long term in-home and community-based care.

Programs of All-inclusive Care for the Elderly (PACE) is a Medicare and Medicaid program available in many states that allows individuals who would otherwise need nursing home care to remain in their homes in the community. To qualify for this program, one must be aged 55 or older, live in the area served by the PACE program, be certified by the state that you need nursing home-level care, and show that you can safely live in the community while receiving PACE services. Many services are covered among these are doctor or healthcare practitioner visits, prescription drugs, hospital or nursing home stays, transportation, psychosocial services at a center, meals, teaching, and much more.

HEALTH AND WELLNESS TECHNOLOGIES

Technology increasingly impacts the maintenance of a healthy lifestyle for older adults. Gerotechnology focuses on creating the tools to facilitate older adult learning. For example, computers offer information related to dietary planning, medications, and healthy living. Gerotechnology also offers the means to enhance health promotion and the management of illness (McCallum et al., 2017). Telemedicine is especially helpful in recording the health status of older adults with chronic diseases and relaying it to healthcare providers. Medical information such as vital signs, blood glucose levels, and spirometry can be transmitted to the professional caregiver at intervals, allowing more close monitoring and intervening, thus averting potential health crises (Nocella, 2014). Healthcare portals allows individuals to access health information such as doctor's notes, test results, and reminders of appointments. "Skype" or "Facetime" is used for telemedicine visits as well as interaction between older adults, their family, and friends (McCallum et al., 2017). More recently, platforms like Zoom and Microsoft Teams have also become very popular. Many apps for smartphones, along with company websites, have been developed that allow older adults to shop for groceries, and other necessities, to order and have items delivered to the home as well as many other services.

BARRIERS TO HEALTH PROMOTION

Some common barriers to older adults' interest and participation in health promotion activities include the following:

- Cost of preventive health services may be prohibitive for those already on a fixed income (Resnick, 2011).
- Difficulty in obtaining adequate treatment for existing illnesses results in a sense of futility about seeking additional services for health promotion and disease prevention.
- Ethnic and cultural influences regarding health-seeking behavior vary greatly and are complex. Diversity has not always been considered or even recognized by many healthcare policymakers or providers.
- Transportation is costly or not available (Resnick, 2011).

For most older adults, promoting and maintaining health means making major changes in lifestyle and habits formed over many years. Behavior modification (which includes motivation and incentives), counseling, and consistent support from others are often necessary for such changes to occur along with an individualized approach.

Sennott-Miller and Kligman (1992) discuss specific strategies to prevent relapses into former unhealthy behaviors. These include locating a successful role model, identifying the health risks of the former behavior, imagining the risks of a relapse, practicing relaxation techniques regularly, deciding what to do in case of a relapse, and developing an attitude of distancing oneself from the urges and cravings leading to the unhealthy behavior. Vigilance in maintaining the new behavior is imperative for success and healthcare professionals can be potent motivators in facilitating behavior change.

EXERCISE

Regular, systematic, appropriate exercise is one of the best known antiaging agents. It offers many health benefits such as improvement in blood pressure, diabetes, osteoporosis, osteoarthritis, and neurocognitive functioning and reduced morbidity and mortality (Allen & Morelli, 2011). Research documents its positive influence on the older population. Physical fitness maintained by a consistent exercise regimen can substantially reduce the behavioral effect of many age-related changes that limit mobility, reduce independence, and affect the enjoyment of life in older age.

Yen (2005) describes *sedentary death syndrome*, a term created by health professionals, that refers to the many diseases and life-threatening problems brought on by an inactive lifestyle. Inactivity is known to be a major risk factor for the development of obesity, diabetes; cardiovascular, respiratory, and musculoskeletal disease; and a host of other illnesses. It is now established that many common physical changes long regarded as an inevitable part of growing older in reality mostly are due to inactivity and a sedentary lifestyle (disuse, or hypokinetic disease). The adage "*use it or lose it*" becomes even more true as attention is increasingly focused on the study of normal aging rather than on disease and pathology.

Age-Related Changes Modified by Exercise

Age-related changes that can be modified by regular exercise include the following:

1. *Aerobic capacity and cardiovascular–pulmonary functioning.* Maximal oxygen consumption or VO_{2max} is the most commonly used measure of exercise capacity. VO_{2max} is an indicator of the ability of the cardiovascular system to deliver blood and oxygen to the muscles and the ability of the muscles to use oxygen in performing work. VO_{2max} declines with age; however, exercise research indicates a 10% to 30% improvement with exercise training (Emery et al., 1991). Similarly, other studies indicate an increase in pulmonary function with systemic exercise. High-Intensity Interval Training (HIIT) for older adults offers a type pf

aerobic exercise. It is composed of groupings of work and rest sessions done at regular intervals on the land or in the water. Individuals considering such an exercise program should obtain a doctors permission prior to exercising (Larkin, 2017).

2. *Muscle strength*. Both muscle mass loss and strength decline with age (sarcopenia), with the decline generally greater in weight-bearing muscles. The amount of loss is usually 1% to 5% after age 50, but it can begin as early as in the 30s (Haber, 2016). These changes can have a significant effect on the ability to carry out daily activities and can increase the possibility of falls and other accidents. Research shows impressive results of muscle strengthening and functioning in older adults after an appropriate program of strength training. The literature on exercise in older age emphasizes the importance of strength training for older adults, including the oldest-old (Fahlman et al., 2007; Schwartz & Buchner, 1999).
3. *Flexibility and balance*. Joint flexibility tends to decline with age, especially in the shoulder, elbow, wrist, hip, knee, ankle, and spine. Flexibility is needed to perform activities of daily living and also for gait and locomotion. Studies report improved flexibility in older adults who participate in exercise programs that include a flexibility component (Kart et al., 1992). Postural control declines with age, as does vibratory sense in the feet. Furthermore, there is a decreased stride, gait changes, and lessened ability to raise foot height, all of which influence balance. Medications also may affect balance and postural control (Baum et al., 2002). Tsang and Hui-Chan (2004) report improved joint proprioception (awareness of the body in space) and balance in older people who took part in tai chi and golf.
4. *Bone mass*. Osteopenia, or decrease in bone mass, is common in older adults, especially women. When osteopenia is severe enough to cause fractures, the condition is known as osteoporosis. Stressing bones with weight-bearing exercises increases bone mass, and muscle-strengthening exercises reduce the risk for osteoporosis (Sebastian, 2007). Vigorous flexion exercises are not recommended for older persons because they may contribute to compression fractures of the vertebrae, but increasing bone mass, muscle strengthening, balance, and gait stability reduce the risk of future falls and fractures.
5. *Metabolic functioning (glucose and lipoproteins)*. Some variability in metabolic functioning in older age has a genetic component; other aspects of age-related metabolic change are influenced by obesity, diet, smoking, medications, physical activity, fluid and electrolyte balance, and central nervous system functioning. A number of these factors are based on lifestyle and therefore can be modified by choice. Strength training positively influences glucose metabolism, and resistance training escalates glucose uptake as well as sensitivity to insulin in the skeletal muscles (Sebastian, 2007). Changes in glucose and lipoprotein metabolism predispose older adults to diabetes and high levels of fats in the blood. These disorders increase the likelihood of coronary artery disease, the leading cause of death in this age group. Regular systematic exercise has been shown to improve glucose tolerance, lipoprotein levels, and weight loss for those who also eat a healthy diet (Fleg & Goldberg, 1990).
6. *Blood pressure*. Older adults who maintain physical fitness through exercise have lower blood pressure and heart rates, improved cardiac output, and more efficient skeletal muscles, all of which reduce the workload on the heart. Both clinicians and researchers (Allen & Morelli, 2011; Goldberg & Hagberg, 1990) agree that older adults with hypertension can safely participate in exercise programs if heart rate and blood pressure are carefully monitored.
7. *Psychological benefits*. Systematic studies regarding the psychological effects of exercise on older adults are sparse and inconsistent. However, numerous research findings report a more positive mood, improvement in cognitive functioning (such as memory), a sense of improved self-esteem and body image, more meaningful social relationships, and a general feeling of well-being resulting from a regular exercise regimen (Langlois et al., 2013). Research studies also conclude that exercise is strongly linked to feelings of happiness.

Types of Exercise

Isometric

In isometric exercise, very specific muscles are contracted for a short time (5–8 seconds) without joint movement. For example, clasp your hands and then try to pull them apart. Considerable effort can be expended, but there is no movement of a joint. Isometric exercises are appropriate for building the strength of specific muscles and are used to rehabilitate muscles after an injury, but they do not provide cardiovascular conditioning. Care must be taken when using isometrics with cardiac patients because the Valsalva maneuver may be initiated; however, care can be taken to avoid this when doing these exercises (Haber, 2016).

Isotonic

Isotonic exercise produces muscle contraction and movement at adjacent joints, but the activity is not maintained consistently enough for significant cardiovascular and respiratory conditioning. Bowling and golf are examples of isotonic exercise.

Aerobic

Aerobic exercise requires the body to use oxygen to produce the energy needed to carry out the activity. It involves rhythmic or repetitive activity using several large muscle groups (as in walking) for 20 to 30 minutes of sustained exercise. Both the pulmonary system and the cardiovascular system are conditioned with aerobic exercise. Ideally, aerobic exercise programs should be tailored to the individual's abilities. Activities involved in the program should be varied to prevent boredom and to encourage cross-training, which uses various muscle groups rather than the same ones over and over again. Aerobic exercises include swimming, walking, rowing, cycling, tennis, aerobic dancing, and jogging (Eliopoulos, 2018).

Strength-training (*resistance*) exercises involve muscles contracting against a type of resistance greater than normally experienced such as using resistance bands or hand or ankle weights. Muscle groups are exercised, resulting in increased muscle strength and muscle building. These should ideally be performed at least two times a week (Miller, 2009; Touhy, 2012).

Endurance exercises are individual movements of large muscle groups in 10-minute segments. Examples of endurance exercises are swimming, dancing, walking briskly, and playing tennis. Initially, these exercises should be started for a short period and gradually increased as tolerated (Struck & Ross, 2006).

Flexibility/stretching exercises put the joints through their full range of motion, keeping the body flexible and limber. Yoga is a type of exercise that stretches and flexes the muscles. Flexibility and stretching exercises are recommended two to four times a week (Miller, 2009; Sebastian, 2007).

Balance exercises are important for maintaining standing and walking. They also help prevent falls and accidents and should be done three times a week (Miller, 2009; Sebastian, 2007).

The U.S. Department of Health and Human Services (USDHHS) regularly publish their physical activity recommendations for all ages, including older adults (2018). It is recommended that adults should do at least 150 to 300 minutes a week of moderate-intensity exercise, or 75 to 150 minutes a week of vigorous-intensity aerobic activity, spread throughout the week (USDHSS). Moderate aerobic exercise results in a noted increase in heart rate and breathing yet being able to converse with someone. In addition, muscle-strengthening activities of moderate or greater intensity which involve all major muscle groups on 2 or more days per week, as this provides additional health benefits (USDHSS). For older adults, the guidelines are the same, with the addition of balance training, though consideration of level of effort relative to level of fitness is emphasized, along with understanding how chronic conditions can affect ability to undertake exercise, but the aim should be to be as physically active as abilities and health conditions allow (USDHSS). For a more detailed discussion of these specific recommendations from the report, go to https://health.gov/sites/default/files/2019-09/Physical_Activity_Guidelines_2nd_edition.pdf

The American College of Sport's Medicine (ACSM, 2011) also stress the importance of variety in exercise and recommends healthy older adults regularly participate in four types of exercise in addition to the usual activities of the day: (a) cardiorespiratory (aerobic), (b) flexibility (stretching), (c) resistance (weight lifting), and (d) neuromotor (motor skills).

Exercise Programs

Walking briskly is considered the most popular and enjoyed exercise among older adults. No special equipment is needed and most people can access it by walking in the neighborhood. Walking groups have become popular in both indoor and outdoor settings. Many malls open early, allowing for walking in a safe, climate controlled environment. Individuals who have been walking regularly may wish to gradually increase their pace. Those who have just begun walking should increase their exercise time by a minute or two per week. Research indicates that exercise does not have to be intense or exceptionally vigorous to be effective. A moderately paced walk for approximately 20 to 30 minutes each day can promote cardiovascular conditioning and be effective for general physical fitness. If dizziness, pain, or shortness of breath occur, they should stop exercising and rest until the symptoms disappear. If symptoms remain or if they reoccur, a medical evaluation is recommended (Tabloski, 2014).

Some walkers use a pedometer to count steps taken during a walk, others wear a pedometer all day to determine the total steps taken during the day. There are also fitness trackers such as Fitbit Flex or BodyMedia armbands that measure the intensity and duration of the exercise as well as total energy expenditure (Haber, 2016).

Group exercise classes are popular among older adults. Haber (2016) suggests including the components of aerobics, strength building, balance and flexibility, and a period of education. The latter may be used to teach fall prevention recommendations, share a special recipe, demonstrate a type of massage, or progressive relaxation and other topics. Group exercises offer the opportunity for socializing with others and joining in a class directed by an exercise leader. Careful attention by the leader to each person in an exercise group is essential (Tabloski, 2014).

Yoga and Tai Chi are also popular exercise programs among older adults. These types of exercise include movements that are graceful and affect the body as well as the mind. Dancing, swimming, water aerobics, cycling, cross-country skiing, kayaking, or canoe paddling and others are also enjoyed by this age group. Of great importance is that the exercise chosen is appropriate for the individual's health status and that a physician's permission is obtained prior to exercising when appropriate.

These suggested exercises would need to be modified for persons who are disabled or who have prohibitive health issues. Such individuals should be encouraged to be as physically active as they are able. Exercise for less than 10 minutes at a time does not benefit a healthy person's heart and lungs, but it may offer some benefit to older adults who are not well conditioned (Rogers & Rogers, 2012).

ACSM (2011) purports that consistent exercise increases an older person's years of independent living and quality of life and decreases the likelihood of disability. Many other research studies further document the benefits of regular exercise. These benefits include a longer period of good health and lessened disability from arthritis, diabetes, osteoporosis, cardiovascular and pulmonary disease, stroke, and some cancers. Additionally, psychological benefits are observed as improved quality of life and cognitive skills as well as lessened incidence of anxiety, dementia, and depression (Rogers & Rogers, 2012).

Motivation to Exercise

Statistics demonstrate the heightened challenge of motivating older adults to exercise regularly. The Kaiser Permanente Health Care System implemented an exercise vital sign (EVS) initiative, monitored annually, for 1.8 million adults and implemented it with 85% of their members on their medical record (Kaiser Permanente, 2012). They found that two-thirds of their patients were not meeting national guidelines for exercising, and one-third did not exercise at all. Angela Smith, MD, past president of ACSM, believes that exercise offers a very low risk in relation to its benefits. She recommends physicians write a prescription for the type and amount of exercise for each patient and that they be a positive role model by exercising themselves. She also proposes approaching exercise as "fitness is fun," a pleasant experience. Rebranding the word *exercise* by emphasizing its ability to enhance "quality of life," such as through an improved mood, stress reduction, and being more energetic, is also suggested (Lazare, 2012).

Allowing the person to select a preferred type of exercise may also be motivating. Leadership of an exercise class by a dynamic, well-informed, positive, fun instructor is a high motivator and encourages adherence to a program (Rogers & Rogers, 2012).

GENERAL RECOMMENDATIONS FOR EXERCISE PROGRAMS FOR OLDER ADULTS

General recommendations for exercise in older adults include the following:

1. Because cardiovascular, pulmonary, and musculoskeletal diseases are so common in older adults, a physical examination is recommended before embarking on an exercise program. In addition to the physical exam, a medical history, pertinent laboratory testing, and possibly a stress test if necessary should be performed. Existing medical conditions should be monitored for stability before and during exercising (Struck & Ross, 2006).
2. It is helpful for the older adult to learn how to monitor heart rate. The maximum heart rate is found by subtracting the person's age from 220. The target heart rate to be reached while exercising is usually considered to be 60% to 75% of the maximum heart rate. However, it is now recognized that any exercise program using the target heart rate to monitor intensity of exercise is not appropriate for older adults. Instead, they can use their perceived level of exertion (as, e.g., when they feel tired) as a guide for regulating the intensity of exercise.
3. Larson and Bruce (1997) advocate teaching guidelines for self-pacing during exercise. Using the "talk test," individuals know they are exercising at a comfortable rate when they can carry on a conversation while exercising. Being too breathless to talk while exercising is an indication that the exercise is too strenuous. An exercise program should be started slowly and the exercise time and challenge gradually increased. Being aware of how they feel during exercise is very important. Signs of overly strenuous exercise include wheezing, coughing, difficulty breathing, chest discomfort, excessive sweating, feeling faint or dizzy, exhaustion, and local joint or muscle pain. Awareness of these or other symptoms is an indication to slow down or stop the exercise.
4. For fitness conditioning and safety, an exercise program should include three components: a warm-up period, an aerobic component, and a cool-down period. The warm-up and cool-down periods consist of gentle stretching exercises to warm muscles up before strenuous activity and to allow them to cool down slowly after strenuous activity. Gentle stretching for approximately 5 minutes is important to prevent injury and also to improve flexibility. Shorter periods of exercise can be helpful in improving strength and endurance and may be preferred by the older person.

5. It is very important to walk in the right manner. Chin should be raised and shoulders back slightly. The heel of the foot should be the first to touch the ground and the toes should be pointed straight ahead with body weight forward when walking and swinging the arms in rhythm with the walking. Footwear is important. Shoes should support and protect the feet. Walking in athletic shoes is advised (Tabloski, 2014).
6. Walking might include a mind–body aspect using chi walking, which joins walking with tai chi routines (https://www.chirunning.com/chiwalking/; Mauk, 2018).

Activities involving straining or breath-holding are dangerous because they may increase the possibility of causing cardiovascular problems. Exercises should always be gentle with no bouncing, twisting, or heavy straining. Injury prevention is paramount because muscle soreness and fatigue usually lead to permanent withdrawal from an exercise program.

Music during exercise can enhance the exercise experience and influence interest in attending class. Additionally, it can encourage socialization and emotional and mental health. Music should preferably be slow at the beginning of the class and at the end, with a faster tempo in between. Choosing music that is relevant to the participant is important. Loud rock music with a fast tempo is not appropriate and is offensive and even harmful. Remember, music is to enhance and support the participant's experience and is not played for the preference of the exercise leader. It is good to involve older adults in selecting music they know and appreciate. Music that has a slower tempo is preferable (Haber, 2016).

Monitoring for dehydration is essential because exercise results in body-fluid loss. Furthermore, because muscle loss in aging results in lower total body water, encourage drinking of water before, during, and after exercise (Tabloski, 2014). Comfortable clothing that is nonrestrictive and dry is also advised.

When starting an exercise program, it is important to begin with moderately intense exercise or lower and progress as time goes on. It is important to always assess the health and ability status of each exerciser and design the program based on their health needs and abilities (Rogers & Rogers, 2012).

Regular exercise classes should be an integral part of every long-term care setting. Staff are responsible for encouraging regular attendance. Scheduling exercise in the dining room before lunch is a good time to gather residents for an exercise session. Exercise classes are also offered in senior centers, churches, and other settings. Classes can also be accessed using the computer, television, CDs, and DVDs, and used in the home setting.

In summary, an effective exercise program for older adults increases conditioning (especially endurance), minimizes risk, improves muscle strength, and promotes enjoyment without excessive fatigue or discomfort (Larson & Bruce, 1997). Wischenk et al. (2016) describe exercise as a powerful medicine for middle-aged and older adults to enhance their health status in later life. The MacArthur studies of successful aging found physical fitness to be the single most important factor in remaining healthy in old age (Rowe & Kahn, 1998).

REFERENCES

Allen, J., & Morelli, V. (2011). Aging and exercise. In V. Morelli & M. Sidani (Eds.), *Successful aging* (pp. 661–671). Saunders.

American College of Sport's Medicine. (2011). Position stand: Quantity and quality of exercise for developing and maintaining cardio-respiratory, musculoskeletal and neuromotor fitness in apparently healthy adults: Guidance for prescribing exercise. *Medicine and Science in Sports and Exercise*, *43*(7), 1334–1359. https://doi.org/10.1249/MSS.0b013e318213fefb

Baum, T., Capezuti, E., & Driscoll, G. (2002). Falls. In V. T. Cotter & N. E. Stumpf (Eds.), *Advanced practice nursing with older adults: Clinical guidelines* (pp. 245–270). McGraw-Hill.

Eliopoulos, C. (2018). *Gerontological nursing* (9th ed.). Wolters Kluwer.

Emery, C. F., Burker, E. J., & Blumenthal, J. A. (1991). Psychological and physiological effects of exercise among older adults. In K. W. Schaie & M. P. Lawton (Eds.), *Annual review of gerontology and geriatrics* (Vol. 11, pp. 218–238). Springer Publishing Company.

Fahlman, M. M., Tapp, R., McNevin, N., Morgan, A. L., & Bradley, D. J. (2007). Structured exercise in older adults with limited functional ability: Assessing the benefits of an aerobic plus resistance training program. *Journal of Gerontological Nursing, 33*, 32–39. https://doi.org/10.3928/00989134-20070601-06

Filner, B., & Williams, R. (2000). Health promotion for the elderly: Reducing functional dependency. In *Healthy people 2010*. U.S. Department of Health and Human Services.

Fleg, J. L., & Goldberg, A. P. (1990). Exercise in older people: Cardiovascular and metabolic adaptations. In W. R. Hazzard, R. Andres, E. L. Bierman, & J. P Blass (Eds.), *Principles of geriatric medicine and gerontology* (2nd ed., pp. 85–100). McGraw-Hill.

Fried, L. P. (1990). Health promotion and disease prevention. In W. R. Hazzard, R. Andres, E. L. Bierman, & J. P. Blass (Eds.), *Principles of geriatric medicine and gerontology* (2nd ed., pp. 192–200). McGraw-Hill.

Fries, J. F. (1980). Aging, natural death, and the compression of morbidity. *New England Journal of Medicine, 303*, 130–135. https://doi.org/10.1056/NEJM198007173030304

Goldberg, A. P., & Hagberg, J. M. (1990). Physical exercise in the elderly. In E. L. Schneider & J. W. Rowe (Eds.), *Handbook of the biology of aging* (3rd ed., pp. 407–428). Academic Press.

Haber, D. (2016). *Health promotion and aging: Practical applications for health professionals* (7th ed.). Springer Publishing Company.

Haber, D. (2018). Health promotion, risk reduction, and disease prevention. In K. Mauk (Ed.), *Gerontological nursing: Competencies for care* (4th ed., pp. 241–269). Jones & Bartlett.

Hansen-Kyle, L. (2005). A concept analysis of healthy aging. *Nursing Forum, 40*, 45–57. https://www.nejm.org/doi/full/10.1056/NEJM198007173030304

Hickey, T., & Stilwell, D. L. (1991). Health promotion for older people: All is not well. *Gerontologist, 31*, 822–829. https://doi.org/10.1093/geront/31.6.822

Jett, K., & Touhy, T. A. (2016). Health and wellness in an aging society. In T. A. Touhy & K. Jett (Eds.), *Toward healthy aging: Human needs and nursing response* (9th ed., pp. 1–12). Mosby Elsevier.

Kaiser Permanente. (2012). *Exercise as a vital sign initiative informs better care*. https://about.kaiserpermanente.org/total-health/health-topics/kaiser-permanente-study-finds-efforts-to-establish-exercise-as-a

Kart, C. S., Metress, E. K., & Metress, S. P. (1992). *Human aging and chronic disease*. Jones and Bartlett.

Langlois, F., Vu, T. T., Chassé, K., Dupuis, G. Kergoat, M. J., & Bherer, L. (2013). Benefits of physical exercise training on cognition and quality of life in frail older adults. *The Journals of Gerontology. Series B, Psychological Sciences and Social Sciences, 68*(3), 400–404. https://doi.org/10.1093/geronb/gbs069

Larkin, M. L. (2017). HIIT: High-intensity interval training can boost health, wellbeing. *Journal of Active Aging, 16*(3), 34–41. https://monumentalresults.com/wp-content/uploads/2018/10/HIIT-low-res.pdf

Larson, E. B., & Bruce, R. A. (1997). Exercise. In C. K. Cassel, H. J. Cohn, E. B. Larson, D. E. Meier, N. M. Resnick, L. Z. Rubenstein, & L. B. Sorensen (Eds.), *Geriatric medicine* (3rd ed., pp. 815–821). Springer Verlag.

Lazare, J. (2012). Rebranding exercise: Physicians should focus on patients' quality of life. *Aging Well, 5*(3), 10–13. https://doi.org/10.1186/1479-5868-8-94

Mauk, K. (2018). *Gerontological nursing: Competencies for care* (4th ed.). Jones and Bartlett.

McCallum, T. J., Agree, E. M., & Coppola, J. F. (2017). Health management, health promotion and disease prevention in gerotechnology. In S. Kwon (Ed.), *Gerotechnology: Research, practice, principles in the field of technology and aging* (pp. 351–368). Springer Publishing Company.

Miller, C. (2009). *Nursing for wellness in older adults* (5th ed.). Wolters Kluwer/Lippincott Williams & Wilkins.

Nakasato, Y. R., & Carnes, B. A. (2006). Health promotion in older adults: Promoting successful aging in primary care settings. *Geriatrics, 61*, 27–31.

Nelson, J. M. (2018). Identifying and preventing common risk factors in the elderly. In K. Mauk (Ed.), *Gerontological nursing: Competencies for care* (4th ed., pp. 291–297). Jones and Bartlett.

Nocella, J. (2014). Technology. In E. A. Capezuti, M. L. Malone, P. R. Katz, & M. D. Mezey (Eds.), *The encyclopedia of elder care* (3rd ed., pp. 731–735). Springer Publishing Company.

Resnick, B. (2011). Health promotion and illness/disability prevention. In S. E. Meiner (Ed.), *Gerontologic nursing* (4th ed., pp. 135–147). Mosby Elsevier.

Rogers, N. L., & Rogers, M. E. (2012). Exercise recommendations for older adults: An update. *Journal of Active Aging, 11*(2), 40–48.
Rowe, J. W., & Kahn, R. L. (1998). Successful aging. *The Gerontologist, 37*, 433–440. https://doi.org/10.1093/geront/37.4.433
Schwartz, R. S., & Buchner, D. M. (1999). Exercise in the elderly: Physiologic and functional effects. In W. R. Hazzard, J. P. Blass, W. H. Ettinger, J. B. Halter, & J. G. Ouslander (Eds.), *Principles of geriatric medicine and gerontology* (4th ed., pp. 149–158). McGraw-Hill.
Sebastian, L. A. (2007). Exercise recommendations. *Advance for Nursing, 8*, 27–29.
Sennott-Miller, L., & Kligman, E. W. (1992). Healthier lifestyles: How to motivate older patients to change. *Geriatrics, 47*, 52–59.
Struck, B. D., & Ross, K. M. (2006). Health promotion in older adults: Prescribing exercise for the frail and homebound. *Geriatrics, 61*, 22–27.
Tabloski, P. A. (2014). *Gerontological nursing* (3rd ed.). Pearson.
Touhy, T. A. (2012). Health and wellness. In T. A. Touhy, & K. Jett (Eds.), *Toward healthy aging: Human needs and nursing response* (8th ed., pp. 21–32). Mosby Elsevier.
Touhy, T. A. (2016). Physical activity and exercise. In T. A. Touhy & K. Jett (Eds.), *Toward healthy aging: Human needs and nursing response* (9th ed., pp. 233–243). Mosby Elsevier.
Tsang, W. W., & Hui-Chan, C. N. (2004). Effects of exercise on joint sense and balance in elderly men: Tai Chi versus golf. *Medical Science Sports Exercise, 36*, 658–667. https://doi.org/10.1249/01.mss.0000122077.87090.2e
U.S. Department of Health and Human Services. (2018). *Physical activity guidelines for Americans* (2nd ed.). https://health.gov/sites/default/files/2019-09/Physical_Activity_Guidelines_2nd_edition.pdf
U.S. Department of Health and Human Services. (2019). *About the Affordable Care Act*. https://www.hhs.gov/healthcare/about-the-aca/index.html
U.S. Department of Health and Human Service. (2020a). *What is the Healthy People 2030 framework*? https://www.healthypeople.gov/2020/About-Healthy-People/Development-Healthy-People-2030/Framework
U.S. Department of Health and Human Services. (2020b). *Healthy people 2030 Older adults, Goal: Improve health and well-being for older adults*. https://health.gov/healthypeople/objectives-and-data/browse-objectives/older-adults
Vita, A. J., Terry, R. B., Hubert, H. B., & Fries, J. F. (1998). Aging, health risks, and cumulative disability. *New England Journal of Medicine, 338*, 1035–1041. https://doi.org/10.1056/NEJM199804093381506
Wischenka, D. M., Marquez, C., & Friberg Felsted, K. (2016). Benefits of physical activity on cognitive functioning in older adults. In B. Resnick, & M. Boltz (Eds.), *Annual review of gerontology and geriatrics: Optimizing physical activity and function across settings* (pp. 103–122). Springer Publishing Company.
World Health Organization. (1946). *Preamble to the Constitution of the WHO as adopted by the International Health Conference, NY* (pp. 19–22). Author.
Yen, P. K. (2005). Physical activity—The "new" nutrition guideline. *Geriatric Nursing, 26*, 341–342. https://doi.org/10.1016/j.gerinurse.2005.09.008

18

Complementary, Alternative, and Integrative Medicine

INTRODUCTION

The popularity and use of complementary, alternative, and integrative medicine is growing rapidly in the United States and worldwide. Most studies and surveys indicate that 30% to 60% of adults use complementary and alternative medicine (CAM; Mackenzie & Rakel, 2006). The National Health Interview Survey 2007 interviewed more than 23,300 adults regarding health and illness experiences. The section on CAM focuses on the use of 36 therapies. Data indicate that 38% of the adults use CAM, with an increase in the use of mind–body therapies such as acupuncture, massage, naturopathy, meditation, and deep-breathing exercises. The more commonly used CAM modalities are natural products such as fish oil, omega-3, glucosamine, echinacea, flaxseed oil/pills, and ginseng. CAM is used primarily for chronic pain or problems in the back, joints, and neck and for arthritic pain. Additionally, it is used for the treatment of anxiety, cholesterol management, and head or chest colds (Bauer, 2013). Individuals ages 50 to 59 are the most frequent users of CAM, but the greatest increase in the use of CAM is among those ages 60 to 85 plus. Data describing the profile of individuals who tend to use CAM show that 50% earn more than $50,000 yearly; 50% or more are college graduates; and 50% or more have chronic pain, poor health, or comorbidities (Faass, 2006). In retail pharmacy, herbal products represent the most rapidly growing area, increasing 20% to 25% each year when compared with conventional drugs (Lilley et al., 2017). Nutritional supplements and herbs are used by individuals in all ethnic groups (Robbins & Burroughs Phipps, 2016).

Previously known as National Center for Complementary and Alternative Medicine, the National Institutes of Health's Institute regarding this topic is now known as the National Center for Complimentary and Integrative Health (NCCIH). NCCIH defines complementary medicine as "health care approaches that are not typically part of conventional care or that may have origins outside of usual Western practice" (2018, para. 1). According to the NCCIH, complementary medicine is when non-mainstream practices are used *together with* conventional medicine. Alternative medicine is when non-mainstream practices are used *in place of* conventional medical care, and it is not widely practiced in the United States (NCCIH, 2018, para. 1) Complementary medicine or complementary integrative medicine (CIM) more specifically describes the use of alternative therapies as an adjunct to, not in place of, conventional medical care. *Integrative medicine* is a relatively new term that describes the combination of conventional high-tech medicine with alternative medicine such as nontraditional therapies, products, and practices, using the best of both approaches in healing (Bauer, 2013). It focuses on a patient-centered holistic approach to healthcare and wellness of the whole person's social, emotional, mental, functional, spiritual, and community aspects. An increasing number of healthcare systems are using complementary therapies. NCCIH organizes alternative and complementary medicine into types (NCCIH, 2018).

1. Natural products such as vitamins, herbs, (botanicals), probiotics, and minerals. These may be sold as dietary supplements.
2. Mind-body practices among which are yoga, osteopathic manipulation, chiropractic, acupuncture, tai chi, guided imagery, meditation, healing touch, and others.

Whole medical systems are included in other approaches including Ayurvedic medicine, homeopathy, naturopathy, traditional Chinese medicine, and others.

The Consortium of Academic Health Centers for Integrative Medicine and Health is composed of Academic Institutions and Health Systems across America whose goal is to advance the principles and practices of integrative medicine. The guiding principles of integrative medicine include prevention, natural healing, active learning, and holistic care (Bauer, 2013). Dr. Andrew Weil has been prominent in the United States in promoting this newer approach to patient care. The Andrew Weil center for Integrative Medicine at the University of Arizona College of Medicine offers a 2-year Fellowship in Integrative Medicine designed for physicians (M.D.& D.O.), Pharmacists, Advanced Practice Registered Nurses and Physician Assistants. Board certification is available from the American Board of Physician Specialties in Integrative Medicine.

Some therapies formerly considered to be alternative or complementary are now mainstream; for example, chiropractic is an acknowledged, reimbursed treatment modality. Gradually, as research supports the effectiveness of various complementary and alternative therapies more will gain acceptance in the healthcare system. Many of these therapies have been used in other cultures for years. In China, conventional medicine is used as a complementary approach to traditional Chinese medicine (Bright, 2002a; Parkman, 2006). Conventional medicine has traditionally been used to treat acute health problems such as trauma and infections, more successfully than chronic health issues. Alternative therapy is particularly helpful for older adults who experience multiple chronic health problems that are not as treatable by conventional medicine (Fontaine, 2011). Older adults tend to respond to a holistic approach that considers physical, emotional, energetic, mental, and spiritual issues. This biopsychosocial approach addresses the cause and treatment of chronic illnesses rather than relying exclusively on various medications that may cause dependency and contribute to polypharmacy (Mackenzie & Rakel, 2006).

HISTORY

The concept of healing through the use of diverse alternative modalities has intrigued individuals from ancient to contemporary times. In ancient Greece and Rome (700 BCE to 300 CE) there were 200 or more temples dedicated to the god of healing. Healing comprised a complex system of symbolism, mythology, and priest healers who surrounded the person with a spiritual, physical, and mental healing environment. Herbs, surgery, art, massage, music, and more were used to treat individuals. Hippocrates wrote 75 volumes describing these treatments. The Catholic Church led the movement from the first to the sixteenth centuries. Christian concepts based on seeing God's image in the person became the foundation of healing through the Works of Mercy. A holistic approach of mind, body, and spirit was integrated into the healing process (Keegan, 2001).

Cultures such as Native American, African, Chinese, and Indian considered the life forms of sun, rain, rocks, and animals as having the human qualities of soul or spirit (animism). Various types of spiritual interventions for healing were offered through rituals, rites, and incantations often through "shamans" (an individual who is associated with the supernatural both as a doctor and priest). Indian (Hindu) culture developed the Ayurvedic system of healing with eight identified branches of medicine. Each specific ailment was considered according to five elements (fire, water, earth, air, and ether). Yoga was another major approach used in healing where chakras (unseen energy fields) were innervated by meditation or concentration and physical body movements. Traditional Chinese medicine focused on chi (or qi). This vital life force or energy is believed to circulate throughout the body following an organization of meridians or pathways. In healthy individuals the chi (qi) flows unrestricted throughout the body, but if there is an impediment to this flow a health problem may arise. Yin and yang, opposing energies, influence chi (qi). Acupuncture involves the use of small needles inserted into acupoints along the meridians, whereas acupressure uses pressure exerted on the acupoints. Through these treatments, the flow of energy is restored, as is the balance between yin and yang to promote health (Ergil, 2006; Hisghman, 2006; Keegan, 2001).

The use of cathartics, emetics, and bloodletting was replaced in the United States in the 18th and 19th centuries by homeopathy, which emphasizes the efficacy of natural cures such as herbs, hydrotherapy, nutrition therapy, and manual manipulation (Frye, 2006). Concomitantly, Thompson founded a therapy using steam baths and botanicals, believing individuals could be their own doctors. Naturopaths promoted "water cures" in tandem with hygienic practices and natural foods. Franz Mesmer's belief in "magnetic healing" held that disease was caused by an imbalance in the magnetic field of the body. His theory of healing was a forerunner to energy-based healing. In the 1870s osteopathy, the use of musculoskeletal manipulation to treat disease, was developed, and in the 1880s Palmer initiated the chiropractic approach to treatment. During this time Mary Baker Eddy introduced Christian Science, a form of mental–spiritual healing (Bright, 2002a; Cuellar, 2006).

Gradually, scientific data were incorporated into the curriculum of both physicians and nurses, and the allopathic approach to healthcare was born. *Allopathy* describes regular medicine in which disease is cured by remedies that overpower it; for example, antibiotics destroy organisms such as bacteria (Whorton, 2006). During this era immigrants flooded the United States and the major foci were hygienic issues, immunizations, and sanitation, with less attention given to the holistic approach to healthcare. In the 20th century, with the discovery of vaccines and antibiotics, the biomedical approach permeated medical and nursing education and the entire healthcare system.

The focus on and use of alternative therapies receded until the 1970s, when they again began to gain popularity. Congress appropriated funds and established the Office of Alternative Medicine (OAM) in 1991. In 1998, the National Center for Complementary and Alternative Medicine was established by Congress in the National Institutes of Health. Its mission is to conduct and support applied research and education, to dispense health information on complementary and alternative medicine, and to function as a clearinghouse for information regarding these therapies (Ignatavicius, 2013).

Allopathic (conventional) medicine is gradually becoming more receptive to the philosophy of complementary and alternative medicine. Because greater numbers of patients believe in and use these modalities, physicians feel more obligated to learn about them. CAM courses are being offered in medical and nursing schools throughout the country. Practitioners of CAM are now better educated and represented by professional organizations, lending greater credibility in the eyes of those in allopathic medicine. Furthermore, several CAM modalities, such as chiropractic, biofeedback, massage, and others, are now covered by Medicare, health maintenance organizations (HMOs), and other types of health insurance plans (Whorton, 2006). Coverage depends on the provider, the length of healthcare coverage, and the state in which the individual resides. It is wise, before using these treatments, to investigate whether a healthcare plan covers the particular therapy, at what level, and for how long, and whether the practitioner accepts coverage by the insurance company (Cuellar, 2006).

Holistic health focuses on the uniqueness and totality of the person physically, emotionally, socially, and spiritually. Holistic healthcare providers, whether conventional, complementary, or alternative, take time to know the individual and family and to view them as equals. Treatment is given within the scope of the person's beliefs and practices. Individuals actively participate in their healing and seek the state of wellness in bringing about a healthy lifestyle on an ongoing basis. Practitioners of any kind are not considered holistic unless the practitioner treats the whole person (Bright, 2002a; Bright et al., 2002; Keegan, 2001).

COMPLEMENTARY AND ALTERNATIVE THERAPIES

Only a selected number of therapies are discussed here, and their use is constantly being evaluated and modified. Modalities introduced here are those better known and most likely to be used by older adults.

Natural Products

Herbs and Dietary Supplements

Dietary supplements are comprised of alternative medicines that are administered orally. They are intended to supplement the diet and include vitamins, herbs, minerals, and other botanicals, enzymes, and amino acids. The Federal Drug Administration (FDA) estimates that there are over 29,000 different dietary supplements currently being consumed in the United States along with about 1,000 new ones being introduced each year (Lilley et al., 2017). Botanical products are used either alone or with conventional medicine to treat various health problems or to maintain health. They are defined and regulated specifically by the Dietary Supplement Health and Education (DSHEA) Act of 1994 and are used primarily to supplement the diet.

Herbs are available in a variety of forms, the most common being extracts, oils, teas, salves, capsules, tablets, and tinctures. They are categorized according to their intended action, such as for comfort and pain relief, digestion, skin care, and elimination (Lilley et al., 2017; Robbins & Burroughs Phipps, 2016).

Herbs have been grown and taken as medicine for thousands of years before the emergence of prescription and over-the-counter (OTC) drug availability. Nearly 50% of adults use dietary or herbal supplements in the United States, and about 70% do not inform their regular healthcare provider because of fear or they believe herbs to be harmless because they are organic and natural (Lilley et al., 2017; Miederhoff, 2006). One-fourth of individuals who consume herbs report experiencing adverse reactions with symptoms such as nausea, diarrhea, headaches, constipation, and even disruption to the kidneys or liver. Used for a variety of ailments they are easily purchased in a variety of settings and their effectiveness is often overly exaggerated (Lilley et al., 2017).

Although certain vitamins and minerals are regulated and approved by the FDA, the majority of the 1,400 or more herbs and nonherbal products are not regulated. They vary widely from company to company and from one batch to another depending on such things as what part of the herb is used or the type of fillers and agents used (Ciocon et al., 2004). Some herbal products may even be contaminated by pesticides, microorganisms, and heavy metals, or they may contain only small amounts of the active ingredient (De Smet, 2002; Robbins & Burroughs Phipps, 2016). The U.S. Pharmacopeial Convention compiled standards that ensure quality products. Herbal products labeled "USP" have been manufactured in compliance with these standards. However, because all companies do not abide by these standards, there remain doubts about the actual composition of some herbal products. Herbs and food supplements may not be labeled for prevention, cure, or treatment of a disease, but the label may indicate the effects it has on the body (Lilley et al., 2017).

The use of herbs and supplements is not always safe and can be potentially dangerous when adverse reactions and interactions occur, especially if ingested with certain prescribed or over-the-counter drugs. It is always wise to discuss the use of herbs and dietary supplements with your healthcare provider before taking them. Do not ingest them if you are pregnant or breast-feeding, if you are younger than 28 or older than 65, if you are taking prescription or nonprescription drugs, or if you are having surgery (Bauer, 2013). Individuals with a history of hypertension, thyroid disease, diabetes, glaucoma, heart disease, and stroke are especially at risk because of the potential for adverse reactions (Miller, 2009). For example, garlic, ginseng, and ginger increase bleeding, and St. John's wort, parsley, dill, celery, and fig cause photosensitivity. Siberian ginseng and hypericum perforatum interact with some cardiac drugs; thus it is imperative older adults inform their medical provider and pharmacist regarding the use of any of these supplements. One week or longer before surgery or a diagnostic procedure, all herbs and supplements should be stopped and the physician informed regarding the supplement or herb used (Ciocon et al., 2004). Several Internet sources, such as the Office of Dietary Supplements (http://dietary-supplements.info.nih.gov/) and Natural Medicines Comprehensive Database (www.naturaldatabase.com), are helpful resources, as is the *Mayo Clinic Book of Alternative Medicine and Home Remedies* (Bauer, 2013) and *Complementary and Alternative Medicine for Older Adults* (Mackenzie & Rakel, 2006).

Garlic

Some earlier studies supported the use of garlic to lower cholesterol and blood pressure and to inhibit the formation of blood clots, but more recent investigations question its benefits (Haber, 2016). However, garlic may increase the incidence of bleeding if taken with aspirin (or other nonsteroidal anti-inflammatory drugs [NSAIDs]) and warfarin. Garlic should be discontinued 7 to 14 days before surgery. It is most effective if eaten in the raw form, but it can be purchased in pill form also; however, the odor-free pills may not contain allicin, the active ingredient in garlic (Lee, 2007; Miller, 2009). When a clove of garlic is chopped, chewed, or crushed, allicin, a powerful antioxidant that is the primary active ingredient, is released (Bauer, 2013; Robbins & Burroughs Phipps, 2016).

Ginseng

There are many species of ginseng, and it is grown commercially in the United States. Ginseng is thought to be helpful in reducing the effects of stress, fatigue, exhaustion, memory, appetite, and sleep patterns. Some studies show it may improve learning and thinking, memory, appetite, sleep patterns, lower blood sugar levels, and reduce certain cancer risks (Bauer, 2013; Vallerand et al., 2017). Ginseng is not recommended for children or individuals with hypertension or those who are acutely ill. Standardized doses of ginseng are not easy to locate; thus, special care should be taken when purchasing ginseng (Haber, 2016). It is not advisable to take ginseng for longer than a 3-month period, and it should always be taken at the recommended dosage (Libster, 2002). It should be used cautiously by individuals with an autoimmune or cardiovascular disease, diabetes, those with a bleeding disorder or receiving anticoagulants (Vallerand et al., 2017).

Echinacea

Echinacea has been heralded as a preventative for upper respiratory infections. The German E Commission supports its use for infections, colds, and urinary and respiratory tract infections. However, some later studies indicate that it may not be as effective in treating or preventing colds as previously thought. It should be used in the short term and not for a chronic condition. It is most beneficial if used as a gargle (Bauer, 2013; Libster, 2002; Miller, 2009). It may interfere with immunosuppressant drugs, and is not recommended in diseases related to the immune response such as HIV, multiple sclerosis, and tuberculosis (Pepa, 2018).

Ginkgo

Ginkgo's origins are more than 200 million years old, and it has been used by the Chinese for centuries in treating such dysfunctions as peripheral vascular disease, reduced flow of blood to the brain, dizziness, and tinnitus. Elders with depression, anxiety, headache, and short-term memory loss may also benefit from it. Ginkgo should be used cautiously with individuals on anticoagulant or antiplatelet therapy and aspirin (and other NSAIDs) and not ingested with monoamine oxidase (MAO) inhibitors (Haber, 2016; Haller, 2006; Pepa, 2018).

St. John's Wort

A valued herbal remedy since the Middle Ages, St. John's wort has been used to enhance healing and as an anti-inflammatory agent. Some studies support its use in treating mild-to-moderate depression and anxiety (Vallerand et al., 2017). It is the second most used herbal drug in the United States and the most used in Germany for depression. Generally it is safe to ingest, but it is not compatible with steroids, anticoagulants, birth control pills, antidepressants, and some asthma preparations (Lee, 2007). It is known to interfere with antibiotics and with medications taken for heart disease, asthma, acquired immunodeficiency syndrome, and depression. Older adults are particularly susceptible to headaches, anxiety, confusion, and dizziness when ingesting both prescribed antidepressants and St. John's wort (Haber, 2016).

Ginger

Used in China for more than 2,500 years, the German E Commission believes ginger is helpful in preventing motion sickness and as an anti-inflammatory treatment. Often found in medications that aid digestion, nausea, flatulence, and gastric acidity, it is also effective in allergy relief, as a cholesterol-lowering agent, and as an arthritis treatment. It is not to be taken casually because it may increase bleeding (Haber, 2016). It should be used with caution for individuals with bleeding tendencies, those on anticoagulant therapies, and those with diabetes (Pepa, 2018). Ginger is not recommended for those with gallbladder disease, and it should be discontinued 7 to 14 days before surgery (Libster, 2002; Miller, 2009).

Glucosamine and Chondroitin Sulfate

Located in and around the cells of the connective tissue and cartilage glucosamine and chondroitin sulfate are natural substances. Glucosamine is thought to stimulate the growth of cartilage, and chondroitin sulfate supports the cartilage in retaining water. They can be purchased separately but are frequently combined in one formulation (Robbins & Burroughs Phipps, 2016). They are primarily used for making cartilage and to reduce osteoarthritic pain, especially in the knees. Numerous studies have been done on the efficacy of these substances, some supporting their use in controlling arthritic knee pain and others not. It is widely used, especially in the United States, being the sixth top-selling dietary supplement (Haber, 2016). Individuals allergic to shell fish or those who have diabetes or asthma should be cautious using glucosamine.

This discussion includes only a small number of herbs taken by older adults. Others include chamomile, hawthorn, saw palmetto, licorice, valerian, black cohosh, cat's claw, horse chestnut, and kava. The Food and Drug Administration offers valuable tips for those who take herbs and homeopathic products. Among their recommendations are the following:

1. Do not to substitute any of these products for prescribed drugs.
2. Inform your healthcare provider if herbs are used.
3. Be aware of prescribed and over-the-counter drugs that interact with herbs.
4. Do not ingest any herb that is contraindicated if you have hypertension, heart disease, stroke, diabetes, glaucoma, thyroid, or bleeding disorders.
5. Be aware that they may cause nausea and diarrhea. Self-diagnosis and treatment with herbs is definitely not advised without seeking advice from a healthcare provider experienced in their use.

Nutraceuticals

Nutraceuticals, or functional foods or drinks, have made a resurgence on the market. Foods that are nutritionally weak (junk foods) are fortified with various additives such as calcium, vitamin D, iron, herbs, and other elements in an effort for companies to make a profit. They have made a profit with sales sky-rocketing to $31 billion dollars in 2008. Fortunately the amounts of these additives are small and quite safe for the consumer. However, the consumer is wasting money consuming junk food (Haber, 2016).

MIND AND BODY PRACTICES

Energy Therapies

Acupuncture

Acupuncture began in China thousands of years ago and is now used widely in the United States. It is based on the Chinese definition of health, which states that vital life energy (qi or chi) flows along 14 primary pathways (meridians) around and through the body in a balanced manner. Blockages of the balanced flow cause disease or dysfunction resulting from outside factors (organisms or injury) or are due to an internal imbalance visible as a functional or structural malfunction (neck pain or heart palpitations; Fontaine, 2011; Rotchford, 2006). Prior to initiating acupuncture, the acupuncturist takes a patient's history, completes a physical assessment, identifies the symptoms, and considers the individual life circumstances and family environment.

Modulation of the flow of chi or qi is the goal of treatment. It is either diminished or intensified as circulation is reinstated and the imbalance corrected by the insertion of fine sterile needles at certain points along the meridians. There are 1,000 to 2,000 acupuncture points on the body. Sometimes heat is applied to the needles or they are gently moved or are stimulated with electricity. The physiological effect of this treatment activates endorphin and serotonin release. Usually, biweekly or weekly sessions last 30 to 60 minutes. Acupuncture is effective in treating fibromyalgia, nausea and vomiting, and pain. It improves blood flow and removes toxic substances from the body. Successful treatment of many other diseases has been documented as well as the regulation of various physiological functions (Bauer, 2013; Korngold & Faass, 2006).

Many medical doctors use acupuncture along with traditional medicine. There are more than 10,000 licensed acupuncturist practitioners in the United States who are required to pass national board examinations. Practitioners should be selected with careful consideration of their credentials and reputation. Other types of pressure point therapies include reflexology, shiatsu, myotherapy, and tuina (Olsen et al., 2006).

Therapeutic Touch

Considered to be an energetic therapy, therapeutic touch is a holistic evidence-based therapy that incorporates the intentional and compassionate use of universal energy to promote balance and well-being. Founded in the late 1960s and early 1970s by Delores Krieger and Dora Kunz, it focuses on the phenomenon of a human energy field within and surrounding the body. It is based on the assumption that illness is a lack of balance in the energy flow or pattern within the person's body (Ignatavicius, 2013). The practitioner increases and reorients the client's energy flow both around and in the body to enhance the process of healing (Bauer, 2013). During treatment, the practitioner places his or her hands usually a few inches off the body as energy is directed toward or away from the body to increase and stimulate the flow of energy and to decrease congestion and dampens areas of increased activity. The client must consciously or unconsciously accept the energy from the practitioner for the healing to occur (Coughlin, 2006). Krieger (1993) outlines three effects noted on the client receiving therapeutic touch: (1) in 2 to 4 minutes a relaxation response occurs when the autonomic nervous system is quieted down, as evidenced by blood pressure, pulse, and respiratory rate decline and a dilation of the blood vessels of the peripheral nervous system; (2) if pain is present, it is reduced, often necessitating a modification in pain medication; and (3) it speeds up the healing process of both psychological and physical wounds. Decreased client anxiety has been reported in a variety of situations. Therapeutic touch may be especially comforting at the end of life. Treatment time is 20 to 30 minutes and may require one or multiple sessions. More than 250,000 health professionals have been taught this therapeutic approach in colleges and universities throughout the world (Bright, 2002b). Other types of energetic therapy include biofeedback, magnetic therapy, Reiki, polarity therapy, and spiritual healing (Keegan, 2001).

Aromatherapy

Aromatherapy involves the use of botanical aromatic oils through a bath, inhalation, or aerial diffusion. The emotion and smell centers bypass the brain's cognitive center allowing the sense of smell to stimulate, regulate, and produce a variety of emotions without one's awareness. Various physiological effects also occur when essential oils impact the body through the skin, influencing the organs and the bloodstream. It is thought as the oils contact our body, harmony is restored between the body, mind, spirit, and the world about us that is so often disrupted by stress, diseases, pollution, and other life stressors. Essential oils have a variety of properties and are used based on the their special scent and healing properties. Many research studies have been conducted related to specific health concerns and diseases. They have been found that aromatherapy reduces the need for pain medications and relieves nausea, vomiting, stress, and anxiety (Libster, 2018).

A license is not required to practice aromatherapy, but educational programs are available and encouraged to be used. Best-practice standards have been developed by the American Holistic Nurses Association with certification through the individual programs offering the education (Libster, 2018).

Hands-On Therapies

Massage

Massage is an ancient modality used to promote healing. It uses mechanical movements that affect soft muscle tissues, assist in body fluid movement, and use reflexive methods that potentiate the chemical, endocrine, and nervous systems. Massage improves circulation, decreases edema, lowers blood pressure, eases tension, causes relaxation, and reduces anxiety, among other benefits (Fontaine, 2011; Mauk, 2018). It is used for a variety of disorders related to the musculoskeletal system, soft tissue injuries, and chronic pain and to relieve tension. While massage is considered relatively risk free, it is contraindicated for individuals with bone fractures, open wounds, deep vein thrombosis, skin infections, and advanced osteoporosis (Pepa, 2018).

The massage therapist uses fingers, hands, arms, or fists to rub, press, or pull tissue such as muscles, tendons, ligaments, connective tissue, and skin in light, deep, staccato, or sustained movements with varying degrees of pressure (Bauer, 2013; Vaughan, 2002). Therapists are educated for 500 to 1,000 contact hours and are licensed in most states. A massage therapist may also be educated in specialized techniques such as trigger-point therapy, manual lymph drainage, craniosacral therapy, deep muscle massage, movement therapies, and others (Rosen & Faass, 2006). A massage usually lasts 30 to 60 minutes, but shorter sessions are also offered using massage chairs.

Chiropractic

Chiropractic practitioners believe in the innate wisdom of the body as the source of all healing and that the functioning of the nervous system is necessary for good health. Misalignments or subluxations of the spinal column contribute to a loss of homeostasis resulting in illness and nerve dysfunction. Subluxations or dislocations are caused by trauma, poor posture, emotional stress, and fatigue (Roman & Callanan, 2002). A thrust is used to correct the subluxation, which then allows the body to heal itself (Lawrence, 2006). Chiropractors treat neuromuscular and skeletal disorders such as bursitis, sciatica, and muscle strain, as well as neck and back pain and other ailments (Roman & Callanan, 2002). Usually a series of treatments are given lasting 30 minutes to 1 hour. Other types of therapy, such as hydrotherapy, heat therapy, massage, and ultrasound, are often used in concert with chiropractic treatments.

Chiropractors complete 4 years of college and chiropractic college and must pass the national board examination. They can become board certified in subspecialties such as sports medicine, nutrition, or orthopedics (Haber, 2016). Medicare and other types of insurances usually reimburse for chiropractic treatments (Chapman-Smith, 2006).

Mind–Body Therapy

Yoga

Yoga is a mind–body practice with roots in India. It integrates physical, spiritual, and mental health to allow the person to be in harmony with the universe. Yoga combines gestures, postures, meditation, and disciplined breathing. Widely available, yoga has been used to reduce stress, relax muscles, improve well-being, and enhance overall fitness. It is also effective in reducing depression, pain, and functional disabilities (Mauk, 2018; Tabloski, 2014).

Animal-Assisted Therapy

In recent years, pets have gained much greater recognition for their work with individuals who are sick or dying. Animal-assisted therapy (AAT) involves the use of selected animals, usually dogs, as a therapeutic modality in treating persons of all ages. Dating back centuries, animals, especially dogs, have been used with the physically and mentally ill. Over the last 10 years, because of the acknowledged value of AAT, animals have been allowed in long-term care facilities, hospitals, and treatment facilities (Fontaine, 2011).

Therapy dog handlers report a transformation often takes place with residents when interacting with a dog. Their presence offers calmness; a sense of acceptance and normalcy; enhanced feelings of self-worth, unconditional love, and trust; and decreased loneliness. Dogs are able to sense emotional states, and they have a heightened sense of smell compared with humans. Many report a dog's ability to recognize pain and illness in a person or other animal (Marshall, 2012). Dr. Sue Saxon, an author of this book, had a therapy dog with whom she had a very special bond (see Figure 18.1). Abigail was a beautiful Pekingese dog with the special ability to sense the needs of others and reach out to them with caring, love, and healing. Together, they spent twelve years comforting and healing thousands of ill and dying people, their loved ones, caregivers, students, and volunteers. They visited in the hospital, nursing homes, assisted living centers, and hospices; spoke to community groups; held university classes; oriented hospice volunteers; and even attended funerals. Over the years, Dr. Saxon kept a diary of these visits. The following are a few of the responses from these visits: "Abigail, God sent you," "Abigail you changed by life, I am now motivated to cooperate with P.T. and get well," "Thank you for bringing your amazing dog, you gave the staff the very best therapy," and a dying woman hugged Abigail as she lay beside her saying "Thank you for being so kind to me". Cats have been known to sense when an individual is near death. Animals are thought to prompt individuals to participate in restorative therapy and to improve their eating. They assist in recalling memories of the past, improving self-esteem, and promoting relaxation. Some individuals even report a lessened need for pain medication when receiving pet visitations (Childers & Scott, 2013).

Therapy dogs are chosen for their unique calm, friendly, tolerant temperament; ability to relate to others; and companionability. They must complete various training programs, such as obedience training and therapy dog training, as well as clinical testing before becoming a certified therapy dog (Fontaine, 2011). The handler usually also must pass a written examination. Several organizations offer training, such as Therapy Dogs International, Pets Uplifting People, and Paws for Friendship. Each has its own training and testing requirements. A veterinarian must complete a form documenting the dog's health status and that they have received the required inoculations. Additionally, the dog must be at least 1 year old and have a personal trainer who accompanies the dog during the therapy visitations. Most healthcare settings require the animal be certified before working with clients.

Figure 18.1 Dr. Sue Saxon and her therapy dog, Abigail.

Settings in which therapy dogs work include nursing homes, older adults living centers, mental hospitals, rehabilitation centers, hospices, schools, prisons, centers for the mentally challenged, hospitals, and clinics, among others. A new handbook *Older Adults and Animal Programing* can be of great help in initiating and maintaining a pet therapy program. The human-animal bond has been shown to positively influence the lives of the sick, the staff, and family (Kaldy, 2018). Therapy dogs are especially effective with older adults who can touch, caress, talk to, and love them. They help to bring back memories of earlier days, when they too had a treasured pet. In addition to therapy dogs, there are companion dogs, AAT with horses and dolphins, animals trained to work with survivors after a disaster, resident animals, service dogs, and screening dogs.

An animal's value in healing, supporting, and comforting is rapidly gaining recognition and acceptance in many arenas. To gain an appreciation for this special therapy, accompany a certified dog and handler on their visits and see firsthand the marvelous effect they have on the sick and dying.

Healing Through Music

Music as a therapeutic approach reaches back 4,000 years. Pythagoras believed music could improve health and that it acted as a mental catharsis. Music has the power to touch, relieve, inspire, and move the person's sense of the universal and individual as it reaches beyond the cognitive way of knowing. Various research studies have shown that music is beneficial in reducing stress, anxiety, and anger. It promotes relaxation, encourages reminiscence, lowers the heart rate, reduces pain, and creates a peaceful environment for the dying (Fontaine, 2011). Research supports the use of music as a therapeutic approach with dementia patients. It tends to relieve apathy, helps to improve verbal communication, maintain cognitive functioning and enhance the quality of life for individuals (Ellis, 2018). Longer sessions tend to be more beneficial to the client (Dossey & Keegan, 2013). Not only does music promote healing for the ill and soothe the dying, it can reduce stress and promote a calm environment for healthcare providers, the family, and friends.

Music therapists are university educated and study, among other topics, performance, psychotherapeutic theories, counseling, and biopsychosocial issues. They use live or taped music and work with ill individuals, with groups, and with the dying (O'Callaghan, 2010). Certified music practitioners provide prescriptive, therapeutic live music to individuals who are ill or dying. They do not work with groups of people. Individuals study for about 2 years to become board certified after passing a clinical and written examination. Organizations that educate music practitioners include the Music for Healing and Transition Program (MHTP) and the International Harp Therapy Association (IHTA), among others. The National Standards Board for Therapeutic Musicians awards national board certification to eligible individuals.

Meditation, Mindfulness, and Prayer

Meditation is described as a variety of practices that help to relax the body and quiet the mind (Fontaine, 2011). The majority of these practices came from the East (India, Japan, Tibet, and China); however, most cultures have meditative therapies. Meditation has been associated with many spiritual belief systems, such as Buddhism, Islam, Judaism, and Catholicism. Before 1970, the focus of meditation was on spiritual or religious issues. Since then, it has been advocated for the relief of stress of both mind and body (Fontaine, 2011).

Many types of meditation approaches are available, such as transcendental meditation, visualization, breath meditation, moving meditation, and others. It may be used alone or in groups and is easy to learn. Classes, books, DVDs, recordings, the Internet, meditation apps for smartphones, and other resources are helpful in learning how to meditate. The individual focuses attention primarily on breathing, or they might repeat a sound, word, or phrase to suspend thoughts in the mind. It is thought to be helpful in reducing stress and anxiety, relieving pain, and improving symptoms of depression, hypertension, fibromyalgia, asthma, and other conditions (Bauer, 2013).

Mindfulness, a form of meditation, has become a popular mind-body therapy with connection to ancient practices of Buddhism. It supports a mindset of being open, accepting and compassionate while being aware of the natural tendency to judge (Cameron, 2018). It is a form of stress reduction that concentrates on being aware of the present moment from one moment to the next in an effort to defer judgement on what is currently happening and substituting discernment. Mindfulness is concerned with being kind to the self accompanied by an openness to what could be possible. Numerous research studies have supported its ability to bring peace, happiness, and a deeper awareness of consciousness as well as potentially relieving pain (Libster, 2018).

Spirituality and prayer promote connectedness with self and others through formation of a personal value system and in our search for life's meaning (Bauer, 2013). Prayer is defined as a communication with a deity (Fontaine, 2011). All cultures give evidence of using some kind of prayer. Many types of prayer exist, such as prayer alone or with others, meditation, and formal religious observance or a belief in a power beyond ourselves. Although considerable research has been done on prayer, the results have been mixed because of the variety of spiritual practices and the diversity of meaning it has for different people. Some research has associated it with improved quality of life and enhanced immune system functioning (Bauer, 2013).

Selected Therapies

Other posture and mobility therapies include tai chi, pilates, dance/movement therapies, Alexander technique, and body–mind centering. Research has been done on some of these, whereas others have remained relatively unstudied. Complementary and alternative therapies also include bibliotherapy, aromatherapy, art therapy, and others.

Chronic illness and disabilities remain an ever-present challenge both for treatment and management. As greater numbers of individuals live to an older age, the challenge of chronic illness becomes paramount. Complementary, alternative, and integrative therapies may hold the key to this challenge as we commit more time and money researching these modalities and their effectiveness.

REFERENCES

Bauer, B. (2013). *Mayo Clinic book of alternative medicine and home remedies.* Time Home Entertainment.

Bright, M. A. (2002a). Paradigm shifts. In M. A. Bright (Ed.), *Holistic health and healing* (pp. 6–30). F. A. Davis.

Bright, M. A. (2002b). Therapeutic touch. In M. A. Bright (Ed.), *Holistic health and healing* (pp. 171–179). F. A. Davis.

Bright, M. A., Andrus, V., & Lunt, J. Y. (2002). Health healing and holistic nursing. In M. A. Bright (Ed.), *Holistic health and healing* (pp. 31–46). F. A. Davis

Cameron, L. J. (2018). The power of mindfulness and compassion. *Journal of Medical Practice Management, 33*(4), 251–253.

Chapman-Smith, D. (2006). Overview of the chiropractic profession. In D. Rakel & N. Faass (Eds.), *Complementary medicine in clinical practice* (pp. 341–347). Jones & Bartlett.

Childers, L., & Scott, P. S. (2013). The healing power of pets. *Arthritis Today, 27*(3), 56–59.

Ciocon, J. O., Ciocon, D. G., & Galindo, D. J. (2004). Dietary supplements in primary care: Botanicals can affect surgical outcomes and follow-up. *Geriatrics, 59*(9), 20–24.

Coughlin, P. (2006). Manual therapies. In M. S. Micozzi (Ed.), *Fundamentals of complementary and integrative medicine* (3rd ed., pp. 111–138). Saunders.

Cuellar, N. (2006). *Conversations in complementary and alternative medicine.* Jones & Bartlett.

De Smet, P. A. (2002). Herbal remedies. *The New England Journal of Medicine, 347*(25), 2046–2056. https://doi.org/10.1056/NEJMra020398

Dossey, B. M., & Keegan, L. (2013). *Holistic nursing: A handbook for practice* (6th ed.). Jones & Bartlett.

Ellis, B. (2018). Music intervention improves apathy in residents with dementia. *Caring for the Ages, 19*(8), 18. https://doi.org/10.1016/j.carage.2018.12.009

Ergil, K. V. (2006). Chinese medicine. In M. S. Micozzi (Ed.), *Fundamentals of complementary and integrative medicine* (3rd ed., pp. 375–417). Jones & Bartlett.

Faass, N. (2006). Who uses complementary medicine? In D. Rakel & N. Faas (Eds.), *Complementary medicine in clinical practice* (pp. 9–17). Jones & Bartlett.

Fontaine, K. L. (2011). *Complementary and alternative therapies for nursing practice* (3rd ed.). Pearson.

Frye, J. (2006). Homeopathy as an aid to healthy aging. In E. R. Mackenzie & B. Rakel (Eds.), *Complementary and alternative medicine for older adults* (pp. 79–96). Springer Publishing Company.

Haber, D. (2016). *Health promotion and aging* (6th ed.). Springer Publishing Company.

Haller, C. A. (2006). Clinical approach to adverse events and interactions related to herbal and dietary supplements. *Clinical Toxicology, 44*(5), 605–610. https://doi.org/10.1080/15563650600795545

Hisghman, V. (2006). Acupuncture and acupressure. In N. Cuellar (Ed.), *Conversations in complementary and alternative medicine* (pp. 177–184). Jones & Bartlett.

Ignatavicius, D. (2013). Introduction to complementary and alternative therapies. In D. Ignatavicius & M. Workman (Eds.), *Medical-surgical nursing: Patient-centered collaborative care* (7th ed., pp. 8–14). Elsevier.

Kaldy, K. (2018). Older adults and animal programing. *Caring for the Ages, 19*(12), 1–23.

Keegan, L. C. (2001). *Healing with complementary and alternative therapies*. Delmar.

Korngold, E., & Faass, N. (2006). Overview of clinical acupuncture. In D. Rakel & N. Faass (Eds.), *Complementary medicine in clinical practice* (pp. 275–284). Jones & Bartlett.

Krieger, D. (1993). *Accepting your power to heal: The personal practice of therapeutic touch*. Bear and Company.

Lawrence, D. J. (2006). Chiropractic medicine. In N. Cuellar (Ed.), *Conversations in complementary and alternative medicine* (pp. 155–163). Jones & Bartlett.

Lee, M. (2007). Herbs and other dietary supplements. In B. Bauer (Ed.), *Mayo Clinic book of alternative medicine* (pp. 125–131). Time Inc.

Libster, M. (2002). *Delmar's integrative herb guide for nurses*. Delmar.

Libster, M. (2018). *Holistic and integrative nursing*. S.C. Publishing.

Lilley, L. L., Rainforth Collins, S. R., & Snyder, J. S. (2017). *Pharmacology and the nursing process* (8th ed.). Elsevier.

Mackenzie, E. R., & Rakel, B. (2006). Holistic approaches to healthy aging. In E. R. Mackenzie & B. Rakel (Eds.), *Complementary and alternative medicine for older adults* (pp. 1–9). Springer Publishing Company.

Marshall, N. L. (2012). Breaking barriers through pet therapy. *Counselor Magazine, 13*(4), 26–27.

Mauk, K. L. (2018). *Gerontological nursing: Competencies for care* (4nd ed.). Jones & Bartlett.

Miederhoff, P. (2006). Herbal medicines and other natural products. In N. Cuellar (Ed.), *Conversations in complementary and alternative medicine* (pp. 145–153). Jones & Bartlett.

Miller, C. (2009). *Nursing for wellness in older adults* (5th ed.). Wolters Kluwer/Lippincott Williams & Wilkins.

National Institutes of Health, National Center for Complementary and Integrative Health. (2018). *Complementary, alternative, or integrative wealth: what's in a name*? https://www.nccih.nih.gov/health/complementary-alternative-or-integrative-health-whats-in-a-name

O'Callaghan, C. (2010). The contribution of music therapy to palliative medicine. In G. Hanks, N. I. Cherny, N. A. Christakis, M. Fallon, S. Kaasa, & R. K. Portenoy (Eds.), *Oxford textbook of palliative medicine* (4th ed., pp. 214–223). Oxford University Press.

Olsen, K., Lowe, W., Ina, V., Chrisman, L., & Faass, N. (2006). Definitions of clinical massage and body work. In D. Rakel & N. Faass (Eds.), *Complementary medicine in clinical practice* (pp. 253–258). Jones & Bartlett.

Parkman, C. A. (2006). Health information, managed care, and complementary medicine. In N. Cuellar (Ed.), *Conversations in complementary and alternative medicine* (pp. 9–20). Jones & Bartlett.

Pepa, C. A. (2018). Pain management and alternative health modalities. In K. Mauk (Ed.), *Gerontological nursing: Competencies for care* (4th ed., pp. 797–828). Jones & Bartlett.

Robbins, J. L., & Burroughs Phipps, J. L. (2016). The use of herbs and supplements. In T. A. Touhy & K. Jett (Eds.), *Ebersole & Hess' toward healthy aging: Human needs and nursing response* (9th ed., pp. 115–129). Elsevier.

Roman, V., & Callanan, A. (2002). Chiropractic. In M. A. Bright (Ed.), *Holistic health and healing* (pp. 239–246). F. A. Davis.

Rosen, S., & Faass, N. (2006). Referring patients to clinical massage. In D. Rakel & N. Faass (Eds.), *Complementary medicine in clinical practice* (pp. 235–240). Jones & Bartlett.

Rotchford, J. K. (2006). Medical acupuncture. In E. R. Mackenzie & B. Rakel (Eds.), *Complementary and alternative medicine for older adults* (pp. 161–173). Springer Publishing Company.

Tabloski, P. A. (2014). *Gerontological nursing* (3rd ed.). Pearson.

Vallerand, A. H., Sanoski, C. A., & Deglin, J. H. (2017). *Davis drug guide for nurses* (15th ed.). F.A. Davis.

Vaughan, V. (2002). Therapeutic massage. In M. A. Bright (Ed.), *Holistic health and healing* (pp. 161–169). F.A. Davis.

Whorton, J. C. (2006). History of complementary and alternative medicine. In N. Cuellar (Ed.), *Conversations in complementary and alternative medicine* (pp. 1–8). Jones & Bartlett.

19

Nutrition

INTRODUCTION

Health, vigor, and quality of life from infancy to old age depend on adequate nutritional intake. Nutritional status plays a paramount role in living life to the fullest over the entire life span by maintaining body structure and function, providing energy, warding off illness, enabling clear thinking, and participating in social activity (Eliopoulos, 2018). Eating patterns of older adults are an outcome of lifelong experiences with foods. The human body needs certain basic nutrients such as carbohydrates, fats, proteins, vitamins, minerals, and water to build and repair tissues, to supply energy, and to regulate vital body processes. Poor dietary habits are linked with 6 of the 10 leading causes of death in the United States: heart disease, stroke, cancer, diabetes and kidney and liver disease. They are also associated with other chronic diseases such as diverticulosis and osteoporosis (Mauk, 2018).

A calorie is the measurement of the quantity of heat energy stored in a food. Calorie requirements usually decrease with age as a result of (a) reduced physical activity, (b) decrease in metabolic rate, (c) altered body weight and composition, and (d) prevalence of multiple disabilities and diseases. With an increase in body fat as we age, calories are not burned as rapidly (Eliopoulous, 2018). Mauk (2010) suggests older people are indeed at greater risk for a poor nutritional state because of the following:

- Dependency or disability
- Social isolation
- Acute or multiple chronic diseases
- Poverty
- Inappropriate or excessive food intake
- Tooth or mouth problems
- Chronic medication use
- Requiring help with self-care

Nutritional needs are also influenced by age-related changes in various body systems such as (a) diminished enzyme production and mucosal changes in the digestive system, (b) loss of nephrons and altered kidney function, (c) blood vessel changes and decreased cardiac output, (d) alterations in lung function, and (e) glucose intolerance and insulin response decline because of altered carbohydrate metabolism. Less total body water and protein, loss of lean body mass, and more fragile temperature regulation are characteristic of the older body. These, along with other age-related changes, the influence of digestion, absorption, utilization, and excretion of food in the older person is also important.

As a society Americans consume more calories than needed, as well as an excess of saturated fats, cholesterol, salt, and sugars. Concomitantly, recommended amounts of vegetables and fruits are not eaten, all of which results in dietary deficits of fiber, calcium, potassium, magnesium, and vitamins A, C, and E (Tufts University Health and Nutrition Newsletter, 2008). Healthy People 2030 (U.S. Department of Health and Human Services, 2020) sets standards that are data driven to help improve the health of Americans of all ages over the next decade. The objectives include nutrition based issues accompanying various disease conditions, including hypertension, diabetes, gastrointestinal disease, alcohol consumption, and physical activity.

About 45 chemical compounds and single elements from foods are required for human cell functioning. They include carbohydrates, fats, proteins, vitamins, and minerals. Additionally, 13 vitamins are essential for healthy human functioning. Vitamins A, D, E, and K are fat-soluble vitamins, whereas the water-soluble vitamins include vitamin C and eight B vitamins: biotin, thiamine, folate, B_6, B_{12}, riboflavin, niacin, and pantothenate. Linoleic and possibly linolenic acid must be obtained from food. Minerals and trace elements complete the required nutrients. Macrominerals include magnesium, calcium, chlorine, potassium, phosphorus, sodium, chromium, and sulfur. Microminerals include iron, copper, cobalt, fluorine, manganese, iodide, selenium, molybdenum, and zinc. Neglecting to ingest any one of these may lead to illness. Ultratrace elements include aluminum, silicon, tin, and nickel, among others (Gallagher, 2012).

Dietary standards are essential to determine which foods to eat and in what amounts to maintain a healthier lifestyle. The recommended dietary allowance (RDA) was initially developed in 1941 to help protect individuals from deficiency diseases and inadequate diets. When observing the outcome of dietary excesses, it was deduced that the focus should not only be on avoiding excesses but on achieving maximal health; thus, the RDAs have been replaced by the dietary reference intakes (DRIs) (Dudek, 2010). Included under the umbrella of the DRIs are the RDAs, adequate intake (AI), tolerable upper intake level (UL), and estimated average requirement (EAR) indices. The DRIs recognize the special nutritional needs of older adults and include recommendations for the 51 to 70 age group and those aged 71 and older.

A second kind of nutritional guide, the *Dietary Guidelines for Americans* published by the U.S. Department of Agriculture and the U.S. Department of Health and Human Services for 2020 to 2025 (USDA/DHHS, 2020), represents the nation's most highly respected nutritional advice for all Americans to help them lead healthier lives. Updated every 5 years, it represents science-based advice on the food and drink we should ingest to promote health, meet nutritional needs, and reduce the risk of chronic disease. These guidelines are the first to address eating healthfully over the life span including older adulthood, plus pregnant and lactating women. It is used as the basis for federal nutrition policy, government food and nutrition programs such as school meals, Head Start, and older adult nutrition programs. These four guidelines which encourage Americans to "Make Every Bite Count" include:

1. Following a healthy dietary pattern at every life stage.
2. Categorizing and enjoying nutrient-dense food and beverage choices to reflect personal preferences, cultural traditions and budgeting considerations.
3. Focusing on meeting food group needs with nutrient-dense food and beverages from five food groups, vegetables, fruits, grains, dairy and fortified soy alternatives, and proteins, and staying within caloric limits.
4. Limiting food and beverages higher in added sugars saturated fats and sodium and limiting alcoholic beverages.

These Dietary Guidelines for American are made more understandable though the use of MyPlate, which helps to clarify and apply them to the individual. The USDA offers the Start Simple with the MyPlate Campaign plus a new MyPlate website to assist individuals, families, and those in the community to better make healthy food choices that are easy, affordable, and accessible. For more information, see www.MyPlate.gov.

PSYCHOSOCIAL CULTURAL ASPECTS OF NUTRITION

Food plays a major role in the lives of humans. From the beginning of time, food not only provided sustenance but was part of religious and cultural rites. Eating is a social event usually shared and enjoyed with others; however, the social aspect becomes increasingly limited for older persons who are disabled, live alone, or are institutionalized. There may be difficulties in purchasing, storing, and preparing food or little incentive to shop or prepare food only for oneself. A sense of well-being and sharing stimulates interest in shopping, preparing, and eating well-balanced meals.

Food has symbolic meaning for individuals and may represent reward or punishment, security, sociability, age, and sex symbolism. A unique and important part of every culture is the particular food that is prepared and enjoyed by family and friends; we seek comfort, belonging, share memories, and thrive on foods culturally relevant to us. Food and drink are served at a variety of gatherings such as meetings and receptions, and throughout the life span satisfying foods are associated with security. Children are often rewarded with food or punished by not being given a treat; such relationships carry over into adulthood when eating becomes a reward for pain, stress, or loss. Foods can be associated with age; for instance, low-salt, low-fat, or low-cholesterol foods are often linked with older persons, whereas fast foods such as hamburgers and hot dogs are associated with the young. Food also symbolizes affection as when a box of candy or fruit basket is given to a friend or we take someone to dinner. Candy, cookies, cakes, and special foods are an integral part of celebrating holidays, birthdays, weddings, and anniversaries.

Psychological states such as feelings of loneliness often influence one's interest in shopping or cooking. Individuals who are depressed and those experiencing loss and grief often have poor eating patterns, whereas people who are neurotic, psychotic, or demented typically modify the kinds and amounts of food eaten.

Limited funds, habits such as alcoholism, drug overdosing, and even smoking negatively affect food intake. Some older people live in restricted space with limited cooking or refrigeration; others may have little or no access to transportation or are not physically able to shop or carry groceries. These and other variables all influence older adults' eating patterns to a greater or lesser degree.

Cultural practices are gradually learned by individuals in childhood through both conscious and unconscious learning. Food preferences and eating habits are among the most deeply rooted aspects of one's culture. Certain foods have deep symbolic meaning within each culture. Older family members are usually the transmitters of these preferences, which have been passed from generation to generation, as special food served at family, community, and religious gatherings. Such foods bind people together, are served at rites of passage, soothe those who are grieving, and are an integral part of celebrations (Guthrie & Picciano, 1995).

Cultural food preferences may also affect methods of handling, storing, and cooking food, as well as the types of food eaten, attitudes toward food, and how food relates to health. Cultures often identify specific staple foods, times for meals, as well as special feasts for various holidays. Certain foods are even identified in the treatment of disease. Meeting specific cultural food needs and individual preferences is a complex challenge that is best accomplished through the services of a registered dietitian. Another resource is the previously mentioned MyPlate, which is available in several languages on the Internet (www.MyPlate.gov). Food preferences of older adults need to be reviewed to better understand how their dietary intake relates to the maintenance of good health. Changing cultural eating patterns may be very difficult and sometimes impossible. If a dietary change is necessary, including familiar foods if possible may increase compliance to a therapeutic diet.

PHYSIOLOGICAL ASPECTS OF NUTRITION

Aging is accompanied by physiological changes as well as the presence of chronic disease. Nutritional status is often associated with the onset of and complications associated with disease states. For example, obesity may result in a diabetic state, and failure to follow a prescribed diabetic diet results in high blood sugar levels, which eventually can cause impaired vision, tingling in the limbs, or even the amputation of an extremity (Collins, 2012). Food is one of the primary sources of satisfaction and contentment in the later years, even though sensory losses associated with aging often influence the ability to gain as much pleasure and satisfaction as before.

Receptor cells located in the taste buds of the mouth are mostly responsible for the ability to experience taste. The sense of smell is closely allied to the ability to discriminate between various flavors. Sweet and sour (crude taste) are closely related primarily to the taste buds. As we age, our sensations of taste change each in their own unique way. Some receptors remain quite intact, but those for salty, sour, and bitter taste decline. Earlier studies reported that smell acuity declines with age, but a more recent study suggested that pleasant odors may be enhanced with age. Even the common experience of a head cold can distort the sense of taste. Certain medications, smoking, periodontal disease, dentures, and decreased salivation may alter both taste and smell sensation (Touhy, 2012). The sensation of touch also declines with age (Wellman & Kamp, 2012). Consequently, methods of enhancing the appearance, taste, and smell of food are needed when cooking for older persons, such as, for example, the liberal use of foods of different colors, allowing cooking odors to permeate the eating area, and preparing foods with more definite taste by using taste enhancers, herbs, and so on. Be aware, also, that spoiled foods are more likely to be ingested by those whose taste and smell acuity has decreased.

Other physiological age-related changes possibly affecting nutritional status are decreased lean body mass, increased fatty tissue, and lower metabolic rate; thus, older adults require fewer calories to maintain their optimal body weight (Siegler & Hark, 1996). RDAs suggest a 10% reduction in the amount of caloric intake for individuals older than age 51 years (Baker, 2007). DRIs for both the older adult and the general populations are the same except for vitamin B_{12}, B_6, D, and calcium. There is some atrophy of tissues in the mouth, decreased salivation causing xerostomia (dry mouth), and reduced sensation of thirst, which may also contribute to less effective processing, chewing, and swallowing and lessened enjoyment of food (Tabloski, 2014; Wellman & Kamp, 2012). Chewing and swallowing are less efficient, and if the older person also has poorly fitting dentures or missing teeth, the initial processing of food in the mouth will be affected (Mazur & Litch, 2019). Decreased enzyme secretion in the mouth, stomach, and intestines tends to reduce the nutritional value of food eaten.

Health problems related to range of motion, coordination, or ambulation influence shopping for food, cooking, and eating. Arthritis, a disease prevalent among older adults, makes handling food in the grocery store, pushing a cart, and transporting food difficult. Removing food from bags, opening boxes and cans, and storing food can all become arduous tasks for those with musculoskeletal limitations. Disabilities related to lung disease, cancer, heart disease, and fractures can also affect food intake. Many chronic diseases require special diets, and changing lifelong eating habits is a challenge some older adults find difficult or impossible. The palatability of food and food preparation methods often determine whether food is appealing to the individual and will be eaten.

Certain drug therapies affect the appetite by altering taste perceptions or by causing an unpleasant aftertaste or dry mouth. Other medications stimulate or decrease appetite. Some cause nausea, diarrhea, or constipation, which can interfere with the absorption of nutrients.

Psychotropic drugs reduce mental acuity, causing drowsiness and ultimately lessened food intake. Lethargy or weakness caused by other drugs may impede the ability to shop for or cook foods. All of this contributes to a tendency for older adults to adopt a "tea and toast" regimen or to consume diets high in refined sugars and fats.

WATER AND BODY FLUIDS

Water is essential in maintaining life; in fact, six to eight glasses are needed daily to maintain stable body temperature, efficient cell metabolism, and digestion; to eliminate waste products; and to give form and structure to the body (Tabloski, 2014). Homeostatic mechanisms regulate fluid supply, and amounts taken in and excreted should be equal. Fluids are excreted through the lungs, skin, kidneys, and intestines and also lost through diarrhea, vomiting, fever, and hemorrhage. Dehydration caused by limited fluid intake is one of the most common fluid and electrolyte imbalances observed in older adults; it impairs homeostasis and disrupts functions in many major body organs such as the circulatory, digestive, and urinary systems and causes mental confusion and elevated body temperature. Sufficient fluid intake is of equal importance as food for older adults (Luggen et al., 2008). MyPlate recommends eight servings of fluid a day for older adults.

Many older adults have a decreased ability to detect thirst and may not even realize they need to drink fluids; others may voluntarily limit their fluid intake to prevent frequent urination. Diminished intake of fluids may be due to altered thirst which occurs when the person does not feel thirsty even when the liquid portion of the blood is too low (Crogan, 2019). Physical limitations may prevent individuals from obtaining needed fluids, and those with altered mental processes may not recognize when they are thirsty, may lack the motivation to drink fluids, or may fear incontinence. Individuals with dementia or ambulation problems and the old-old are especially vulnerable to fluid imbalances (Eliopoulos, 2014). Warm temperatures can also increase older adults' vulnerability to dehydration. Signs of dehydration include constipation, weakness, thromboembolism, dizziness, agitation or confusion, dark concentrated urine, and dry mouth (Touhy, 2012). These factors may necessitate increasing fluid intake to 1,500 to 2,500 mL daily unless medically contraindicated. It is extremely important that caregivers monitor fluid intake, especially for disabled older adults, and make a variety of fluids available to prevent fluid imbalance. Use of air conditioners, fans, and shades all help to diminish fluid loss. Rubbing the skin with creams, olive oil, or vegetable shortening, especially after bathing, aids in hydrating the skin and providing comfort.

PROTEIN

Protein is essential to regulate body processes, preserve lean body mass and structure, build cells and maintain organ system performance, maintain blood pressure and volume, and for adequate functioning of the immune system (Lutz & Przytulski, 2011). The RDA for men is 56 g, and for women is 46 g (Touhy, 2012). This amounts to 20% to 30% of total caloric intake, with preference given to complete protein foods. Amino acids are the structural units of proteins, and both essential (those supplied by food) and nonessential (those produced by the body) amino acids are necessary to maintain health. Nitrogen and both types of amino acids are provided by protein of animal origin, including meat, fish, poultry, eggs, milk, and cheese. Grains and vegetables, though, are deficient in one or more of the essential amino acids. Protein is necessary for growth and maintenance of body tissue, as well as for other physiological and metabolic activities, and must be continually replenished in the body by an adequate intake. Unfortunately, the high cost of meat and other animal products often prohibits those on restricted incomes from purchasing protein-containing food. However, substituting chicken, fish, soy products, or nuts for red meat may be more healthy choices. Older persons are at risk for protein-calorie malnutrition because many eat fewer calories and less protein and a low-protein diet is likely to contain reduced minerals and vitamins.

With age, atrophy of the mucosa in the stomach results in a decrease in the ratio of somatostatin cells to gastric-secreting cells. Thus, older adults are more likely to become anemic because reduced hydrochloric acid in the stomach and loss of the intrinsic factor (a protein in the gastric juice) lead to poor iron and vitamin B_{12} absorption (Ignatavicius, 2013). Foods rich in iron and vitamin B_{12} such as liver, fortified cereal, and red meats should be included in most diets to avoid the iron-deficiency anemia caused by a reduced meat intake, quite common in this age group. Individuals with high cholesterol are encouraged to eat lean meat, egg whites, and low-fat desserts and breads, and drink fat-free milk (Lutz & Przytulski, 2011).

Older adults with infections, trauma, burns, fever, and malignancies, as well as those under stress or undergoing surgery, require greater protein and calorie intake because these conditions can produce a negative nitrogen imbalance resulting in lowered body resistance and slower wound healing (Dudek, 2010). Skin breakdown and the formation of pressure ulcers are especially related to low protein ingestion. Overall, the protein intake of older adults should reflect individual needs at any particular time. Dietitians can greatly assist in assessing and recommending the best food to ensure adequate protein levels, and sometimes liquid protein supplements are prescribed to maintain adequate protein ingestion.

CARBOHYDRATES AND FIBER

Carbohydrates are the most preferred source of energy for the majority of body functions. Older adults should obtain about 50% to 60% of total calorie intake from carbohydrates, which are necessary for various physiological activities such as contraction of muscles, transmission of nerve impulses, and brain and lung functioning (Lutz & Przytulski, 2011). Complex carbohydrates found in whole-grain cereals and breads, fruits, and vegetables are rich in vitamins, fiber, and minerals. Refined carbohydrates such as crackers, cookies, candies, and pastries are said to contain "empty calories" because they only contribute calories to a diet and may cause malnutrition if eaten at the expense of other nutrient- and fiber-rich foods.

Fiber, an indigestible complex carbohydrate, has limited nutritive value in itself, but cereal fiber (fiber in grains) absorbs many times its weight in water and helps to move food through the digestive system more rapidly, aiding in the elimination of wastes. Soluble or gel-forming fiber found in peas, beans, and some fruits, though, actually slows transit time and may contribute to constipation. There is increasing evidence that cereal fiber is helpful in preventing constipation, cancer of the colon, hiatal hernia, appendicitis, hemorrhoids, and diverticular disease and in lowering serum lipoproteins (fatty proteins in the blood). Prudent increases in fiber are recommended as a substitute for laxatives and as a means of improving intestinal musculature; however, individuals must drink sufficient quantities of water or the fiber may actually cause constipation. Eating too much fiber is not recommended because it may impede the absorption and digestion of other nutrients. Recommended daily dietary fiber intake is 20 to 35 g, which ought to include ample fresh fruits and vegetables, high-fiber cereals and whole grains, and legumes (Chernoff, 2006; Lutz & Przytulski, 2011).

Older adults have a reduced tolerance for glucose and are more likely to experience fluctuations of high or low blood sugar. Elevated blood sugars usually decline more slowly than in the young. These changes are thought to be due to the "secondary aging phenomena" of physical deconditioning caused by decreased activity, obesity, improper diet, reduced muscle mass, and the use of various medications, all of which possibly influence reduced glucose tolerance and insulin action (Goldberg et al., 1990). Decreasing intake of refined sugars and substituting complex carbohydrates are suggested to avoid such sudden fluctuations and high blood sugar levels. Increasing dietary fiber can lower blood sugar levels and reduce or even eliminate the need for insulin or oral diabetes medications in some people with diabetes. Moderating the intake of carbohydrates is especially important for older people with diabetes. Those who are overweight should reduce calorie intake, monitor blood sugar, and carefully regulate carbohydrate, fat, and protein intake.

FATS

Fats are a member of the class of compounds commonly called lipids. As a concentrated form of energy, fats yield twice as many calories as equivalent amounts of carbohydrates and proteins. In addition to energy, fats form an integral part of the cell membrane; they help the body absorb vitamins A, D, E, and K and promote healthy body functioning. Fat serves a variety of other purposes in the body; it cushions and protects the body and insulates it from extremes of heat and cold. Fats are also a major source of flavor in food and contribute to feelings of fullness and satiation. Oils in the skin and scalp facilitate a healthy look.

Older adults may have a reduced ability to utilize fats, which is reflected in high cholesterol levels. Serum cholesterol levels peak between 50 to 59 years of age for men and 60 to 69 years in women. Serum triglycerides, however, continue to rise, possibly due to a lessened ability to remove dietary fat from the blood. Unsaturated fats from vegetable sources are likely to lower cholesterol levels, whereas saturated animal fats tend to raise cholesterol levels. Twenty to thirty percent of calories ideally should come from fats, but trans fats are to be avoided and preference given to polyunsaturated and monounsaturated fats from nonanimal sources such as olives, peanut butter, and nuts. Cholesterol ingestion of less than 300 mg daily is advised (Biggs, 2007).

Long implicated as a potential cause of obesity, researchers have documented the association of fat ingestion with high cholesterol levels and coronary artery and cardiovascular disease. Likewise, a high-fat diet has been linked to cancer of the colon and breast. Individuals with high blood pressure, diabetes, or obesity, or those who smoke should adhere to a low-cholesterol diet, as should those with elevated cholesterol levels or who have a family history of atherosclerosis.

VITAMINS AND MINERALS (MICRONUTRIENTS)

Adams and Urban (2013) and Eliopoulos (2014) state that greater than 50% of Americans take vitamins and nutritional supplements regularly. Billions of dollars are spent on multivitamin and multimineral supplements each year in the United States as individuals hope to prevent heart disease, cancer, bone loss, and other chronic diseases. Vitamins are necessary in small amounts for the physiological functioning of the body by their coenzyme (enzyme-activating) activity in the metabolic process as they promote biochemical reactions in the cells. Vitamins and minerals together are micronutrients the body needs or specific diseases will result. Most of these deficiency-caused diseases can be cured when appropriate amounts of micronutrients are restored. Both vitamins and minerals must come from food or supplements because the body cannot usually manufacture them.

Vitamins and minerals likely to be deficient in the diets of older persons include vitamins C, B_6, B_{12}, folic acid, calcium, and zinc. There is some evidence that with age, C, D, B_6, B_{12}, folic acid, and zinc may be less well absorbed and utilized in the body (Rowe & Kahn, 1998). Older adults in acute or long-term care settings are especially vulnerable to vitamin and mineral deficiencies because of acute or chronic illnesses, eating insufficient amounts of food, or taking various prescribed medications. Vitamin deficiency in older adults results from a lack of meat, fish, fresh fruits, vegetables, milk, and eggs. Adequate vitamin intake can only be ensured if the required foods from each food group are eaten daily. A daily multivitamin is recommended for older adults, especially those who do not eat a balanced diet. Multivitamins enhance the immune system and decrease the likelihood of developing various infections. Ingesting too many and a high dose of vitamins can result in serious adverse side effects. Toxic levels of vitamins have been noted with vitamins A, C, D, E, B_6, and folic acid (Adams et al., 2019).

Vitamins are either fat soluble or water soluble. The water-soluble vitamins, B and C, are readily eliminated from the body through urine and perspiration, whereas fat-soluble vitamins A, D, E, and K are eliminated only when used up by the body. Because they remain in the body much longer before depletion occurs, vitamin toxicity is more likely with fat-soluble vitamins.

WATER-SOLUBLE VITAMINS

Vitamins B and C, the water-soluble vitamins, are primarily located in the watery portions of food and distributed to the body's cells, tissues, and organs. They are readily absorbed into the bloodstream and excreted if their levels in the blood become too high.

Vitamin B

Vitamin B, important in preventing deficiency diseases, also serves vital control-agent roles in building tissue and in energy metabolism reactions as coenzyme partners with critical cell enzymes. There are eight B vitamins: thiamin (vitamin B_1), riboflavin (vitamin B_2), niacin (vitamin B_3), pyridoxine (vitamin B_6), folic acid (vitamin B_9), cyanocobalamin (vitamin B_{12}), pantothenic acid, and biotin. These vitamins do not provide energy per se, but they help burn carbohydrates, fats, and proteins. Deficiencies in B vitamins are reflected in skin changes such as flaking, dermatitis, or roughness. Mucous membranes may atrophy and become painful. Anemia, convulsions, constipation, diarrhea, anorexia, heart abnormalities, irritability, seizures, depression, and confusion progressing to psychosis have all been attributed to a lack of vitamin B.

Age-Related Changes to Vitamin B

Thiamine (vitamin B_1), riboflavin (vitamin B_2), and pyridoxine (vitamin B_6) deficiencies have been noted among older adults, even those taking supplements. Decreased hydrochloric acid secretion in the older adult's stomach may inactivate thiamine, causing a thiamine deficiency. Reduced levels of riboflavin usually accompany decreases in the other B vitamins and is also linked to protein metabolism. The need for riboflavin is based on protein need. Lessened niacin levels are more prevalent among those who are chronic alcoholics, have low incomes, or are institutionalized. Symptoms of niacin depletion include diarrhea, dementia, and dermatitis.

Vitamin B_6 is found in many foods, yet many older adults are deficient in this vitamin, which can lessen the ability to ward off disease, increase homocysteine levels, and increase the risk for stroke or heart disease (Rowe & Kahn, 1998). The RDA for vitamin B_6 is 1.7 mg daily for older men and 1.5 mg per day for older women.

Folic acid (vitamin B_9) is not as readily accessible in foods as other B vitamins and is more readily excreted from the body. Individuals taking anticonvulsant medications and those who are alcoholics may have reduced levels of folic acid. Atrophic gastritis in older adults results in reduced folic acid and vitamin B_{12} absorption (Rowe & Kahn, 1998). Mental confusion, anemia, fatigue, and apathy can result from folic acid deficiency as well as increased homocysteine levels predisposing one to stroke or heart disease.

Reduced vitamin B_{12} seems to be observed more in those older than age 60. This may be due to (a) minimal intake of red and organ meats and green leafy vegetables; (b) lessened intrinsic factor, a protein secreted by the stomach that makes absorption of vitamin B_{12} possible; (c) ingesting certain medications; and (d) the presence of intestinal diseases. Symptoms of vitamin B_{12} deficiency include a lemon-yellow skin tint; smooth, beefy red tongue; anemia; depression, confusion, and psychosis; and reduced pain and temperature sensations. Individuals are usually given monthly injections of vitamin B_{12} as replacement therapy, to be continued for life. Nutrition experts advise that individuals older than age 50 should take vitamin B_{12} supplementation (Lehne, 2010).

Food sources of vitamin B_{12} and folic acid (vitamin B_9) are leafy green vegetables, yeast, some fruits, legumes, liver, red meats, soy, and fortified breads. The RDA recommendation for vitamin B_{12} is 2.4 mcg daily for both older men and women. Increased dietary intake of vitamin B_{12}, antioxidants, carotenoids, lutein, and zeaxanthin may slow the risk and progress of macular degeneration (Workman, 2013).

Vitamin C

Vitamin C (ascorbic acid), an antioxidant, plays an important role in building and maintaining tissues, in overall body metabolism, in strengthening resistance to infections, and in helping in the absorption of iron. It must be replenished daily, and the body's stores of vitamin C can become depleted from smoking, stress, fever, hemorrhage, infection, burns, wound healing, and inadequate intake (Dudek, 2010). Suter (2006) reports there are no age-related changes in the metabolism of vitamin C, but an adequate intake is necessary because vitamin C deficiency may play a potential role in the development of certain diseases, such as cancer of the esophagus, stomach, and colon, all quite common in older adults. The RDA for vitamin C is 90 mg daily for older men and 75 mg per day for older women. Major sources of vitamin C are citrus fruits, tomatoes, potatoes, cabbage, cantaloupe, and peppers. Generally, with aging, increased vitamin C is needed for adequate body functioning because of its importance in tissue healing, resisting infections, collagen repair, and aiding in response to stress.

FAT-SOLUBLE VITAMINS

Fat-soluble vitamins include vitamins A, D, E, and K. Most often, they occur together in oils and fats in foods and are absorbed by the body from the gastrointestinal tract. They are not readily excreted and can build up to toxic levels. Deficiencies in fat-soluble vitamins are linked to diets low in fats; diseases interfering with transport, absorption, and storage of these vitamins; and overingestion of laxatives such as mineral oil.

Vitamin A

Vitamin A is necessary for healthy epithelial tissues in the skin, eyes, and gastrointestinal, genitourinary, and respiratory systems. It is also needed for visual light and dark adaptation, reproduction and growth, bone growth, and energy regulation (Adams & Urban, 2013). Major food sources are liver, beef, dark green leafy vegetables, yellow or orange vegetables, milk, cheese, and eggs. Symptoms of deficiency include night blindness, sensitivity to glare, corneal ulceration, and rough, dry skin.

Absorption of vitamin A does not seem to be appreciably impaired in healthy older adults and for most, levels of vitamin A appear to be adequate (Suter, 2006). Vitamin A deficiency may occur as a result of normal age-related changes, alcohol use, nutrition malabsorption, and medication interactions (Adams & Urban, 2013). Some research, however, does indicate that an increased absorption of vitamin A in older adults could produce toxicity if excessive amounts are taken as a supplement (Lipschitz, 1997). Vitamins A, E, and C are antioxidants and serve to neutralize free radicals in the body. The RDA of vitamin A is 900 mcg for older men and 700 mcg for older women.

Vitamin D

Vitamin D is essential for calcium and phosphorus absorption and for bone mineralization. The principal sources of vitamin D are sunshine, fortified milk, yeast, deep-sea fish, and fish-liver oils. Because only minimal amounts of vitamin D are found in most foods, fortified sources such as milk and margarine, as well as eggs, liver, and fish, are advised. Six hundred IU are recommended daily or up to 1,000 IU if individuals are not exposed to sun. Older adults may not be exposed to sun, especially if living in northern climates or if they live in nursing homes or other long-term care settings. Recent studies found that older women are more likely to have decreased vitamin D levels, which may even increase the risk of death (Keller, 2012). Exposure to sunlight must be twice as long for an older person as for the young to produce equivalent amounts of vitamin D in the skin. To maintain necessary blood levels of vitamin D, the back, shoulders, and arms must be exposed to sunlight without sunscreen between 11 a.m. and 2 p.m. for 15 minutes during the summer and 20 minutes in the spring and fall (Lutz & Przytulski, 2011). Individuals with inadequate intake of vitamin D or with chronic digestive diseases, as well as alcoholics and those exposed to little or no sun, may be at high risk for developing osteomalacia or osteoporosis.

Vitamin E

Vitamin E maintains cell membrane structure and integrity by safeguarding fatty acids and other lipids from the damage of oxidation and by protecting red blood cell membranes and protecting against blood clot formation (Adams et al., 2019). Vitamin E is an important antioxidant. Vitamin E deficiency may occur in older adults as a result of age-related changes, malabsorption of nutrients, or medication interaction (Adams & Urban, 2013). Major sources of vitamin E are wheat germ and soybean oil, vegetable oils, nuts, whole grains, legumes, milk, eggs, fish, leafy vegetables, and fortified cereals.

The major symptoms of vitamin E deficiency include anemia, reduced blood clotting time, neuromuscular degeneration, weakness, leg cramps, difficulty walking, and fibrocystic disease. A vitamin E overdose may result in fatigue, muscle weakness, reduced thyroid hormone concentrations, general gastrointestinal discomfort, and an enhanced effect of anti–blood-clotting medication. The RDA for vitamin E is 15 mg daily for both older men and women.

Vitamin K

The major function of vitamin K is in the liver, where it helps speed up the synthesis of several blood-clotting factors such as prothrombin as well as playing a role in bone metabolism (Adams et al., 2019). Food sources of vitamin K include green leafy vegetables, milk, cheese, eggs, and liver. Diseases interfering with fat and bile absorption impede vitamin K absorption, resulting in a greater tendency to bleed. Anticoagulant drugs also inhibit vitamin K action. The AI for vitamin K is 120 mcg daily for older men and 90 mcg for older women.

MINERALS

Minerals are classified as macrominerals, microminerals, and ultratrace minerals. Major minerals include calcium, sodium, potassium, phosphorus, magnesium, sulfur, and chloride. Macrominerals are needed at a daily level of 100 mg or more and microminerals at a few mg daily (Gallagher, 2012; Mazur & Litch, 2019). Minerals are responsible for a variety of metabolic and regulatory processes, such as building bone mass, and nerve and muscle functioning.

Adequate absorption of minerals is essential if the body is to utilize them properly. Absorption is often impaired in older adults because of (a) diarrhea; (b) excess or deficiency of one nutrient that diminishes another's absorption time; and (c) certain minerals, such as iron and calcium, in combination with chemical compounds contained in some foods, which become insoluble compounds and are excreted from the body. The minerals having the greatest influence on body functions in older adults are calcium and iron, but sodium, chloride, and potassium are also important in maintaining electrolyte balance. Medications such as diuretics (non-potassium sparing, i.e., those that deplete potassium) are often implicated in causing electrolyte imbalance.

Calcium

Calcium balance is necessary throughout life because calcium is a significant mineral in maintaining bone structure and is metabolically vital for certain enzyme activities. The most abundant mineral in the body is calcium. Ninety-nine percent of all calcium, in the form of calcium salts, is found in bones and teeth, where it also serves as the body's calcium bank in case blood calcium levels drop. Once deposited in bone, calcium does not remain there forever because bones are continually in a state of flux with ongoing building-up (bone deposition) and tearing-down (bone resorption) processes. Both hormones and vitamin D promote deposition of calcium in bone. About 30% of the calcium taken in daily is retained in the body, with the remainder excreted in feces. Only 500 mg should be ingested at one time since larger amounts are not absorbed well (Eliopoulos, 2018).

Calcium absorption decreases with age, especially in postmenopausal women. Absorption is decreased by insufficient vitamin D intake, excessive fiber intake, large amounts of phosphorus and magnesium, and a sedentary lifestyle. Osteoporosis is more common among women than men because body mass is usually less in women, and men generally eat twice as much calcium-containing foods as women. Both smoking and long-term excessive alcohol intake increase the risk of osteoporosis. Several lifestyle modifications can prevent or slow calcium loss:

- Increasing dietary calcium
- Taking calcium supplements (calcium citrate is most readily absorbed in the gastrointestinal [GI] tract)
- Getting regular exercise
- Taking estrogen replacement therapy or other medications
- Maintaining health status
- Taking adequate amounts of vitamin D, protein, phosphorus, lactose, magnesium, and fluoride

Milk and milk products are the best sources of calcium. Adequate intake of 1,200 mg per day is recommended, or 1,500 mg for postmenopausal women.

Phosphorus, Potassium, Sodium, Chloride, Magnesium, Sulfur

Phosphorus is found in all body cells and is needed for all growth processes. The second most abundant mineral in the body, it is important for energy transfer in cell metabolism and in the development of bones and teeth. Eighty-five percent of phosphorus is found in bones and teeth. Fats (lipids) also contain phosphorus (phospholipids) and aid in carrying lipids in the blood and in the transport of nutrients in and out of cells. Milk is the best source of phosphorus; other sources are eggs, meat, fish, and carbonated beverages. The RDA is 700 mg daily for both women and men.

Potassium, *sodium*, and *chloride* are involved in primary body functions such as fluid, electrolyte, and acid–base balance, as well as muscle irritability. These are all vitally important in maintaining health in older age. Loss of *potassium* through the use of non–potassium-sparing diuretic drugs, surgery, vomiting, injury, or diarrhea may have serious consequences such as weakness, heart irregularities, or muscle impairment. Potassium sources are fruits, milk, vegetables, and meat. The recommended AI is 4.7 g daily.

Sodium levels are often too high among Americans, but certain sodium-restricted diets, diuretic medications, vomiting, diarrhea, or excessive perspiration may result in sodium depletion. Neither extreme is desirable for good health. Estimated minimum requirement for sodium is approximately 1/4 teaspoon daily. The recommended AI is 1,200 mg per day.

Chloride is found in body fluids and makes up a portion of gastric secretions. RDA is 750 mg.

Magnesium is necessary for more than 300 chemical reactions in our body (Adams et al., 2019). It is essential in the bones, where it combines with calcium and phosphorus, and also in body tissues and fluids as an agent to control metabolic activity. Available in many green vegetables and whole grains, deficiencies are rare except in certain intestinal disturbances and in alcoholism. Older adults may develop magnesium deficiencies because of disease and poor diet (Linderman, 2006). The RDA is 320 mg daily for older women and 420 mg for older men.

Sulfur is present in the protein of all body cells, and sulfur maintains the structure of nails, skin, and hair. If diets are adequate in protein, sulfur levels will usually be sufficient.

Iron, Copper, Iodine, and Zinc (Trace Elements)

Iron is important in the formation of hemoglobin in the red blood cells because hemoglobin distributes 98.5% of the blood's oxygen. Iron levels may be low in the older population because of inadequate intake of iron-containing food such as meats, altered absorption of iron, or blood loss. Nutritional anemia results from lack of iron, ascorbic acid, protein, folic acid, or vitamin B_{12}; diminished acidity of the stomach; or combinations of these factors. Iron deficiency may also result from using aspirin, nonsteroidal anti-inflammatory drugs (NSAIDs), or anticoagulants, all of which cause GI bleeding. Stomach ulcers, hemorrhoids, pressure sores, infection, surgery, and cancer also deplete iron levels in the body. Symptoms of low levels of iron are weakness, fatigue, anemia, pallor, atrophy of the tongue, and spoon-shaped nails. Poor nutrition seems to account for some of the unexplained anemia often observed in older adults; however, much is attributable to blood loss (Lipschitz, 1997). Increased intake of meat, enriched grains, and green vegetables are recommended and iron supplements may be necessary. The RDA is 8 mg daily.

Copper is an important element in body functioning because it operates synergistically with iron in the iron absorption process. Copper deficiency in older age, however, is rare. Food sources include seafood, organ meats, whole grains, and nuts. The RDA for copper is 900 mcg daily.

Iodine is most highly concentrated in the thyroid gland, with varying amounts in other body tissues. Iodine is important in regulating vital metabolic activities. Individuals deficient in iodine often develop a goiter. Food sources of iodine include iodized table salt, seafood, milk and milk products, and bread. The RDA for iodine is 150 mcg daily.

Zinc, present in all body tissues, is important in growth and tissue repair, in metabolic activities such as collagen formation, and as a complement to critical enzymes. Lutz and Przytulski (2011) report that zinc is in at least 70 or more of the 200 enzymes in the body. It enhances the ability to taste and smell, assists in immune system efficiency, and is significant in healing, in carbohydrate metabolism, and in the growth process. Deficits in zinc are found in older adults with inadequate diets. Stress, diabetes, alcohol consumption, surgery, and burns increase zinc excretion and may produce abrupt zinc losses. Food sources include meats, shellfish, milk, eggs, and whole-grain foods. RDA for zinc is 11 mg for older men and 8 mg for older women.

Other trace elements such as fluorine, selenium, nickel, and chromium are necessary in very small amounts to maintain normal body functioning. Selenium is an antioxidant that together with vitamin E collaborates in the prevention of cell and lipid membrane damage by free radicals. It also plays a role in thyroid functioning. Some evidence suggests that those with low levels of selenium seem to be at greater risk for cancer. The RDA for selenium is 55 mcg daily.

MALNUTRITION

Malnutrition is defined as deficiencies in dietary intake (undernutrition) or overconsumption of food that increases the risk of developing disease or an imbalance of necessary nutrients (DiMaria Ghalili, 2017). Nutrient imbalances result that interfere with normal functioning in cells, tissues, and organs, setting the stage for illness. Diagnosis of malnutrition is especially important in older adults because losing weight and being underweight increase both morbidity and mortality in this age group (Lipschitz, 1997). Malnutrition is a commonly found condition especially among older adults in long-term care settings.

Malnutrition is often overlooked in older adults who seek medical care. This leads to incorrect diagnoses or to assuming that ailments such as headaches, skin rashes, insomnia, fatigue, confusion, pallor, depression, hair loss, delirium, delayed wound healing, debilitation, and general malaise are part of the aging process when in actuality these symptoms reflect malnourishment. Adequate dietary intake greatly influences dental health. Decreases in vitamin C increases the risk of gum disease while lack of vitamin D and protein promotes bone and tooth loss. Deficiencies in B vitamins can cause inflammation, irritation, and even cracking of the tongue and lips (Thompson & Manore, 2018). Morley (1991) believes malnutrition is often not given high priority by physicians because (a) they may lack the knowledge to diagnose malnutrition and recognize those at risk; (b) they seem unaware that protein–energy malnutrition may be the first symptom of a treatable disease; or (c) they generally are not aware of the best methods to manage individuals who have protein–energy malnutrition. Protein–energy malnutrition (PEM) is a metabolic response to stress in which there are increased requirements for protein and calories to meet the demands of the day (Wallace, 2009).

About 50% of those older than age 65 have dietary intakes of less than the daily recommended levels. Factors thought to result in malnutrition include the following (Luggen et al., 2008):

- Poverty or near poverty
- Obesity
- Polypharmacy
- Lack of ability to shop for or prepare food because of physical or mental impairments
- Social isolation caused by loneliness, depression, or apathy
- Alcohol abuse
- Ignorance about adequate dietary requirements
- Poor teeth or ill-fitting dentures
- Digestive system disease
- Bereavement

There are nutritional risks associated with many medications. Careful monitoring of older adult's medication is advised, particularly if there are changes in nutritional intake (Eliopoulos, 2018).

Adequate dietary intake greatly influences dental health. Decrease in vitamin C intake contributes to gum disease, while lack of vitamin D and protein promotes bone and tooth loss. Furthermore, deficiencies in the B vitamins can cause inflammation, irritation, and cracking of the lips and tongue (Thompson & Manore, 2018).

UNDERNUTRITION

Older adults are especially prone to undernutrition (less than body requirements) for a variety of reasons such as living alone, eating empty calories rather than nutritious food, poor fluid intake, medications, depression, anorexia, malignancy, and bereavement. Although many factors, such as medications, mood, lack of socialization, and diminished sense of taste and smell, alter appetite, psychological aspects also inhibit appetite. Appetite plays a significant role in causing malnutrition. Lessened motivation to eat, oral dental problems, chronic diseases, and certain medications may also decrease appetite. Careful and daily assessment of the type and amount of food eaten is essential; consultation with a dietitian is also recommended. Caregivers are not always alert to preventing malnutrition by monitoring food intake, feeding older adults, and developing effective strategies to ensure adequate nutrition. Institutional malnutrition is also a common occurrence. Approximately 17% to 65% of older adults in hospitals and long-term care settings experience undernutrition (Luggen et al., 2008). Individuals in acute care settings may be NPO (nothing by mouth), receiving intravenous fluids, or have medical or surgical conditions that interfere with nutritional intake. Orexigenic medications are now available to stimulate appetite and prevent loss of appetite and, consequently, malnutrition.

Undernutrition can result in agitation, depression, dementia, anemia, inadequate wound healing, weakness, fatigue, increased incidence of pressure sores, impaired elimination and immunological functions, and rehospitalization. Those malnourished experience 90% longer stays in the hospital, require 40% longer to recovery from illness, and experience two to three complications (Haber, 2016). High mortality and morbidity rates are associated with inadequate nutritional levels (Champion, 2011).

Silver (1993) believes the key to treating malnourished states is to intervene as early as possible by doing the following:

- Increasing protein intake during periods of stress
- Promoting exercise to enhance appetite
- Encouraging fluid intake to normal levels
- Increasing the diet to a minimum of 2,000 calories per day
- Avoiding constipation by ensuring adequate fluid intake and drinking prune juice instead of taking cathartics
- Reviewing the drug regimen, including over-the-counter (OTC) drugs, vitamins, minerals, and alcohol
- Screening for depression
- Promoting a more independent and active lifestyle
- Routinely examining the person's mouth for lesions or dental needs

Specific dietary management techniques to promote eating such as environmental modifications to enhance the pleasantness of eating, focusing on methods to stimulate the appetite, and ongoing monitoring of nutritional status are useful and effective. Underweight older adults may be given a high-fat diet, including milk, cream, red meats, and ice cream (unless other health issues prohibit these choices); frequent small meals; and high-calorie and protein supplements. Regular nutritional screening is highly recommended along with biochemical assessments, especially for those who are most vulnerable. It is important to keep in mind that undernutrition may be confused with the normal aging process; thus, continued vigilance is essential (Furman, 2006).

OVERNUTRITION

Obesity is associated with the imbalance of energy occurring when taking in more calories from food than expended in carrying out activities of daily living or physical exercise (Birn, 2013).

The U.S. Federal Interagency Forum on Aging-Related Statistics (USFIF, 2020) stated in 2018 that among individuals age 65 and older, 40% were obese. This compares with just 22% in 1994 (USFIF, 2020). Among the general population, 68% of adults are overweight, or obese with those obese outnumbering those individuals who are in the overweight category (Haber, 2016). This represents a real epidemic of obesity with all its concomitant health issues. Patterns of overeating developed in childhood often continue into old age; more often, though, eating habits leading to obesity are related to sedentary lifestyle. Other reasons for overeating include anxiety, a sedentary occupation, difficult life situation, mental illness, glandular imbalance, cultural food preferences and grief or loss. Metabolism (the process in our bodies that build and destroy tissue) slows down as we age; therefore, there is a need for fewer calories to maintain our weight. Consuming the usual number of calories will actually increase our weight (Haber, 2016). Obesity is so prevalent that manufacturers have had to design and produce larger chairs, wheelchairs, beds, caskets, and other items. The travel industry too is challenged as airlines offer seat belt extenders to obese travelers, or they must purchase two seats.

The dangers from overeating are many. Among these are strokes, heart disease, and type 2 diabetes, which are leading causes of death. These conditions greatly increase medical costs and have become a huge burden on the medical delivery system in the United States. Some professionals believe it may not be harmful for older adults to be slightly overweight. A tendency to lose weight after age 75 has been observed; however, this is attributed to muscle loss, not fat loss, resulting in a thinner body but also one that is weaker and not as functional (Haber, 2016).

Although crash or extreme diets are not recommended, reduction in calorie intake, not eating empty calories, and including all the food groups in the diet are appropriate. Older adults who need to lose weight should have specific and realistic dietary guidance. Weight loss programs of every kind are available, but it is prudent to seek a primary care practitioner's advice before embarking on any specific program because some have proven to be harmful. A physical examination is advised, as is consultation with a dietitian and pharmacist if appropriate, before starting a formal diet program (Flood & Newman, 2007). Diets that call for less than 800 calories a day should be avoided; the recommended weight loss is from 0.5 to 1 pound per week, but fluctuating weight gain and loss patterns are not desirable. Exercise is important for weight reduction because it not only burns calories but also enhances feelings of well-being, and improves cardiovascular and lung function, as well as having many other benefits. Monitoring weight within a healthy norm should be a major goal for everyone throughout their lifetime (Touhy, 2016).

Since the number of obese people in the population has increased, bariatric surgery has become a more popular option. This surgery modifies the gastrointestinal tract, allowing for appreciable weight loss. Several types of bariatric surgery are available, but the majority are performed laparoscopically. Bariatric surgery may be an option for selected older adults who are grossly overweight, whose weight-loss program has been unsuccessful, and who have a body mass index (BMI) of 40 or greater. The survival rate has improved particularly since the laparoscopic procedure has been available. Medicare Part B will pay for counseling sessions for individuals whose BMI is 30 or higher (Haber, 2016).

ANOREXIA OF AGING

Anorexia affects those individuals who have no appetite or who have lost their appetite. An estimated 20% of old-old persons suffer from anorexia. They are more disabled, more frail, and have a higher mortality rate (Morley, 2010). Unintentional weight loss is due to multiple factors that contribute to "the anorexia of aging" and sarcopenia (loss of muscle mass). Causes include inadequate fluid and food intake, iatrogenic issues (adverse condition because of the effects of medical treatment), increased loss of various nutrients, and hypermetabolism (Tabloski, 2014).

Anorexia is an increasing problem in the older age group and may be caused by physiological, psychological, sociological, and pathological factors. Both mortality and morbidity rates are increased, as well as infections and pressure ulcers among those who are anorexic (Champion, 2011).

Anorexia as a result of any cause warrants unique approaches, such as the following (Haber, 2013):

- Eating when the person is hungry, no matter what the time
- Eating higher calorie foods
- Eating smaller meals more frequently
- Eating in a pleasant environment
- Sharing meals with others
- Eating or drinking nutritional supplements
- Requesting a change of medications or the prescribing of other medications that may relieve indigestion, nausea, or other symptoms while eating

FAILURE TO THRIVE

Failure to thrive (FTT) has long been associated with infants who do not gain weight. In frail older people, it is a syndrome defined as a gradual decline in physical or mental functioning, along with weight loss, decreased appetite, depression, and withdrawal from social interactions in the absence of an explanation for these symptoms (Rourke, 2006; Wellman & Kamp, 2012).

FTT is a syndrome often used as an admission diagnosis to hospitals and nursing homes. Verdery (1997) found that more than 50% of adults older than 65 hospitalized with a history of weight loss continued to lose weight after discharge and 75% of them died within 1 year. The most common causes of FTT in older adults are depression, delirium, dementia, drug reactions, chronic inflammation, and disease. Economic factors often contribute to the problem (Marcus & Berry, 1998; Sarkisian & Lachs, 1996). Normal aging changes such as impaired sensory systems or lessened homeostatic reserves also cause FTT. Individuals display multiple problems, including physical, social, mental, and environmental difficulties along with severely diminished coping abilities and functional capacities. Because it begins gradually, family members often fail to recognize its presence. FTT is not considered to be a normal age-related change nor is it necessarily exhibited in all older individuals who have a chronic disease.

Crogan, (2019), Newbern (1992), and Rourke (2006) recommend an in-depth evaluation, including assessments of nutritional and mental status and relational attachments and the use of a genogram to assess the behavioral, cultural, and social development of the older person's family. Other useful evaluations include a medical history; physical assessment (especially of the special senses: vision, hearing, taste, and smell); pulmonary, musculoskeletal, cardiovascular, and neurological examinations; interviews with family members or caregivers; and selected laboratory studies. Osato et al. (1993) believe that when curing the underlying cause of an individual's FTT is not possible, concern should be directed to their symptoms and the prevention of complications. Furthermore, prevention of complications, providing comfort, decreasing symptoms, and restoring or preserving functioning as much as possible, is paramount. Attention should focus on dietary needs, education, functional status, special equipment necessary, and development of a coordinated plan for discharge and aftercare. Priority should be given to older high-risk individuals in an effort to prevent FTT.

FOOD LABELS

Food labeling provides a major source of information for consumers when choosing a healthy diet. The Nutrition Labeling and Education Act of 1990 allowed the Food and Drug Administration (FDA) to develop and enforce specific labeling. Uniform and mandatory nutrition labeling is now required for most prepared food but is voluntary for eggs, milk, fish, poultry, and produce.

Food packages must list the total calories, total energy from fat, total fat (both saturated and trans fats) plus the amounts of cholesterol, total carbohydrates, fiber, sugar, and protein, sodium, vitamins A and C, calcium, and iron. Major nutrients must be listed in grams or milligrams and also as a percentage of the total recommended intake of an individual consuming 2,500 calories a day, along with the number of servings based on a serving size. The FDA also regulates health claims (relationship between the food and a specific disease or illness). Furthermore, they regulate structure and function claims (those claims that the nutrients or ingredients are intended to influence structure and function in human beings) such as "helps to lower cholesterol." Descriptions such as "free," "high," "low," "light," and "lean" used on food must meet the specific legal definition of the words (Lutz & Przytulski, 2011). The above regulations change from time to time regarding the labeling of foods; thus, updating on the latest advisements is recommended. Older adults should be instructed on how to read labels and to choose and shop for healthy foods based on their content.

OLDER ADULTS AND INSTITUTIONAL DIETS

About 5% of adults older than 65 are in nursing homes, with many having health problems necessitating special diets. Most states require consulting dietitians and trained food service managers to plan and prepare diets served to institutional populations. Policies also exist that control the frequency with which the same foods are served, and usually weekly menus must be posted. Because food is extremely important to most older adults, facilities should offer opportunities for varied food selection. The quantity and quality of food served in these settings is highly variable; some facilities serve nutritious and tasty meals, whereas meals at others need decided improvement.

All nursing home residents are assessed on entrance into the nursing home setting and periodically thereafter using the Minimum Data Set (MDS). The MDS includes a nutritional assessment portion in which percentage of meals eaten, disabilities, age, weight status, and so on are assessed and appropriate interventions initiated. Other useful information includes current diagnoses, sex, height, appetite, dietary history, pattern of weight gain or loss, food preferred or disliked, types and amounts of food eaten at every meal, pattern of snacking, ethnic and religious food preferences, food intolerances, special therapeutic diets, current medications, diagnostic laboratory test results, and emotional states. It is also important to determine the person's ability to feed himself or herself and to chew and swallow. Using these data, preferences and problems can be identified and a dietary plan, including goals, is developed to ensure a pleasant and adequate diet for each resident. Periodic dietary evaluations are essential because both physical and mental status change over time.

Various health problems of many older adults in long-term care settings necessitate feeding by others. Suggestions for caregivers include the following:

- Allow sufficient time to eat; do not hurry the person.
- Attempt to offer a diet as close as possible to the person's accustomed diet.
- Encourage the person to eat the breakfast meal because the appetite is usually best at this time of day.
- Inquire whether the family or significant others (when available) might like to feed the person, then instruct them on proper feeding techniques.

For severe undernutrition, using feeding tubes in conjunction with high-calorie and high-protein diets or hyperalimentation may help stabilize the person's nutritional state. Some older adults may require modified diets because of chronic or acute diseases. For instance, people with diabetes require a reduced carbohydrate diet and those with cardiovascular problems often need a low-fat, low-cholesterol, or low-sodium diet. Individuals with cancer may require vitamin and nutritional supplements to ensure adequate dietary nutrition. Certain drugs such as diuretics and antidepressants also make dietary modifications necessary. Regimens of strict therapeutic diets such as low cholesterol or low salt may contribute to a decreased food intake and are not advised for frail older adults (Morley, 2003).

MYPLATE

In 2005, the U.S. Department of Agriculture developed MyPyramid, which included recommendations for a healthy diet along with recommended activity. In 2011, the USDA replaced MyPyramid with MyPlate (www.myplate.gov). The current MyPlate includes the food groups and how much of each type food one should eat depending on the calorie allowance. It is based on age, height, weight, sex, and the individual's level of activity. The plate is divided into fruits, vegetables, grains, and proteins with dairy needs on the side. Fruits and vegetables make up half of the plate, with the other half consisting of grains, mostly whole, lean protein, dairy, and fortified soy alternatives. Half of the grains eaten should be whole grains, and milk should be 1% or skim milk. Portion control is recommended as well as reducing sodium and sugar intake. Alcohol intake may be one drink per day for women and two for men. .

Tufts University's Jean Mayer USDA Human Research Center on Aging (USDAHRCA; Tufts University, 2016) introduced the latest model of MyPlate for Older Adults (see Figure 19.1). This model addresses the special nutritional needs and physical activity for older adults. Basic nutritional requirements stay the same in older age or may increase for some individuals. MyPlate for Older Adults pictures types of foods that yield high levels of vitamins and minerals per serving of food. It recommends 50% of the diet be comprised of fruits and vegetables, 25% grains and 25%

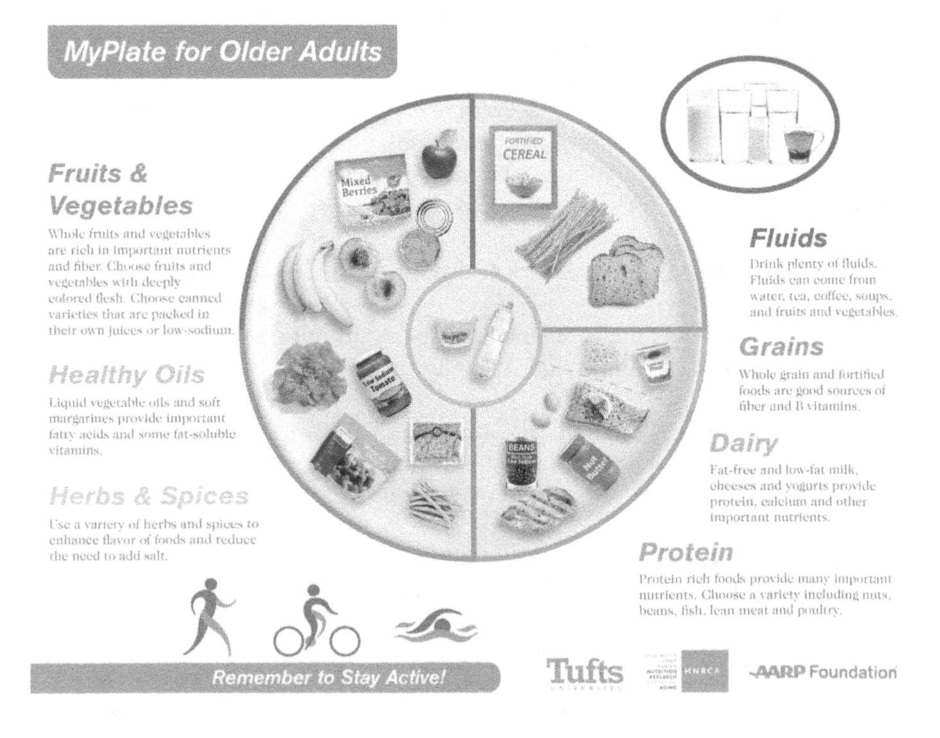

Figure 19.1 MyPlate for Older Adults
Copyright © 2015 Tufts University. For details about the MyPlate for Older Adults, please see https://hnrca.tufts.edu/myplate/.

protein-rich foods. Guidelines recommend limiting salt, added sugars, and saturated and trans fats and advise eating whole grains.

For each section of MyPlate for Older Adults, the best types of food in this group are identified. The fruit and vegetable sections encourage deep-colored fruits such as plums and berries and brightly colored vegetables such as sweet potatoes and carrots. The protein section includes fish, poultry, eggs, lean meat, and nuts. Dairy foods such as fat free cheese, low-fat milk, and yoghurts are encouraged. The grains section recommends whole or fortified and enriched grains such as 100% whole wheat bread and brown rice. Spices and flavoring should replace salt, and soft spreads low in trans fats and saturated fats and liquid vegetable oils are advised. Lastly, fluids such as water or tea should be drunk. Physical activity such as walking, cycling, and swimming is advised.

Registered dietitians or nutritionists are available to develop appropriate nutrition and dietary planning. A useful feature of www.myplate.gov is the interactive design to access dietary information related to each food group. Based on the Dietary Guidelines for Americans, it assists in planning wholesome well-balanced meals and useful resources. Using the MyPlate App initially allows one to develop simple daily food goals, see real-time progress, and to even earn badges to reward success.

NUTRITIONAL RECOMMENDATIONS FOR OLDER ADULTS

General nutritional recommendations for older adults include the following:

1. Overall consume fewer calories since they tend to have reduced metabolism and lower levels of activity.
2. Consume a variety of food daily.
3. Reduce fat intake to 20% to 30% of total calories consumed; substitute unsaturated fats for saturated ones and limit cholesterol intake to 300 mg a day or less.
4. Daily protein consumption should total 0.8 g per kg of body weight for a healthy older person and 20% to 30% of the total caloric intake.
5. Carbohydrate consumption should make up about 50% to 60% of the total caloric intake, with the major part from complex sugar sources such as fresh fruits, vegetables, cereals, and breads. Few calories should come from simple sugars such as sugar, candy, preserves, and syrup.
6. Consume six or more servings of grain products, especially whole grains, daily.
7. Vitamin intake should be adequate, especially vitamins A, B complex, C, D, E, and K.
8. Ensure intake of adequate minerals, including 1,200 to 1,500 mg of calcium daily plus sufficient amounts of phosphorus, potassium, zinc, iron, and other minerals.
9. Consume 3 servings of milk daily.
10. Food such as fresh fruits and vegetables provide important roughage in the diet. Preventing constipation depends on ingesting adequate amounts of fiber-rich food and water. Three or more servings of fruits and two or more of vegetables daily are necessary.
11. Regularly planned daily physical activity will help prevent constipation and assist in the digestive process. Thirty minutes a day for 5 days each week is advised.
12. Eating in moderation helps to maintain ideal weight and prevent or decrease obesity.
13. Avoid junk foods.
14. If alcoholic beverages are consumed, they should be used only in moderation.
15. Smaller, more frequent meals help to prevent snacking and they also serve as a source of greater satisfaction for some individuals.
16. Salt intake should be moderated or decreased.
17. Be aware of harmful food and drug interactions.

18. Foods should be stored at the proper temperature and hands washed frequently when handling food.
19. Instructions for specialized therapeutic diets should be followed.
20. Diet planning should take into account individual preferences for ethnic and other foods, specific nutritional needs, and idiosyncratic intolerances of various foods.
21. Drink 8 glasses of fluid each day unless contraindicated.

EDUCATION

Educating older adults and their caregivers is one of the most important factors in promoting adequate nutrition. Teaching should focus on the meaning of food for people, its various functions in the body, food groups, MyPlate, major vitamins and minerals, food supplements, food selection and preparation, drug–food interactions, methods of food selection, healthy cooking, food storage, specific therapeutic diets, diet preparation that takes into consideration ethnic and individual food preferences, over- and undernutrition, selected methods for ensuring adequate food intake for individuals with disabilities, and available dining sites. Most congregate nutrition sites offer some type of nutrition education to their participants. The purposes of a nutritional education program are as follows: (a) assist the individual in selecting the required food for good health from the best sources, and for the least money; (b) explain methods of identifying and obtaining various nutritional services such as home health aides or homemaker services; (c) increase the older person's awareness of nutrition programs such as Meals on Wheels or congregate dining in the community or availability of food stamps; and (d) provide information about special diets or menus needed for good health.

Teaching should also include how to maintain a healthy mouth and proper dental care. Regular dental visits with teeth cleaning, as well as mouth, natural teeth, and denture assessment are recommended. These are often neglected when one becomes disabled or is institutionalized. Reduced nutritional intake is closely associated with alteration in adequate mouth and teeth care. Using creative approaches to nutrition teaching in an environment that promotes enjoyment, socialization, and support should help ensure improved dietary intake as well as increased compliance with various necessary dietary modifications.

SUPPLEMENTAL NUTRITION

If the older adult cannot or refuses to eat the necessary food to sustain adequate nutritional intake, supplementation with vitamins, minerals, and herbs may be necessary. Nutritional supplements are a source of compensation for nutrients and deficiencies as a result of inadequate food intake or the effects of medications (Eliopoulos, 2018). A dietary supplement contains vitamins, minerals, amino acids, herbs, or enzymes. They are meant to supplement not usually become the total dietary intake. They are available in several form, such as liquids, powders, capsules, tablets, and soft gels (Thompson & Manore, 2018). Although supplementation can treat an individual with less than needed nutritional intake, it is difficult to provide all the needed nutrients found in food (Yen, 2005). Supplemental feedings are available for those who cannot or will not ingest a balanced diet. Four types of supplements are available and used as oral feedings: (a) modular supplements containing only one nutrient, such as a carbohydrate or protein; (b) intact or "polymeric" formulas used when the older adult needs all nutritional requirements within a specific amount (e.g., Ensure, Meritene, and Sustacal); (c) elemental or "predigested formulas" that are easier to digest, such as Vivonex or Flexacal; and (d) disease-specific formulas designed for individuals with specific metabolic problems, such as a lung or kidney disorder; examples of these are

Hepatic-Aid and Pulmocare. When individuals are unable to eat in the usual manner or in sufficient amounts, enteral nutrition (tube feedings) may be administered via a tube inserted through the nose into the stomach or through a surgically inserted tube into the stomach. Intermittent or continuous formulated liquid foods help meet the necessary nutrition for each person. In the case of an individual with more serious and immediate nutritional requirements, an intravenous (IV) infusion (hyperalimentation) of high-caloric nutrients in a central vein may be administered (Adams & Urban, 2013).

COMMUNITY-BASED NUTRITION PROGRAMS FOR OLDER ADULTS

A variety of programs aimed at promoting improved and accessible nutrition at minimum cost are available. In 1972, the Older Americans Act (OAA) of 1965 was amended, establishing the National Nutrition Program for Older Americans. A national network of programs for home-delivered and congregate dining was made available to the states and U.S. territories through Title 111c of the OAA. It requires food be provided 5 days a week with at least one meal per day that meets a third of the RDA. It also provides nutrition screening, counseling, and education plus other health support services. Seniors' Farmer Market Nutrition Program offers coupons to purchase fresh locally grown fruits and vegetables to those age 60 and older. The Supplemental Food Commodity Program is available in some states and distributes commodity foods such as cereals, cheese, flour, fruits, and vegetables to low-income seniors (Wellman & Kamp, 2012). Meals on Wheels nutrition program for the elderly is a national program providing food to the needy at specific locations or delivered to the home of older adults regardless of income. Various other services may be offered to serve frail older adults such as weekend meals, multiple daily meals, or liquid supplemental snacks. Even diets to meet specific health needs may be offered along with ethnic or culturally preferred foods.

The 1964 Food Stamp Act offers the USDA food stamp program now known as Supplemental Nutrition Assistance Program (SNAP) to individuals in low-income households by issuing an electronic benefit card to purchase food. Certain third-party payers such as Medicare, Medicaid, Veterans Affairs, and insurance carriers offer and reimburse the cost of nutritional support and teaching. Nutritional counseling is offered in acute and long-term care settings, home health, and hospice care, as well as in ambulatory care settings (Wellman & Kamp, 2012).

Volunteers play important roles in delivering food to the homebound or those eating at congregate dining sites. In both settings, they offer socialization and someone to show concern and caring. A registered dietitian develops the dietary plan, including specific therapeutic diets. Sites such as churches, schools, social halls, and other settings are used for congregate dining and also offer opportunities to participate in various recreational, educational, and counseling programs. Similar programs are offered at adult day care centers. In addition, homemaker services are available through governmental and private sources to assist older persons with shopping, food preparation, and even light housekeeping. Homemakers are specially trained in meal preparation and diet modification, and a dietitian or nutritionist is usually available for consultation. Other specific programs are offered by county extension services throughout the country.

SUMMARY

The importance of an adequate diet throughout life cannot be overestimated. We are indeed what we eat. The prevention of chronic and life-threatening illnesses, rapid recovery from surgery or disease, and optimal physical and psychosocial functioning depend on the ingestion of necessary amounts of carbohydrates, proteins, fats, vitamins, minerals, and fluids. Many older adults do not have a nutritionally sound diet because they are unable to shop for, pay for, or prepare and eat proper food. All who work with older persons have responsibility to assess dietary intake, teach about nutrition, and refer individuals to the various nutritional programs and to qualified professionals in the local community.

REFERENCES

Adams, M. P., & Urban, C. Q. (2013). *Pharmacology connections to nursing practice* (2nd ed.). Pearson.

Adams, M. P., Urban, C. Q., & Sutter, R. E. (2019). *Pharmacology: Connections to nursing practice* (4th ed.). Pearson.

Baker, H. (2007). Nutrition in the elderly: An overview. *Geriatrics, 62*(7), 28–31.

Biggs, A. J. (2007). Nutritional considerations. In A. D. Linton & H. W. Lach (Eds.), *Matteson & McConnell's gerontological nursing* (3rd ed., pp. 161–197). Saunders.

Birn, C. S. (2013). Adult obesity in the United States: A growing epidemic. *Nurse.com, 4*(3), 36–41.

Champion, A. (2011). Anorexia of aging. *Annals of Long Term Care, 19*(10), 18–24. https://www.hmpgloballearningnetwork.com/site/altc/articles/anorexia-aging

Chernoff, R. (2006). Carbohydrate, fat, and fluid requirements in older adults. In R. Chernoff (Ed.), *Geriatric nutrition: The health professional's handbook* (2nd ed., pp. 23–30). Jones & Bartlett.

Collins, N. (2012). Food as medicine. *Caring, 35*(5), 40–41.

Crogan, N. L. (2019). Nutrition. In S. E. Meiner & J. J. Yeager (Eds.), *Gerontologic nursing* (6th ed., pp. 145–158). Elsevier.

DiMaria Ghalili, R.A. (2017). Nutritional problems. In Lewis, S.L. , Bucher, L.. Heitkemper, M.M. & Harding, M. (Eds.), *Medical-surgical nursing* (10th ed., pp. 854–875). Elsevier.

Dudek, S. (2010). *Nutrition essentials for nursing practice* (6th ed.). Lippincott.

Eliopoulos, C. (2014). *Gerontological nursing* (8th ed.). Wolters Kluwer Health/Lippincott Williams & Wilkins.

Eliopoulos, C. (2018). *Gerontological nursing* (9th ed). Wolters Kluwer.

Flood, M., & Newman, A. M. (2007). Obesity in older adults. *Journal of Gerontological Nursing, 33*(12), 19–34. https://doi.org/10.3928/00989134-20071201-04

Furman, E. F. (2006). Undernutrition in older adults across the continuum of care. *Journal of Gerontological Nursing, 32*(1), 22–27. https://doi.org/10.3928/0098-9134-20060101-11

Gallagher, M. L. (2012). Intake: the nutrients and their metabolism. In L. K. Mahan, S. Scott-Stump, & J. L. Raymond (Eds.), *Krause's food and nutrition care process* (13th ed., pp. 32–128). Elsevier-Sanders.

Goldberg, A. P., Andres, R., & Bierman, E. L. (1990). Diabetes mellitus in the elderly. In W. R. Hazzard, R. Andres, E. L., Bierman, & J. P. Blass (Eds.), *Principles of geriatric medicine and gerontology* (2nd ed., pp. 739–758). McGraw-Hill.

Guthrie, H. A., & Picciano, A. (1995). *Human nutrition*. Mosby.

Haber, D. (2013). *Health promotion and aging* (6th ed.). Springer Publishing Company.

Haber, D. (2016). *Health promotion and aging* (7th ed.). Springer Publishing Company.

Ignatavicius, D. D. (2013). Gastrointestinal system. In D. D. Ignatavicius & M. L. Workman (Eds.), *Medical-surgical nursing* (7th ed., pp. 1177–1191). Saunders Elsevier.

Keller, M. (2012). Physicians as supplement supervisor. *Aging Well, 5*(2), 22–25. https://www.todaysgeriatricmedicine.com/archive/031912p22.shtml

Lehne, P. A. (2010). *Pharmacology for nursing care* (7th ed.). Saunders.

Linderman, R. D. (2006). Mineral requirements. In R. Chernoff (Ed.), *Geriatric nutrition: The health professional's handbook* (2nd ed., pp. 77–93). Jones & Bartlett.

Lipschitz, D. A. (1997). Impact of nutrition on the age-related declines in hematopoiesis. In R. Chernoff (Ed.), *Geriatric nutrition: The health professional's handbook* (pp. 271–287). Aspen Publications.

Lipschitz, D. A. (1997). Nutrition. In C. K. Cassel, H. J. Cohen, E. B. Larson, D. E. Meier, N. M. Resnick, L. Z. Rubenstein, & L. B. Sorensen (Eds.), *Geriatric medicine* (3rd ed., pp. 801–813). Springer Publishing Company.

Luggen, A. S., Bernstein, M. J., & Touhy, T. A. (2008). Nutritional needs. In P. Ebersole, P. Hess, T. A. Touhy, K. Jett, & A. S. Luggen (Eds.), *Toward healthy aging* (7th ed., pp. 194–221). Mosby.

Lutz, C., & Przytulski, K. (2011). *Nutrition and diet therapy: Evidence-based applications* (5th ed.). F.A. Davis.

Marcus, E. L., & Berry, E. M. (1998). Refusal to eat in the older adult. *Nutrition Review, 56*, 163–171. https://doi.org/10.1111/j.1753-4887.1998.tb06130.x

Mauk, K. L. (2010). *Gerontological nursing: Competencies for care* (2nd ed.). Jones & Bartlett.

Mauk, K. L. (2018). *Gerontological nursing: Competencies for care* (4th ed.). Jones & Bartlett.

Mazur, E., & Litch, N. C. (2019). *Lutz's nutrition and diet therapy* (7th ed.). F.A. Davis.

Morley, J. E. (1991). Why do physicians fail to recognize and treat malnutrition in older persons. *Journal of the American Geriatrics Society, 39*(11), 1139–1140. https://doi.org/10.1111/j.1532-5415.1991.tb02884.x

Morley, J. E. (2003). Anorexia and weight loss in older persons. *Journal of Gerontology: Medical Sciences, 58A*, 131–137. https://doi.org/10.1093/gerona/58.2.m131

Morley, J. E. (2010). Anorexia, weight loss and frailty. *Journal of the American Medical Directors Association, 11*(4), 225–228. https://doi.org/10.1016/j.jamda.2010.02.005

Newbern, V. B. (1992). Failure to thrive: A growing concern in the elderly. *Journal of Gerontological Nursing, 18*(8), 21–25. https://doi.org/10.3928/0098-9134-19920801-06

Osato, E. E., Stone, J., Phillips, S. L., & Winne, D. M. (1993). Clinical manifestations: Failure to thrive in the elderly. *Journal of Gerontological Nursing, 19*(8), 28–34. https://doi.org/10.3928/0098-9134-19930801-07

Rourke, K. M. (2006). Nutrition. In S. E. Meiner & A. G. Lueckenotte (Eds.), *Gerontologic nursing* (3rd. ed., pp. 210–228). Mosby.

Rowe, J. W., & Kahn, R. L. (1998). *Successful aging*. Pantheon Books.

Sarkisian, C. A., & Lachs, M. S. (1996). "Failure to thrive" in older adults. *Annals of Internal Medicine, 124*, 1072–1077. https://doi.org/10.7326/0003-4819-124-12-199606150-00008

Siegler, E., & Hark, L. (1996). Older adults. In G. Morrison & L. Hark (Eds.), *Medical nutrition and disease* (pp. 142–155). Blackwell Science.

Silver, A. J. (1993). The malnourished patient: When and how to intervene. *Geriatrics, 48*(7), 70–73.

Suter, P. M. (2006). Vitamin metabolism and requirements in the elderly: Selected aspects. In R. Chernoff (Ed.), *Geriatric nutrition: The health professional's handbook* (2nd ed., pp. 31–76). Jones & Bartlett.

Tabloski, P. A. (2014). *Gerontological nursing* (3rd ed.). Pearson.

Thompson, J., & Manore, M. (2018). *Nutrition: An applied approach* (5th ed.). Pearson.

Touhy, T. A. (2012). Nutrition and hydration. In T. A. Touhy & K. J. Jett (Eds.), *Ebersole and Hess' toward healthy aging* (8th ed., pp. 240–264). Mosby Elsevier.

Touhy, T. A. (2016). Nutrtition. In T. A. Touhy & K. J. Jett (Eds.), *Ebersole and Hess' toward healthy aging* (9th ed., pp. 170–190). Elsevier.

Tufts University Health and Nutrition Newsletter. (2008). *26*(5), 1–4.

U.S. Department of Agriculture and U.S. Department of Health & Human Services. (2020). *Dietary guidelines for Americans, 2020–2025* (9th ed.). https://www.dietaryguidelines.gov/sites/default/files/2020-12/Dietary_Guidelines_for_Americans_2020-2025.pdf

U.S. Department of Health and Human Services. (2020). *Healthy People 2030: Building a healthier future for all*. https://health.gov/healthypeople

U.S. Federal Interagency Forum on Aging-Related Statistics. (2020). *Older Americans 2020: Key indicators of well-being*. https://www.agingstats.gov/docs/LatestReport/OA20_508_10142020.pdf

Verdery, R. B. (1997). Failure to thrive in old age: Follow-up on a workshop. *Journal of Gerontology, 53*, M333–M336. https://doi.org/10.1093/gerona/52a.6.m333

Wallace, J. I. (2009). Malnutrition and enteral/paraenteral alimentation. In J. B. Halter, J. J. Ouslander, M. E. Tinetti, S. Studenski, K. P. High, & S. Asthana (Eds.), *Hazzard's geriatric medicine and gerontology* (6th ed., pp. 469–482). McGraw.

Wellman, N. S., & Kamp, B. J. (2012). Nutrition in aging. In L. K. Mahan, S. Escott-Stump, & J. L. Raymond (Eds.), *Krause's food and nutrition care process* (13th ed., pp. 442–460). Saunders Elsevier.

Workman, M. L. (2013). Care of patients with eye and vision problems. In D. Ignatavicius & L. M. Workman (Eds.), *Medical-surgical nursing* (7th ed., pp. 1052–1076). Saunders Elsevier.

Yen, P. K. (2005). Food and supplement safety. *Geriatric Nursing, 26*(5), 277–280. https://doi.org/10.1016/j.gerinurse.2005.08.001

20

Medications

INTRODUCTION

An increase in health issues is often concomitant with aging; those older than 65 are likely to have one to three chronic diseases involving major body systems. The use of prescribed medications then is most often necessary to manage the various disease entities, with many older adults taking five or more medications simultaneously. Polypharmacy, the use of one or more medications concurrently, especially multiples of the same drug is commonplace in this age group. It is estimated one in three older persons ingest greater than eight various drugs each day, with many taking 10 to 15 drugs daily (Lilley et al., 2017). The more medication ingested, the higher the risk for adverse for drug reactions, interactions, and prescription cascading (Barclay et al., 2018). In addition, many take readily available over-the-counter (OTC) drugs for a variety of ailments such as headache, colds, arthritis, constipation, or indigestion. Another significant issue is the use of herbal products or nutritional supplements.

Older adults, who constitute about 15% of the population, take about one-third of all prescription drugs and 40% of all nonprescription drugs used by the general population (Lilley et al., 2017). It is not unusual for some older adults to ingest 10 to 20 pills each day. As the number of prescription and non-prescription drugs, herbal remedies, and nutritional supplements ingested increases, so does the likelihood of experiencing adverse reactions. Furthermore, the incidence of medication errors escalates as inappropriate dosages are taken at the wrong times for the wrong ailment, often causing hospitalizations (Mauk, 2018). A national study of older adults (average age 75) on Medicare revealed that 40% did not take medications as prescribed over the prior 12 months (Wilson et al., 2007). Reasons cited were prohibitive cost, belief they did not need them, or a belief that they were taking more than necessary (Wilson et al., 2007). At times, contraindicated medications, outdated or duplicated medications from different physicians, or even those shared with a neighbor are ingested. These behaviors often result in a litany of adverse reactions and interactions such as gastric irritation and bleeding, electrolyte imbalance, heart irregularities, orthostatic hypotension, nausea, altered mental states, constipation, movement disorders, falls, and urinary retention (Kane et al., 1999). The likelihood of adverse drug reactions is further enhanced by alterations in pharmacodynamics, pharmacokinetics, lower body fat and body mass, reduced liver size and kidney functioning, and lessened blood flow, as well as multiple and increasing numbers of severe health problems (Rolita & Freedman, 2008).

Drug side effects are responsible for at least 40,000 deaths in the United States every year, with older adults much more likely to experience toxic effects from drugs than younger people (Locklear, 2017). Before 1989, drug trials traditionally had been done on young white men.

Pharmaceutical companies are now including a more representative sample of the population (Adams et al., 2019). Since 1989, the FDA has required that new drug applications show evidence that they were studied using individuals who represented the drugs' targeted age group. New drug packet inserts now include a section called "Geriatric Considerations." Although older individuals are now included in clinical drug trials, the majority of those studied seem to belong to the young-old group—those younger than 75 years. This age group may not reflect the old-old (85+) persons' situation since they tend to have more chronic health problems (Tabloski, 2014). The standard adult dosage of a drug easily can be an overdose for an older person with various age-related changes in body composition.

Not adhering to a prescribed medication regimen is common among older adults. Noncompliance is influenced by a variety of factors such as complex medication schedules, lack of doctor–patient discussion, memory impairment, inability to hear or see well, loss of hand and finger dexterity, and not being given adequate information about the drug. Other issues leading to noncompliance include uncomfortable drug side effects such as dry mouth or frequency of urination, lack of literacy skills, and the inability to organize the drug regimen and clearly understand how they are to be taken. It is essential to consider these and other reasons for noncompliance when teaching and assessing the ways older adults take their prescribed medications. Lehne (2010) states that 75% of noncompliance among older adults is intentional because the dosage is too high, the drugs are expensive, and the side effects not tolerable.

CULTURAL RESPONSES TO DRUGS

Our country includes increasing numbers of individuals from diverse cultural backgrounds. Each culture brings with it beliefs and practices influencing the health and illness of its members. Treatment of illness through the use of medications and other practices is deeply rooted in each tradition and passed on from one generation to the next. A variety of regimens are used by cultural groups to treat disease, including (a) use of herbal and home remedies exclusively, (b) use of Western medicine and medical practices exclusively, (c) combining prayer and certain rituals along with various drugs, (d) use of Western medicine as an adjunct to usual folk practices, and (e) following certain environmental and dietary guidelines. Because the older generation often has closer ties to past generations and cultural practices, it is important to determine the various treatments individuals use because some of these may influence the effectiveness of prescribed medications and have the potential to cause adverse reactions.

OLDER ADULTS' RESPONSES TO DRUGS

Older adults' unique responses to drugs and their incidence of drug misuse place them at high risk for impaired physical and psychological states, accidents, and even institutionalization. Karch (2013) indicates that older adults respond quite differently in regard to every aspect of pharmacokinetics because of reduced absorption, distribution, and perfusion and altered metabolism, which results from age-related changes in the kidneys and liver. Furthermore, highly variable individual responses to medications appear to increase with aging. It is necessary for drugs to reach their target cells in sufficient amounts to be therapeutically effective (Adams et al., 2019).

Pharmacokinetics is the study of the time it takes for drugs to be liberated, absorbed, distributed, metabolized, and excreted (LADME) from the body and the correlation between where they are distributed in the body and the duration of intensity of therapeutic effects (Tabloski, 2014). Normal aging influences each drug's pharmacokinetics somewhat differently. Individual responses to drugs vary widely, and sensitivity to drugs may either increase or decrease with age. Factors such as age, disease, the presence of other medications or food in the body, smoking, alcohol ingestion, body weight and composition, genetics, and environment all influence the processing of drugs in the body (Kane et al., 1999; Shorr, 2007).

The effectiveness of a drug depends on its concentration at the site of action. The response of the drug is also determined by pharmacodynamics, which are processes in which the particular drugs change in the body. Pharmacodynamics is associated with the therapeutic and biologic effects of drugs at the drug receptor sites on the target organ. Some research indicates that there may be diminished numbers of receptors and possible changes in receptor sensitivity which may alter the older adults response to drugs (Adams et al., 2019; Eliopoulos, 2018). Thus, the rate at which liberation, absorption, distribution, metabolism, and excretion occur influences the speed at which the drug works, how long it remains in the body, and the blood concentration of the drug.

Liberation

Liberation occurs when the coating of a pill or capsule of a medication dissolves in the mouth, thus liberating the active drug ingredient.

Absorption

Absorption occurs when the medication is ingested and absorbed in the mouth, stomach, or intestinal tract. Drugs must be absorbed in solution into body systems to be effective. Most drugs are absorbed through the gastrointestinal (GI) tract into the general blood circulation. Generally, there is a slowing of drug and nutrient absorption. Possible age-related impediments to drug absorption, though, do exist. A higher pH in the stomach can reduce the absorption and solubility of drugs such as tetracycline and iron preparations, or may inactivate penicillin. Furthermore, a delay in stomach emptying, diminished gastrointestinal blood flow, and changes in the number, structure, and functioning ability of the absorbing cells' surfaces may also influence absorption of drugs. Decreased intestinal motility slows the passage of nutrients and unabsorbed drugs through the intestines, increasing the chance that drugs will become inactive or not be completely absorbed, which increases the risk of adverse side effects (Adams et al., 2019). Other factors that may modify drug absorption include the route of administration, the solubility and concentration of the drug, and the diseases and symptoms experienced by the individual (Eliopoulos, 2018).

Some medications are best absorbed on an empty stomach, whereas others need the presence of food to reduce gastric irritation. However, when taken together, a drug may interact with food; for example, orange juice increases and tea decreases iron absorption, laxatives containing mineral oil reduce the absorption of fat-soluble vitamins in food, and carbonated beverages and fruit juices tend to lessen the action of penicillin. Even though age-related changes in the gastrointestinal tract have minimal influence on drug absorption, the GI tract is a common site for both mild and severe reactions that may lead to hospitalization.

Distribution

The process by which drugs in the bloodstream are sent to various parts of the body is called distribution. Depending on their chemical characteristics, drugs absorbed from the intestinal tract pass into the portal vein (which carries blood to the liver) and are partly metabolized by the liver before entering the bloodstream where they are transported to various body sites. The majority of drugs are attached or bound to proteins in the blood, a process that is both reversible and variable. Other drugs are not bound to blood proteins but are "free" drugs in the blood. Bound drugs serve as a reserve supply of drugs, which are released into the bloodstream as the unbound or "free" drugs are metabolized and excreted.

Plasma proteins decline up to 13% in the older person, reducing the total number of usable binding sites (Adams et al., 2019). Older persons, then, tend to have increased amounts of "free" drug in the body, which can result in elevated drug levels in the blood. Certain drugs such as warfarin (an anticoagulant) and nonsteroidal anti-inflammatory drugs (NSAIDs) are highly protein bound. They may displace each other by competing for available protein-binding sites. Free drug molecules may then rapidly enter body tissues, causing dangerously high drug concentrations (Adams et al., 2019; LeFever Kee & Hayes, 2000).

Reduced cardiac output and diminished blood flow to various organs decrease the amount of blood reaching body tissues and affect the speed of drug distribution. Drugs are more rapidly transported to organs with a rich blood supply, whereas it may take hours for drugs to reach fatty tissue. The aging process may also cause a greater permeability of the blood–brain barrier, allowing certain drugs to enter the central nervous system and cause unexpected neurological reactions.

Metabolism (Biotransformation)

Most drugs are metabolized in the liver into metabolites (substances produced during metabolism). This process enhances drug excretion through the kidneys. An older person's liver function is reduced because of lessened blood flow in the liver, lessened enzyme activity, and smaller liver mass. Diminished liver function can influence the rate at which drugs are metabolized, creating a potential for drug toxicity. These changes may cause increased blood and tissue concentrations of some drugs or may prolong the half-life of others (Adams et al., 2019). For example, in older adults a specific cardiac medication such as propranolol (Inderal), a bronchodilator such as theophylline (Elixophyllin), certain antidepressants, and narcotics such as meperidine (Demerol) may produce higher blood levels of the drug as a result of altered liver metabolism. Individuals with liver diseases such as cirrhosis or hepatitis or those with decreased blood flow to the liver are especially sensitive to drugs metabolized by the liver.

Excretion

The kidneys, a major route for excretion of drugs from the body, eliminate metabolites from the liver into the urine as well as drugs not metabolized by the liver. Drugs are also eliminated through feces, exhalation, perspiration, and saliva. With age, there is reduced blood flow to the kidneys, fewer functional nephrons, reduced glomerular filtration rate, and less efficient tubular secretion and reabsorption. Despite these changes, older adults' elimination of waste and fluid is usually adequate for health.

Medications primarily eliminated through the kidneys are thought to have a longer half-life in older persons. Half-life is the time needed for the concentration of the drug in the blood to decrease by 50%. Some medications have short half-lives, others have long half-lives; the longer the half-life of a drug, the longer it will remain in the body. For instance, chlordiazepoxide's (Librium's) half-life is from 5 to 30 hours, and the half-life of digitalis is 36 to 48 hours. Drugs with a long half-life should be taken with longer intervals between doses. If a drug is taken more frequently than 1.5 of its half-life, accumulation will occur. For example, if a drug's half-life is 10 hours, it should not be taken more than every 15 hours to prevent accumulation. Knowledge of the specific half-life of each drug prescribed is very important information to use in preventing adverse reactions. Concomitant kidney diseases further impair drug elimination and increase the likelihood of drug toxicity, thus necessitating careful monitoring of drug dosages. Individuals taking medication primarily excreted through the kidneys should have renal function assessed periodically.

Tissue Sensitivity

Pharmacodynamic interactions (how drugs affects the body) are due to the additive, synergistic, or antagonistic effects of drugs (LeFever Kee & Hayes, 2000). Greater therapeutic effects or likelihood of toxicity can result from age-related changes in pharmacodynamics (Tabloski, 2014). Age-related changes at the site of action determine the individual's responsiveness to the drug. Some evidence indicates older adults have lessened numbers of receptors and possible changes in receptor sensitivity. Older adults are more sensitive to certain drugs, although sensitivity to other drugs may decrease with age. There is limited information on the actual effects of age-related pharmacodynamics of certain drugs for older persons (Adams et al., 2019; Kane et al., 1999).

Pharmacogenetics

Pharmacogenetics reflects genetic variations that may influence the body's reaction to a drug. An individual's response to medications may depend on ethnicity as well as genetic background. There are drugs for which an individual may be tested to determine whether that drug would cause an adverse effect or be effective or ineffective. Although these tests are not done routinely because of their expense, some are available (Tabloski, 2014). Such testing may assist primary caregivers in prescribing more appropriate medications for older adults.

OVER-THE-COUNTER DRUGS

OTC drugs, those purchased without a prescription, are considered by the FDA to be effective and safe therapy if used as indicated on the packaging of the drug. Dosages are usually lower than for prescription drugs and therefore may not be as effective. Some are single-drug products, whereas others are combined with other ingredients. Approximately 40% to 80% of older adults take one or more OTC drugs regularly, with 5.8% of them ingesting five or more OTC products each day. There are about 300,000 OTC products available, which make up 60% of all the drugs used in the United States (Lilley et al., 2017). About $32 billion is spent on OTC drugs annually (Mikulic, 2020). The average medicine cabinet contains 24 various OTC drugs (Lehne, 2010).

Newer, more strict "drug facts" are used for labeling OTC drugs. These include the use and purpose of the drug, specific warnings about its use, when to contact the pharmacist or physician, active and inactive ingredients, storage information, warnings regarding activities and substances to avoid, instruction about dosage, and side effects that may occur (Lilley et al., 2017).

Professionals have a responsibility to alert older persons to possible problems associated with OTC drugs, including interactions with other drugs, alcohol, or food. They should be encouraged to become enlightened consumers by using the many reliable sources of information about drugs available in books, health newsletters, and from the Internet. Healthcare providers, particularly pharmacists, are excellent resource persons for up-to-date and accurate information regarding a drug regimen.

Over the last 10 years, consumer groups have advocated for certain prescription drugs with a high margin of safety to be reclassified from prescription to OTC drugs. About 700 prescription drugs have since been reclassified to OTC drugs. Included among these are omeprazole (Prilosec), loratadine (Claritin), famotidine (Pepcid AC), and others. Special care should be taken when ingesting these drugs to make sure they are compatible with other medications and food that is eaten (Adams et al., 2019).

Some individuals believe OTC medications are harmless or do not even consider them to be drugs, but both OTC and prescription medications are drugs that have the potential for overuse and interaction with other drugs, alcohol, or foods. OTC drugs may also interfere with the accuracy of laboratory tests, alter nutritional states, mask symptoms of a disease, or even delay a diagnosis. Weight loss, diarrhea, confusion, depression, toxicity, or change in appetite could also result from OTC drug use. Reasons for their overuse include lack of money to visit a healthcare professional, lack of a personal physician, a desire to self-medicate, and the presence of an acute or chronic health problem such as a cold, arthritis, indigestion, insomnia, or constipation. OTC drugs most commonly used by older adults include analgesics, laxatives, antacids, cough medications, milk of magnesia, vitamins, and nonsteroidal topical preparations.

GENERIC DRUGS

Generic drugs are commonly prescribed and used today, but each state has specific regulations regarding how they may be dispensed. A generic name is the common chemical name of the active ingredient in a drug product. A drug product includes not only the active ingredient but also several other elements such as coatings, dyes, fillers, and binding agents. The latter cause the equivalency and quality of drug products to vary widely between manufacturers. "Generic equivalent" means that chemically two drugs are the same, but they may not be therapeutically the same because of differences in the pharmacokinetics, such as how they are absorbed. Medications are designated as therapeutically equivalent if they have identical active ingredients; are in the same concentration, strength, and dosage form; and are administered via the same route. To be bioequivalent, the extent and route of absorption of a generic drug may not be 20% or less or 25% greater than the drug with the brand name. If a tested drug earns an AB rating, this indicates that the drug is therapeutically equivalent to other drugs given the AB rating and has the same active ingredients. The FDA publishes *Approved Drug Products with Therapeutic Equivalence Evaluation*, which rates the therapeutic equivalency of drugs (Ditrapano & Peoples, 2008).

When a drug is discovered, it takes about 12 years before it is approved for human use and costs millions of dollars to develop (Cuozzo, 2007). Drug patents last for 17 years. After this period, other drug companies may apply to the FDA to manufacture, market, and sell the drug as a generic drug under a different brand name (Ditrapano & Peoples, 2008). The drug manufacturer obtains FDA approval by showing that the drug is safe, effective, and therapeutically comparable to the brand-name drug. However, some physicians recommend that generic drugs not be used by individuals with diagnoses such as congestive heart failure, diabetes, and depression.

The majority of commonly prescribed drugs are available in generic form at about 30% to 40% less than the cost of brand-name drugs. Consumers should ask health professionals about the advisability of using a generic drug and know when generics are dispensed to them. Currently, many HMOs or insurance companies pay primarily for generic drugs; some states permit substitution of a generic drug when filling a prescription. The increasing availability of generic drugs saves money for Medicare, Medicaid, insurance companies, and managed care plans.

Trademark laws prevent generic drugs from having the same appearance as the brand-name drug. Therefore the size, shape, and color of generic drugs look different from the brand-name drug. It is most important then to teach older adults to observe the characteristics of each of their drugs when prescriptions are refilled. If the pill looks different from the pill they had been taking, they should verify this change with the pharmacist (Bhimji & Pavlovich-Danis, 2012).

ADVERSE DRUG REACTIONS

Adverse drug reactions (ADRs) are a leading cause of morbidity and mortality in the United States. ADRs are unexpected and undesirable or excessive pharmacological responses to drugs that (a) result in serious disability or harm that is temporary or permanent; (b) cause the individual to be hospitalized, needing more intense care or prolonging their care; and (c) cause death. There are four types of ADRs: (a) dose-related reactions (toxic reactions), (b) drug–drug reactions, (c) idiosyncratic reactions, and (d) hypersensitivity reactions (Vallerand et al., 2017).

Tache et al. (2011) report about 16% of older adults in ambulatory care settings experience adverse drug reactions. Thirty-five percent of older adults who live in noninstitutionalized settings experience adverse drug reactions yearly, with 29% needing medical attention. Frequently reported adverse reactions include dry mouth, blurred vision, delayed voiding, constipation, and tardive dyskinesia (involuntary muscle movements caused by long-term use of antipsychotic drugs). Other commonly reported problems include electrolyte imbalances, a decrease in potassium levels when using some diuretics, gastric bleeding as a result of NSAIDs, and falls. Prescribing appropriate medications is imperative in preventing adverse drug reactions. As each group of drugs is considered, adverse reactions will also be discussed.

The Beers Criteria for Potentially Inappropriate Medication (PIM) for older adults (American Geriatrics Society, 2019) was first developed 30 years ago by Dr. Mark Beers and has undergone several revisions since then. The newest edition (2019) is useful for research, training, clinical decision, and policymaking to improve the safety and quality in dispensing medications to older adults. The Beers Criteria is a specific list of PIM's recommended to be avoided by older adults under most circumstances or in special situations such as in certain conditions or diseases because of their potential negative side effects. Beers has five criteria: (a) medications that are potentially not appropriate for most older adults, (b) medications that should be avoided in older adults who have certain conditions, (c) specific drugs to use with caution, (d) drug–drug interactions, and (e) drug-dosage adjustment related to kidney function. The 2019 revision of the Beers Criteria include 25 medications or classes of medications to be outright avoided or in a specific disease condition were dropped from the (American Geriatrics Society) Beers Criteria. A few were modified or moved to a new category. Some drugs or drug-disease combinations were omitted. Four new medications or medication classes to use cautiously were added to the list. Several drugs were added to the drug–drug and drug–disease interaction table.

The incidence of polypharmacy and adverse reactions to drugs among the older age group is startling; thus, the use of the Beers Criteria in prescribing medications for older adults is paramount. Using the criteria in concert with good clinical judgement and a patient-centered approach is paramount.

Consideration should be given to allergic drug reactions because these account for from 6% to 10% of all unusual or unexpected reactions to drugs. Drugs may interact with each other, increasing or decreasing their expected actions. Drugs prescribed for certain diseases, such as liver or kidney diseases, may cause adverse reactions such as confusion, lethargy, agitation, or even seizures.

ALCOHOL AND SUBSTANCE ABUSE

Alcohol and substance abuse among older adults is becoming a national epidemic often over looked by healthcare providers (Lilley et al., 2017). In addition opiates, cocaine and cannabis (marijuana) use is becoming more prevalent as the Baby Boomers age (Yeager, 2019).

Over the years, alcohol ingestion has become an accepted part of social interaction. Alcohol ingestion may result in chronic and progressive addiction, causing irreparable physical damage and psychosocial pain to the alcoholic as well as their family. Symptoms may often mimic those experienced as one grow older and may go unnoticed. Men aged 65 and older are more likely than women to be alcoholics; however, women who are alcoholics are not as likely to be diagnosed (Chou et al., 2011). Merrick et al. (2008) found that 1 out of 10 older Medicare beneficiaries drank more than the recommended amount of alcohol. Age-related changes enhance the likelihood of older drinkers incurring more social, physical, and psychological problems. With the decrease in lean body mass, there is also a decrease in total body water. Because alcohol is dispersed almost entirely in the body's water compartments, there results a greater blood alcohol level per dose of alcohol in older adults than for younger individuals. Additionally, older adults ingest more prescribed and OTC drugs than their younger counterparts. Metabolism may be slower, and most experience chronic disease, thus enhancing the potential for greater drug–alcohol interactions.

Older alcoholics are categorized into two groups: two-thirds represent "early-onset" drinkers who somehow have survived into a usually unhappy, unhealthy old age, and one-third are those older than age 60 who have begun drinking in response to aging or loss. Factors contributing to alcoholism that begins in the older years are (a) recent retirement; (b) lessened control over one's life and less recognition by others; (c) bereavement after the death of a significant person, usually a spouse; (d) impaired physical or mental health; and (e) relocation to a new environment. Any or all of these factors should alert those working with older adults to the possibility of alcoholism and the dangers of mixing alcohol and drugs.

Concurrent ingestion of medications and alcohol can result in central nervous system (CNS) depression, change in sleep habits, unsteadiness in walking, falls, gastric irritability, and impaired metabolism of various medications (Lilley et al., 2017). Major groups of drugs, both prescription and OTC medications, that adversely react with alcohol include tranquilizers, antidiabetic drugs, sleeping pills, antibiotics, anti-infectives, barbiturates, antianginal and antihypertensive agents, blood thinners, diuretics, pain medications (both narcotic and nonnarcotic), antidepressants, gout medications, muscle relaxants, allergy medications, cough- and cold-suppressing products, motion sickness drugs, vitamins, antihistamines, central nervous system stimulants, anticonvulsants, and antialcoholic preparations. Vigilance on the part of caregivers is necessary because the older alcoholic is not easily detected unless actually seen drinking. Older adults and their caregivers should become acquainted with the potentially lethal effects of simultaneously ingesting prescription and nonprescription drugs, alcohol, and cannabis.

Cannabis (Marijuana) is sometimes used by older adults as treatment for chronic conditions. Though little research supports its use, individuals who use it report a reduction in nausea and vomiting and muscle spasticity. Adverse effects are not predictable and may even cause marijuana toxicity, which could be manifest as hypertension, cardiac symptoms, panic attacks, and flashbacks (Harding, 2017). Several states now allow marijuana to be used for medicinal and recreational purposes even though it is a schedule one drug.

Interaction of Drugs With Other Substances

A drug interaction occurs when one substance increases or decreases the action of the drug. This can occur with another drug, food, dietary supplement, and herbal substances (Adams et al., 2019). Older adults, because of possible multiple chronic health problems, subclinical malnutrition, excessive use of drugs, poor cooking methods, and alterations in how the body utilizes drugs, represent an at-risk group with respect to drugs, caffeine, food, and tobacco interactions. Such interactions depend on body size, age, drugs, drug dosages, and current health problems.

Nicotine is also a substance abused by older adults. Tobacco use disorder (TUD) is considered to be the highest cause of disability and preventable disease in the United States. It is a risk factor among 6 of the 13 leading causes of death (Yeager, 2019). The therapeutic action of medication can be adversely affected by nicotine, which (a) alters liver enzyme metabolism, (b) causes vasoconstriction, (c) results in greater amounts of gastric acid secretion in the stomach, and (d) stimulates the central nervous system. To attain the desired therapeutic action of a drug, higher doses may be required because, for example, nicotine reduces the effectiveness of diuretics, heparin, and analgesics (Miller, 2009).

Caffeine, a stimulant, is present in many drinks, foods, OTC, and other medications. Caffeine–medication interactions mostly affect the medication action not the caffeine's action on the body. Such interactions may reduce the absorption of iron, diminish the effect of antiarrhythmic medications, and increase gastric irritation when taken with corticosteroids or analgesics (Miller, 2009).

The action of medications and nutrients may be affected by medication–nutrient interactions for individuals of all ages. Older adults are especially vulnerable because of age-related changes and disease states. A major way that nutrients influence medications is by altering their absorption in the stomach (Miller, 2009). Nutrients can either increase or decrease drug absorption, which could either be serious or beneficial to the individual. Grapefruit can increase the blood levels of a number of drugs such as lovastatin and verapamil (Lehne, 2010). Depleted nutritional states are further aggravated by certain medications that alter vitamin and mineral absorption. For example, mineral, oil when used regularly, impairs absorption of vitamin A, D, E, and K. Vitamin B complex absorption is impaired by digitalis, anti-inflammatory drugs, aspirin, and oral hypoglycemic drugs. Vitamin C levels are negatively influenced by both aspirin and alcohol. Many diuretics cause potassium loss, and aluminum-containing antacids reduce phosphate and calcium absorption.

Drugs can cause nutritional deficiencies by (a) appetite suppression, (b) changes in nutrient absorption, (c) changes in utilization and metabolism of nutrients, and (d) changes in elimination of nutrients. Certain drugs are known to suppress appetite. These include cancer chemotherapy drugs, alcohol, antacids, antihistamines, narcotics, digitalis, cough medicines, amphetamines, and caffeine. Conversely, some medications stimulate the appetite—for example, tranquilizers such as phenothiazines and antidepressants.

Recommendations to prevent drug–food interactions include the following:

1. Drugs and alcohol should never be taken together.
2. Medications should be ingested with sufficient water (preferably a full glass) to allow for complete swallowing.
3. Medications should not be taken with juices, tea, coffee, or soft drinks.
4. The cautions on OTC and prescription drugs regarding food–drug incompatibilities should be read and followed carefully.
5. Persons obtaining a new medication should ask a healthcare professional or pharmacist about any restrictions regarding food intake and medication use.

Drug–Drug Interactions

Drug–drug interactions are possible whenever the individual takes two or more drugs. When these interactions take place, the following may occur: Drug A may enhance or lessen the action of drug B or cause a response not observed with either drug and/or the larger the number of drugs ingested the greater likelihood of serious drug reactions. When considering the number of prescribed and OTC drugs taken by older adults, is it any wonder there are a large number of identified adverse drug reactions, much less those that are never detected (Lehne, 2010)?

LABORATORY VALUES

As individuals age, they become more unique physiologically. Declines in organ function vary from person to person, as does the aging process. Laboratory values may differ from younger to older adults, and there are few established norms for older persons. Some test values increase with aging, others decrease, and some stay the same. Various medications routinely taken by older adults alter laboratory test results. When interpreting these test results, consideration of all these issues is important. Recommended laboratory testing by drug monographs should be ordered routinely for this age group (Mauk, 2018).

PROMOTING ADHERENCE

Adherence to a prescribed drug regimen can be a major concern with older adults. Issues that influence adherence include drug expense, which may result in not taking the medication or skipping some doses. Lack of understanding of the drugs' purpose and how and when to take it is another issue. A low literacy level, inability to read or understand English, and cultural and ethnic beliefs may all influence compliance. Individuals with multiple chronic health problems are often faced with complex medication schedules that can prove too demanding and interfere with life style demands (Metzger, 2017). The situation can be especially challenging when memory or cognitive problems are present. Individuals with low vision or those who cannot manipulate and open pill bottles are also not as likely to take their medications (Mauk, 2018; Rosenjack Burchum, 2011).

Measures useful in promoting adherence include assessing the reasons for noncompliance such as family or financial issues, the recent death of a loved one, and physical or psychological issues. It is essential to establish a positive relationship with the individual (Metzger, 2017). Simplifying a complex drug schedule and offering information to help decrease drug costs can be very valuable. Teaching and learning sessions repeated over time may also be helpful. Encouraging the use of pill containers that separate medications by time and day or the use of a computer-automated pill dispenser and reminder may decrease noncompliance. All prescribed, OTC, and herbal medications should be reviewed regularly (Tabloski, 2014).

Prescriptions are also available online from various pharmacies. Make sure the pharmacy is licensed and located in the United States. These pharmacists must have a prescription from the person's healthcare professional before filling a prescription. It is possible to check the authenticity of the pharmacy by accessing the National Association of Boards of Pharmacy. Be sure that the pharmacy protects your privacy (Tabloski, 2014).

Drug Misuse

The following are some of the many factors responsible for misuse of drugs by older adults:

Psychological Factors

- Cognitive impairments influence the ability to take the correct drug at appropriate times.
- Depression or other psychiatric conditions result in less interest and motivation to take medications.
- Being required to take many medications, especially on complex schedules, may be difficult to understand and manage.
- Needing to take medications is perceived as a threat to self-esteem and independence.
- Some may have a fear of becoming addicted to drugs or fear of taking high-risk drugs.
- A lack of knowledge about medications and medication management may reduce compliance.
- The person may have a low literacy level.

Physiological Factors

- Visual impairment affects the ability to read labels and directions on how to take the drugs.
- Hearing impairment may hinder understanding instructions regarding how and when to take medications.

- Health problems such as arthritis and degenerative diseases often cause weakness and pain, making it difficult to open bottles and manipulate medications.
- Pain, especially chronic pain, may contribute to ingestion of too much medication.
- Using hypnotic drugs to combat insomnia may result in overmedication.
- Individuals may be allergic to medications.
- A weakened, debilitated, fragile state may make it difficult to take the correct medicine at the right time.

Social Factors

- Reduced financial income may restrict individuals from purchasing needed medications.
- Religious or cultural belief may influence use, nonuse, or disuse of some medications.
- Using medications may imply to others that the individual has assumed the "sick role."
- Inability to read or understand directions on the bottle.
- Lack of a social support group that may assist with medications.
- Use of imported drugs.

Other Factors

Research and clinical observations have cited other patterns of drug misuse, such as the following:

- Failure to have prescriptions filled or refilled
- Skipping doses
- Ingesting medications at the wrong time
- Swapping medications with friends
- Stopping medications prematurely
- Overdosing
- Underdosing
- Using outdated medications
- Consuming various drugs prescribed by several physicians and purchasing the drugs at different pharmacies
- Using alcohol, caffeine, smoking, or certain contraindicated foods with medications
- Not understanding the directions on how to take the medication
- Using OTC drugs along with prescription drugs
- Lack of periodic evaluation and follow-up by the primary care practitioner to assess the need to continue using medication
- Having too easy access to prescription drug refills via the telephone or computer
- Being unable to open tamper-proof, child-proof drug packaging
- Using herbal preparations with prescribed medications
- Ordering and using imported medications

All of these possibilities should be considered when monitoring medications used by older adults. Medication misuse has the potential to create mental problems such as confusion or lethargy, physical problems such as unsteadiness and falling, and social impairments such as the inability to engage in everyday activities of daily living (Miller, 2009).

DRUG THERAPY

The treatment of choice for many older persons who have multiple chronic diseases is pharmacological. Drugs, however, can be both helpful and harmful and should be used with caution.

Since limited drugs prescribed for older adults are discussed in this chapter, the use of reliable, up-to-date resources on medications is suggested for both professionals and nonprofessionals interested in geriatric pharmacology. The U.S. National Library of Medicine's MedlinePlus website has an extensive section on drugs, including side effects, dosage, and special precautions (www.nlm.nih.gov/medlineplus/druginformation.html). The reference book *Davis's Drug Guide for Nurses* (Vallerand & Sanocki, 2020) is also recommended for reliable and accurate information. The internet, too, can be a valuable resource of drug information through the use of drug company information sites, online pharmacies, and research information.

MEDICATION STORAGE AND DISPOSAL

Medications should be stored in a safe, dry environment, not in the bathroom medicine cabinet. While medications safety caps help prevent children from ingesting medications, they should also be stored out of the child's reach. Medications should be reviewed periodically and disposed of by the expiration date on the bottle. Disposing of medications properly is important. The Environmental Protection Agency no longer advocates flushing medications down the toilet. Some drug stores have programs that assist in safe medication disposal, and many communities feature special days and places to dispose of medications safely (Tabloski, 2014).

COMMONLY PRESCRIBED DRUGS FOR OLDER ADULTS

Cardiovascular Drugs

Digitalis

Digitalis slows and strengthens the heartbeat and is the drug of choice in treating congestive heart failure and other heart disorders. The half-life of digitalis may be as long as 36 to 48 hours; thus, the likelihood of causing digitalis toxicity and the necessity of carefully monitoring serum drug levels is essential. Those taking this drug are taught to count their pulse rate regularly and to stop the drug if the pulse is less than 60 beats per minute. Nausea, vomiting, and visual disturbances are signs of acute digitalis toxicity. Other symptoms include irregular heart rate, heart block, headache, confusion, agitation, and even psychosis.

Digitalis may depress the appetite and impair nutritional states. Some drugs and foods reduce the absorption of digitalis when taken by mouth. These include aluminum or magnesium antacids, laxatives, large amounts of bran, and the antibiotic neomycin. Older adults may not be able to tolerate digitalis well because of loss of lean body mass and possible impaired kidney functioning. They are also prone to electrolyte imbalances, especially low potassium levels; thus, it may be necessary to prescribe smaller dosages for this age group.

Nitroglycerin

Nitroglycerin is prescribed to dilate the blood vessels. Available in extended-release tablets or capsules, sublingual tablets, lingual spray, transdermal ointment, or intravenously, it helps to improve blood flow through the coronary blood vessels. The sublingual tablets should be stored in the original container and kept tightly closed in a dry, cool place; a fresh supply should be purchased every 3 months. It is advisable not to ingest alcohol, antivasodilators, or antihypertensive drugs with nitroglycerin and not to suddenly discontinue its use. Side effects of nitroglycerin include headache, rash, flushing of the face and neck, nausea, vomiting, low blood pressure, and visual disturbances.

Diuretics

Diuretics (HydroDiuril, Aldactone, and Lasix, for example) are used to treat acute and chronic cardiovascular disease as well as hypertension and heart failure. There are five major groups of diuretics. Thiazide and loop diuretics deplete the body of potassium and require the ingestion of potassium-enriched food such as bananas, oranges, green leafy vegetables, or potassium supplements. Diuretics should be taken at the same time each day, preferably in the morning because they increase voiding and may interfere with sleep if taken in the evening. Side effects related to potassium loss include thirst, erratic or weak heartbeat, mental changes, nausea, vomiting, weakness, fatigue, and muscle cramps. Other reactions to the drug include orthostatic hypotension (lowering of the blood pressure when moving from a lying to a standing position), bleeding, bruising, rash, increase in blood sugar levels, and excess uric acid in the blood. Electrolytes should be monitored regularly and weight taken daily. Furosemide (Lasix), for example, may be toxic to the ear (ototoxic) and result in deafness that is usually transient but can be permanent. Special care should be taken not to use these drugs if hearing is already impaired. Diuretics require close monitoring because they enhance the action of digitalis, lithium, oral diabetes medications, antihypertensives, and skeletal muscle relaxants and reduce the action of anticoagulants.

Beta-Adrenergic Blocking Drugs

Beta-adrenergic blocking drugs, such as atenolol (Tenormin), propranolol (Inderal), and others, lessen oxygen requirements of the heart by reducing the heart's workload with a resultant decrease in heart rate, force of contraction, cardiac output, and blood pressure. Older African Americans and cigarette smokers are often resistant to these drugs' intended actions. They have, however, been proven effective for most older persons with hypertension, angina, arrhythmias, or postmyocardial infarction. Individuals with chronic lung disease or diabetes are not likely candidates for them because of their influence on blood sugar and lung functioning. Side effects include fatigue, insomnia, sexual dysfunction, slow heart rate, and decreased blood flow to the periphery of the body. Certain side effects particularly common in older adults include hypotension, cardiac failure, low blood sugar, thyroid dysfunction, arthritic symptoms, and depression. They should not be withdrawn suddenly because angina, heart attack, or death could result.

Calcium Channel-Blocking Drugs

Diltiazem (Cardizem, Cardizem CD) and nifedipine (Procardia) dilate blood vessels, and other calcium channel-blocking drugs dilate coronary vessels, increasing coronary blood flow. They are used to lower blood pressure, decrease angina attacks, and control cardiac arrhythmias. Side effects are usually mild, with headache and constipation being the most common. Individuals who have asthma, peripheral vascular disease, or diabetes are given these drugs because they may not be able to tolerate beta-adrenergic blockers. They are often prescribed for older adults and African Americans who may not respond to other classes of hypertensive drugs.

Angiotensin-Converting Enzyme Inhibitors

The angiotensin-converting enzyme (ACE) inhibitors such as ramipril (Altace), benazepril (Lotensin), and captopril (Capoten) are used to treat hypertension, myocardial infarction (MI), heart failure, and diabetic neuropathy and to prevent strokes, MI, or death in individuals who are at high risk for developing cardiovascular attacks. Blood pressure may drop suddenly as a result of vasodilation after taking the first dose of an ACE inhibitor; thus, caution should be observed when starting this drug and blood pressure levels should be carefully monitored. Approximately 5% to 10% of those taking ACE inhibitors experience a dry, irritating persistent cough necessitating its discontinuance. It can also cause the kidneys to retain potassium; therefore, those taking it should be advised not to take potassium supplements or use salt substitutes that contain potassium (Lehne, 2010).

Angiotensin Receptor Blockers

Angiotensin receptor blockers such as losartan (Cozaar), irbesartan (Avapro), and valsartan (Diovan) are prescribed for hypertension, heart failure, MI, diabetic neuropathy, and to prevent strokes. They are usually well tolerated and often prescribed for individuals who cannot tolerate ACE inhibitors because they do not cause cough or high potassium levels (Lehne, 2010).

Aldosterone Antagonist

Aldosterone antagonists such as eplerenone (Inspra) and spironolactone (Aldactone) are used for hypertension and heart failure. Their major side effect is high potassium levels, so they should be monitored frequently and the older adult instructed not to take potassium supplements or use a salt substitute that contains potassium (Lehne, 2010).

Antihypertensive Drugs

Hypertension is a commonly occurring health problem among older adults. Aerobic exercises, weight loss, reduced fat and cholesterol intake, salt reduction, smoking cessation, and reducing stress all help to decrease blood pressure or maintain it within normal limits. The stepped-care approach is used in which one drug is prescribed with various others added as time goes on to meet the goal of a normal blood pressure. The first drug of choice is usually a diuretic, with others added as necessary, such as beta-adrenergic blockers, calcium channel blockers, ACE inhibitors, angiotensin II receptor blockers, or direct renin inhibitors.

The greatest risks with antihypertensive drugs involve reduced cardiac output, low blood pressure, dehydration, and orthostatic hypotension. Individuals on antihypertensive drugs should be instructed to rise slowly from a sitting or lying position and to monitor their blood pressure and pulse frequently. If the blood pressure is less than 90/60 mmHg or the pulse less than 60, or if there is dizziness, confusion, weakness, lethargy, or hypotension, a healthcare professional should be contacted immediately.

Drugs for Disorders of Coagulation

Because older persons are at higher risk for blood clots, anticoagulant therapy may be prescribed. Warfarin (Coumadin or Panwarfin) and dicumarol are oral medications used to prolong bleeding time and prevent formation of blood clots. Heparin is given subcutaneously and intravenously to attain a high level of anticoagulation or prevent the formation of postoperative blood clots. Enoxaparin (Lovenox) given subcutaneously is used to prevent deep vein thrombosis, especially after knee or hip surgery. This age group has a higher risk for increased bleeding when given anticoagulants. Blood coagulation time should be regularly monitored because a major side effect is hemorrhage. Because of this dangerous side effect, individuals may be anxious about taking the drug, so careful monitoring and teaching are imperative. Doses should never be omitted, and they must be taken specifically as prescribed.

Aspirin should not be taken with, or 7 days before or after, warfarin therapy. Any drug containing aspirin can increase the effectiveness of anticoagulants and cause bleeding. Because eating foods high in vitamin K promotes blood clotting and decreases the effectiveness of anticoagulants, food containing this vitamin is usually limited to 4 ounces per day. When on anticoagulants, older adults should be advised never to take any new drug, by prescription or OTC, unless first checking with the primary care provider or pharmacist. It is wise to inform caregivers and family if the individual is taking an anticoagulant.

Older adults on anticoagulants should be cautioned about falling, cutting, or bumping themselves because these accidents may cause bleeding. Nose or gum bleeding, blood in the urine, stool, or phlegm may indicate an overdose, and a severe headache might signal intracranial bleeding. All must be reported immediately to the primary care provider or the closest medical service. The use of a medical alert identification bracelet or card is highly recommended for those on anticoagulant therapy.

Antiplatelet Agents

These drugs cause anticoagulation by interfering with a variety of processes involving platelet formation, especially platelet aggregation. They are used to prevent the formation of clots in the arteries. Aspirin, available over the counter, is often prescribed in a small dose such as 80 to 325 mg daily to prevent strokes or coronary heart attacks. Other drugs such as clopidogrel (Plavix), dipyridamole (Persantine), and abciximab (ReoPro) are antiplatelet agents. They increase bleeding time and may cause gastrointestinal (GI) or other types of bleeding or bruising. Anyone taking these drugs should not ingest any medications containing aspirin. If bleeding or bruising occurs, a healthcare practitioner should be contacted immediately.

Thrombolytic Drugs

Anistreplase (Eminase), alteplase (Activase), reteplase (Retavase), and tenecteplase (TNKase) are among the drugs used to dissolve blood clots in the coronary vessels and to treat acute pulmonary embolism and deep vein thrombosis. Because these conditions demand immediate attention, they are usually administered in the emergency department. If an individual is suffering from any of these acute conditions, it is imperative that they be taken to the hospital immediately because each drug must be administered within a strict time frame. Bleeding and its complications are among the most dominant side effects, but streptokinase may also cause an allergic reaction. These medications are being used to treat older individuals but the cost is quite high. Contraindications include recent surgery, serious trauma, cerebrovascular disease, active internal bleeding, recent GI bleeding, and severe hypertension (Lehne, 2010). The use of these drugs have been replaced in hospitals where interventional cardiologic procedures are performed (Lilley et al., 2017).

Cholesterol-Lowering Drugs

Most cholesterol-lowering drugs, which are used to treat hyperlipidemia (greater than normal amounts of plasma cholesterol or plasma triglycerides), lower low-density lipoprotein (LDL), cholesterol, and triglycerides and raise high-density lipoprotein (HDL). Drugs include HMG-CoA (3-hydroxy-3-methyl-glutaryl-CoA) reductase inhibitors (statins), atorvastatin (Lipitor), lovastatin (Mevacor), bile-acid sequestrants such as cholestyramine (Questran), nicotinic acid, and cholesterol-absorption inhibitors such as ezetimibe (Zetia). Statins are the most effective in lowering cholesterol and are widely prescribed, but if withdrawn, serum cholesterol will return to pretreatment levels; thus, they must be taken for a lifetime. Adverse effects include myopathy (injury to the muscle tissue) with symptoms of muscle weakness or aching or joint pain. Such symptoms should be reported immediately to a primary care provider because they could progress to myositis (muscle inflammation) and to even more fatal muscle disintegration. Other adverse effects include liver toxicity, headache, constipation, abdominal pain, rash, and dyspepsia (Adams et al., 2019; Lehne, 2010).

Drugs for Joint Disorders

Two of the most annoying and painful diseases associated with older age are osteoarthritis (degenerative arthritis) and rheumatoid arthritis. Although drugs of choice for these conditions vary, only those commonly prescribed are discussed here.

Nonsteroidal Anti-Inflammatory Drugs

Aspirin (acetylsalicylic acid) is the oldest and most widely used nonsteroidal anti-inflammatory drug (NSAID). It helps to relieve mild-to-moderate pain, decrease inflammation, lower body temperature, and decrease platelet aggregation. Although a commonly used OTC drug for arthritis, it is not without serious side effects such as GI bleeding and iron deficiency anemia. According to Clark et al. (1990), about 10 to 30 mL of blood can be lost daily from long-term aspirin use. Enteric-coated aspirin is sometimes recommended because it is released in the small intestine rather than in the stomach and is less irritating to the stomach. Other adverse effects of aspirin include ringing in the ears, rashes, nausea, vomiting, confusion, and deafness. Salicylate toxicity–induced confusion is often undiagnosed and may occur even with normal therapeutic dosage. It is also important to recognize that many other OTC drugs contain aspirin—for example, Alka-Seltzer, Anacin, Excedrin, Pepto-Bismol, and Doan's pills. Alcohol should be avoided by those taking aspirin. The actions of both oral diabetes and anticoagulant drugs are increased by aspirin; however, it lowers the blood levels of other NSAIDs. The blood levels of aspirin can be measured through blood tests. Side effects are not an accurate indicator of overdose.

Nonaspirin Nonsteroidal Anti-Inflammatory Drugs and Acetaminophen

NSAIDs are most commonly chosen for treating mild to moderate arthritis. Additionally, they may lower body temperature and reduce inflammation. Drugs in this category include indomethacin (Indocin), piroxicam (Feldene), tolmetin (Tolectin), and naproxen (Naprosyn). They are likely to cause gastric irritation, bleeding, peptic ulcers, and water retention. Indocin may result in headaches, dizziness, nausea, vomiting, rash, and difficulty breathing. This group of drugs has a high incidence of adverse effects in older persons and should be used with caution. They should always be taken with milk or food, and careful monitoring is advised for those with heart, kidney, or liver disease.

Ibuprofen (Motrin, Advil, Nuprin) usually causes fewer gastrointestinal symptoms, but they do occur in some individuals. Adverse effects include nausea, vomiting, abdominal pain, gastrointestinal bleeding, headache, dizziness, and skin eruptions. Alcohol, oral anticoagulants, and aspirin should not be taken concurrently with ibuprofen.

Cyclooxygenase 2 (COX-2) inhibitors such as celecoxib (Celebrex) are prescribed for severe arthritic conditions and for those who need a more effective anti-inflammatory drug. Gastric bleeding and peptic ulcers are possible when high dosages are used. There is considerable evidence that COX-2 inhibitors increase the likelihood of developing a stroke, myocardial infarction (MI), or serious cardiovascular episode. Celebrex should be prescribed in the lowest dose possible for the shortest possible time and care taken not to prescribe this drug for individuals who have a cardiovascular disease (Vallerand et al., 2017).

Other drugs are also prescribed for arthritis, such as acetaminophen (Tylenol), which is useful in controlling pain and lowering temperature, but it is not an antirheumatic or anti-inflammatory medication. Tylenol does not cause stomach irritation, but overdosing can cause liver damage (Eliopoulos, 2018).

Disease-Modifying Antirheumatics and Other Drugs

Disease-modifying antirheumatic drugs (DMARDs) slow tissue damage of rheumatoid arthritis, psoriatic arthritis, and other inflammatory conditions. Among them are gold salts and drugs that modify the inflammatory and immune response. They reduce symptoms and mortality rates while improving the patient's quality of life. These medications include adalimumab (Humira), etanercept (Enbrel), azathioprine (Imuran), and others. Because of their toxic nature, patients require close observation (Adams et al., 2019).

Central-acting drugs such as clonidine (Catapres) or tramadol (Ultram) may also be prescribed. Other antirheumatic drugs commonly prescribed include methotrexate (Trexall), leflunomide (Arava), azathioprine (Imuran, Azasan), and others. Individuals with arthritis may experience temporary relief from injections of intra-articular glucocorticoids. A newer type of drug, hyaluronate (Hyalgan), is injected into the knee joint to replace lessened hyaluronic acid because of arthritis. It coats the articulating surface of the cartilage and creates a barrier that prevents future inflammation and friction in the joint (Adams et al., 2019).

Steroids

Adrenal corticosteroids such as prednisone (Deltasone, Meticorten) and hydrocortisone (Cortef) may be prescribed for individuals with rheumatoid arthritis but are to be used for the shortest time in the lowest possible dose and only after other less potent drugs have been ineffective. They should never be discontinued abruptly but must be tapered off gradually. Side effects include nausea, gastric distress, ulcers, hemorrhage, sodium and water retention, euphoria, thinning skin, easy bruising, depression, impaired wound healing, and increased incidence of osteoporosis. Always ingest oral steroids with food (Vallerand et al., 2017).

Drugs for Gout

Individuals with gout have higher than normal levels of uric acid, which crystallizes in joints, tendons, or bursae, causing extreme pain. Approximately 1% to 3% of the population in the United States develop gout. Therapeutic goals are to terminate the acute attack and prevent further episodes and complications such as kidney stones (Adams et al., 2019). Colchicine (Colcrys) is used both for acute attacks and to prevent attacks. This drug requires careful monitoring, especially with older persons who have heart and kidney problems. Alcohol should be avoided when taking it. Side effects include nausea and vomiting followed by diarrhea, stomach pain, mental confusion, numbness, tingling, bleeding, bruising, weakness, and skin rash. Allopurinol (Zyloprim), also prescribed for gout, usually causes minimal side effects, such as skin rash or itching, bruising, bleeding, weakness, or drowsiness, and should be taken with meals or a snack to avoid gastric irritation. Probenecid (Benemid) and sulfinpyrazone (Anturane) are also prescribed but should be avoided by those with kidney stones or kidney failure. Occasionally, GI upset or allergic reactions may occur. It is advisable that those who take these medications drink 10 to 12 eight-ounce glasses of water daily and avoid aspirin or aspirin-containing substances. Regular monitoring of serum uric acid levels and renal function testing are also recommended.

Drugs for Osteoporosis

As ovarian function decreases during menopause, estrogen production falls gradually, or it ceases abruptly after surgical removal of the ovaries, a major cause of osteoporosis. In the United States it is estimated that about 8 million women have osteoporosis. Of these about 40% will incur a fracture with an annual expenditure to society of around 11 billion dollars (Lilley el al., 2017). Because osteoporosis contributes to falls, disability, or death of many older women, considerable research focuses on this health issue. Regular exercise, preferably walking, together with a diet high in calcium, are thought to reduce the likelihood of developing osteoporosis. Postmenopausal women need a calcium intake of about 1,500 mg daily, especially if they are not taking estrogen supplements. When purchasing calcium, it is important to be aware that not all brands deliver the optimal amount of calcium. In addition an intake of 400 to 800 IU of vitamin D is recommended to aid in the absorption of calcium.

Estrogens such as Premarin and estrogen/medroxyprogesterone (Prempro) may be prescribed to alleviate the symptoms of menopause and to slow postmenopausal bone loss. They are available in several forms: oral, transdermal patch, injection, and intravaginal. Adverse effects include breast cancer, fluid retention, MI, deep vein thrombosis, stroke, pulmonary embolism, and dementia.

The Women's Health Initiative and the Heart and Estrogen Progestin Replacement Study research studies have led to major modifications in regard to hormone replacement therapy during menopause. This resulted in large numbers of women discontinuing or not beginning the use of hormone replacement therapy (HRT). Recommendations now discourage the use of HRT for years and advise consulting the primary care provider regarding the use of HRT (Lehne, 2010). Generally when used to treat menopausal symptoms, estrogens are usually prescribed on an individual basis with the lowest dosage for the shortest period, and for women without a history of cancer or cardiovascular disease (Adams & Urban, 2013). Because of the risks involved in taking this drug, product labels must have the highest level of warning (boxed warning) describing its many risks.

Another group of drugs available are the selective estrogen receptor modulators (SERMs), such as raloxifene (Evista). Evista mimics estrogen in certain tissues of the body and its action reduces bone resorption, but it may cause hot flashes and, in some cases, leg cramps and venous thrombosis.

Bisphosphonates, another category of osteoporosis medications, slows bone metabolism and bone turnover, even building new bone and thus increasing bone density or bone mass by 50%. Examples are alendronate (Fosamax), risedronate (Actonel), ibandronate (Boniva), alendronate (Fosamax), and denosumab (Prolia). Alendronate (Fosamax) is also used to treat osteoporosis in men. There is a slight tendency to develop osteonecrosis (bone destruction) of the jaw when taking Fosamax. The first two drugs are taken once a week, Boniva once a month, and Prolia every 6 months. They should be taken in the morning on an empty stomach with a full glass of water. Other food or drink must not be ingested for a half hour and the individual must remain sitting or standing for at least 30 minutes. These drugs are usually well tolerated, but esophagitis, nausea, vomiting, and musculoskeletal pain can result. Zoledronic acid (Reclast) is the only FDA-approved treatment for osteoporosis that is administered once a year by intravenous (IV) injection. Daily calcium and vitamin D supplements are also encouraged to strengthen the bones.

Drugs for Parkinson's Disease

Parkinson's disease is a movement disorder caused by a lack of adequate amounts of dopamine in which the individual develops muscle rigidity, difficulty initiating any movement, and tremor. Certain drugs such as reserpine (Serpasil), norepinephrine (Levophed), and antipsychotic drugs also cause symptoms similar to those of Parkinson's disease. Early treatment involves the use of anticholinergic drugs such as benztropine (Cogentin) or biperiden (Akineton), but possible side effects include dry mouth, urinary retention, constipation, blurred vision, disorientation, insomnia, restlessness, and impairment of recent memory. Individuals with mental problems, narrow-angle glaucoma, urinary or intestinal obstruction, and those with a rapid heartbeat should not take these drugs because they may worsen these conditions.

Dopaminergic drugs potentiate the individual's ability to carry out activities of daily living and are the first line of treatment for Parkinson's disease. They include levodopa (L-dopa), amantadine (Symmetrel), and carbidopa-levodopa (Sinemet). Possible side effects include aggressiveness, dizziness, restlessness, nausea, vomiting, anorexia, agitation, orthostatic hypotension, GI bleeding, darkened urine, confusion, and blurred vision. Carbidopa-levodopa seems to be better tolerated by older persons and has fewer side effects. Pramipexole (Mirapex) is prescribed early in the disease and used in combination with levodopa in the more advanced stages of Parkinson's disease (Lehne, 2010).

Psychotropic Drugs

Included among the psychotropic drugs are sedative-hypnotics, anxiolytics, antidepressants, and antipsychotic drugs, all of which are routinely prescribed for older adults. Care must be taken when prescribing and using these drugs because they have the potential to be toxic. Some are even misused as "chemical restraints" (Meiner, 2015).

Antianxiety (Anxiolytic) Drugs

Anxiety disorders are included among the more commonly diagnosed psychiatric illnesses (Lehne, 2010). Before being prescribed any antianxiety medications, older adults should be carefully assessed through a physical examination, including a history of personal losses and other stresses as well as drug, alcohol, OTC drug, nicotine, and caffeine use. It is preferable to counsel and teach the individual improved coping strategies before prescribing antianxiety medications.

Benzodiazepines are the drugs most commonly used in the treatment of anxiety. Those with a short half-life include alprazolam (Xanax) and lorazepam (Ativan). They are eliminated from the body fairly quickly but may cause rebound anxiety and insomnia. Long-acting antianxiety drugs include chlordiazepoxide (Librium), clorazepate (Tranxene), and diazepam (Valium). These drugs are highly addictive yet effective in treating anxiety. Short-acting benzodiazepines are usually preferable in treating older adults; however, older adults tend to metabolize benzodiazepines less quickly, resulting in the drugs remaining in the bloodstream for prolonged periods, causing toxic effects. They are prescribed for anxiety and insomnia. Side effects may continue for several days after the drug has been discontinued and include dizziness, headache, daytime sedation, motor incoordination, confusion, memory impairment, agitation, cognitive impairment, and drug dependency. Older debilitated individuals require careful monitoring regarding safety issues, and some may be misdiagnosed because side effects can mimic dementia.

Buspirone (Buspar) is a nonsedating, nonbenzodiazepine antianxiety drug that does not cause dependence or withdrawal symptoms. Prescribed for chronic anxiety, it usually takes several days to weeks to become fully effective and is recommended only for short-term use. Adverse effects include lightheadedness, nausea, headache, blurred vision, dizziness, and nervousness, but it is usually well tolerated (Lilley et al., 2017).

Sedative-Hypnotic Drugs

A disorder of sleep is defined as an interruption in the normal pattern of sleep as well as in the quantity and quality of sleep. About 40 million people have a sleep disorder (Adams et al., 2019). Sedatives function by initiating sleep more rapidly and reducing the number of short awakenings during the night. Usually the effectiveness of these drugs is brief, and when they are stopped, former patterns of sleep resume. Certain characteristics of hypnotics such as depression of the central nervous system limit their use with older adults. Tolerance usually develops over a few days, making it necessary to increase the dosage. The half-life of many hypnotics is very long, and some, especially barbiturates, have a very narrow margin of safety between what constitutes a therapeutic dose and a toxic or fatal dose.

Barbiturates, used to induce sleep, are not considered appropriate for older persons because of the likelihood of dependence, their narrow margin of safety, and considerable interactions with other drugs (Rosenjack Burchum, 2011). Benzodiazepines such as flurazepam (Dalmane) and temazepam (Restoril) produce only limited, suppressed REM sleep with no rebound (reoccurrence of the medication's effects after it has been stopped). On the other hand, it suppresses deep sleep stages 3 and 4, which are already appreciably reduced in the elderly. There is danger of accumulation in the body because the half-life of Dalmane may be 120 to 160 hours in older adults. It may be useful for individuals who require prolonged therapy, but side effects such as oversedation, dizziness, and excitement are not unusual in this age group. Chloral hydrate (Noctec), a long-used hypnotic, does not seem to affect REM sleep, rarely causes a hangover, and is excreted from the body fairly rapidly. Zolpidem (Ambien) is used to treat insomnia but has potential for causing sedation and confusion. Melatonin, a natural hormone and OTC drug may improve the onset of sleep as well as the duration. It could however result in headaches, nightmares, mental impairment, and even disturb homeostasis (Adams el al., 2019).

Adverse effects of hypnotics include confusion, ataxia (uncontrolled movements), gastric irritation, excessive drowsiness during the day, and a severe "hangover" effect and falls. They generally account for the stuporous behavior often observed in nursing home and hospital patients. Some sleeping medications contain scopolamine and should not be given to anyone with glaucoma. Acute poisoning can result when certain sedatives are ingested with alcohol. Rather than relying on sedatives, it is best to first use nonpharmacological approaches. Initially determine prior sleep patterns and the causes for not sleeping. Causes might be a full bladder or rectum, anxiousness, loneliness, fear, or grief. Discussing the tendency toward increased short awakenings in older age and emphasizing that little overall sleep is actually lost might alleviate concerns over not sleeping. A glass of warm milk, a back rub, exercise, soothing music, and showing concern can often accomplish better results than administering hypnotics. Daytime napping should be discouraged as much as possible.

Antidepressant Drugs

Depression is a major health issue among older adults (Eliopoulos, 2018). A large array of antidepressant drugs are available that are quite effective in treating depression; however, it takes from 2 to 4 weeks for their action to be fully realized. Selective serotonin reuptake inhibitors (SSRIs) are most often the first choice for older adults and have replaced the tricyclic antidepressants and monoamine oxidase (MAO) inhibitors. The tricyclic antidepressants sometimes cause serious adverse reactions such as cardiac toxicity or memory problems, and the half-life of several is longer than 24 hours, which could cause serious life-threatening side effects. SSRIs may result in GI upset, drowsiness, dizziness, headache, confusion, agitation, and nervousness. Mirtazapine (Remeron) may cause greater sedation than other SSRI antidepressants. Sertraline (Zoloft) may be helpful for an older adult who tends to be sleepy, less social, and showing depressive symptoms because it has a more stimulating effect (Adams et al., 2019). These drugs should not be taken along with OTC or prescription drugs, vitamins, or herbal products without the permission of the primary healthcare provider (Vallerand et al., 2017). These drugs warrant monitoring for enhanced depressive symptoms especially when starting them or when the dosage is changed. The FDA issued a warning of a higher risk for suicide among individuals taking these medications (Lilley et al., 2017).

Mood Stabilizers

Lithium carbonate (Carbolith, Eskalith) is prescribed for the treatment of bipolar affective disorders to level out extreme shifts in emotions. It should be used with caution in older persons because of its narrow margin of safety, and serum levels must be continually monitored to prevent drug overdose. Side effects are multiple and involve major body systems. Among them are confusion, drowsiness, stupor, restlessness, muscle weakness, hypotension, aphasia, and many others. It should be used with caution in older adults and those with diabetes, renal, thyroid, or cardiac disease. It is advisable to take lithium with meals and at specific times each day with no omissions or changes in dosage without contacting a primary care practitioner (Vallerand et al., 2017). Many potential reactions to lithium can be serious because various other drugs, foods, and herbs influence its action on the body (Adams et al., 2019).

Antipsychotic Drugs (Neuroleptics)

These drugs are usually prescribed for older adults with psychosis, agitation, and delirium. Antipsychotic drugs are often quite effective, yet most produce serious side effects. Specifically, they are used to control symptoms such as paranoia, disordered thinking, hallucinations, delusions, aggressive, disordered behavior, and mood and affect issues. There are two major classes of these drugs: first-generation and second-generation drugs. Among the first generation are haloperidol (Haldol), a high-potency medication that should be used cautiously with older individuals. Others include phenothiazine (Thorazine, Sparine) and thioridazine (Mellaril). Side effects and adverse reactions of neuroleptics can be severe depending on the drug used and the dosage prescribed. The lowest possible dosage for the shortest length of time is recommended. An array of adverse reactions is possible when taking these drugs, including tardive dyskinesia (chronic involuntary movements of the tongue, lips, and face, as well as agitated movements of the feet, hand, fingers, and toes). Extrapyramidal effects have been found in 50% of older adults who use antipsychotics and include tremor, agitation, and a shuffling gait, all of which mimic Parkinson's disease. Other side effects include akinesia (a fixed, flat expression), apathetic manner, dulled speech, hypotension, restlessness, photosensitivity, constipation, jaundice, difficult urination, cardiovascular effects, cognitive impairment, delirium, and dementia (Vallerand et al., 2017).

Neuroleptic drugs are to be taken with food or milk and carefully monitored regarding dosage and times taken. Individuals taking these medications should be advised to rise slowly from a lying or sitting position because hypotension may occur. They are not to be ingested with alcohol or other CNS depressants, or within an hour of antacid or antidiarrheal medication. Haldol has a narrow margin of safety regarding dosage, and blood levels must be carefully monitored.

Newer second-generation atypical antipsychotic drugs have been developed that influence the serotonin and dopamine receptors. Among these are clozapine (Clozaril), risperidone (Risperdal), and quetiapine (Seroquel). They have fewer extrapyramidal side effects than others, but they may cause diabetes, weight gain, elevated lipids, and stroke, especially among older individuals (Adams et al., 2019).

Of all the drug classes, neuroleptic drugs are the most overused and misused with older adults. They are not to be prescribed for relatively minor behavior problems. Careful assessment is paramount in determining the cause of the presenting behavior because it could well mask the presence of a urinary tract infection, electrolyte imbalance, dehydration, or adverse drug reactions.

Antibiotics

Antibiotics are a major group of drugs commonly prescribed for older adults to treat infections. Drug–drug interactions and adverse reactions to antibiotics are more common in older adults than in younger persons. Included among the anti-infective drugs are penicillins, cephalosporins, tetracyclines, aminoglycosides, macrolides, fluoroquinolones, sulfonamides, and selected others.

There are also antiviral and antifungal drugs. Broad-spectrum antibiotics are effective against many different pathogens, whereas narrow-spectrum antibiotics are appropriate for only one or a special group of pathogens. Identifying the pathogen (culture and sensitivity testing) before initiating anti-infective therapy is important because taking an inappropriate drug may result in the development of antibiotic resistance bacteria or causing the pathogen to progress further in the body.

Care must be taken to avoid allergic or adverse reactions between medications and foods that may either enhance or reduce medication effectiveness The tetracyclines (such as Vibramycin or Terramycin) enhance the effects of anticoagulants, and aminoglycosides. Gentamicin sulfate (Garamycin) may cause deafness as well as damage the kidneys, and cephalosporin (Cephalexin) may result in bleeding disorders. Because these drugs are mostly excreted in an unchanged form by the kidneys, kidney functioning must be carefully monitored. Doses should be far enough apart to allow for delayed excretion. Other side effects include diarrhea, nausea, or vomiting, which can precipitate fluid and electrolyte imbalance which should be reported to the healthcare provider (Eliopoulos, 2018). It is estimated that more than 17% of individuals who take antibiotics develop antibiotic-associated diarrhea (AAD), which may have a high mortality rate. Probiotics are recommended to prevent and treat AAD (Rogers et al., 2013). It may be necessary for the healthcare provider to prescribe more than one antibiotic. Taking the full course of medications is essential to effective treatment, as is taking the correct dosage at the right time, avoiding certain identified foods or other medications, and drinking sufficient fluids.

Diabetic Drugs

Diabetes mellitus, a chronic illness, increases in prevalence after age 50. It is characterized by elevated blood glucose levels with a decrease in the body's ability to respond to insulin or the absence or decrease of insulin produced in the pancreas. Although some older individuals have type 1, or insulin-dependent, diabetes (IDDM), most have type 2, or non–insulin-dependent, diabetes (NIDDM). Insulin injections act as a replacement for insulin normally secreted by the beta cells in the pancreas. Insulin is available in several forms: rapid, short, intermediate, and long duration, as well as combinations of these and as insulin analogs. Formerly the primary sources for insulin were the pancreas of cows and pigs. Today, most is human insulin. All insulin products essentially contain 100 units of insulin in each ml (Lilley et al., 2017). The type of insulin prescribed is based on blood sugar levels and other factors. Unopened insulin should be refrigerated, but insulin currently being used may be stored at room temperature for up to 1 month. Insulin may be administered by injection intravenously or subcutaneously, by inhalation (Exubera), and via an insulin pump.

Careful instructions are necessary when teaching older persons how to check blood sugar levels and give themselves insulin injections. Understanding the signs and symptoms of hypoglycemia and hyperglycemia as well as appropriate corrective actions is most important. Demonstrations, practice sessions together with oral, audiovisual, computer, and written information are helpful teaching techniques. Special types of injectors to administer insulin are available as well as other assistive devices for those who are visually or otherwise disabled.

People with type 2 diabetes manage their diabetes on a regulated regimen of diet and regular exercise. Others require oral hypoglycemic agents to lower the blood sugar. Individuals who continue to produce some insulin may take sulfonylureas. They use various mechanisms to lower blood sugar. Failure of two drugs to control blood sugar levels usually necessitates a need for insulin. Some drugs are combined into one tablet to achieve the desired effect. Several newer drugs have been approved to treat diabetes, including sitagliptin (Januvia), exenatide (Byetta), and linagliptin (Tradjenta). Diabetic medications may cause hypoglycemia (low blood sugar) and sensitivity to direct sunlight. Careful monitoring of blood sugar levels is necessary as well as attention to possible complications. It is advisable to carry some form of carbohydrate in case of a hypoglycemic reaction and to wear a medical alert bracelet or carry a medical alert card. Older adults are especially prone to hypoglycemia; thus, monitoring food intake, blood sugar levels, the medication dosage, and when taken are very important (Lilley et al., 2017).

Short-term complications of diabetes include hyperglycemia (high blood sugar levels) and hypoglycemia (low blood sugar levels). Ketoacidosis results when hyperglycemia is allowed to continue and could be fatal. Long-term complications develop over many years. It is estimated that more than 90% of people with diabetes die of these complications. Among them are sensory and motor neuropathy, kidney damage, blindness, cardiovascular disease, delayed stomach emptying (gastroparesis), and amputation as a result of infection, especially of the lower limbs. Maintaining blood sugar levels within the normal range can reduce injury to the blood vessels, as can eating the prescribed diet, relieving stress, exercising regularly and maintaining appropriate weight, and avoiding alcohol (Lehne, 2010). Regular visits to a healthcare provider as well as various blood tests are essential in monitoring this disease.

Laxatives

After viewing television and reading the many advertisements for laxatives, it might seem that individuals cannot function without them. Older individuals often abuse laxatives and stool softeners. Laxatives cause dependency, dehydration, loss of muscle tone in the intestines, loss of important salts and minerals, and reduced absorption of vitamins A and D. Anyone using laxatives regularly should consult a primary care practitioner to determine the actual cause of the problem. Keep in mind some drugs taken by older adults also cause constipation. Rather than resorting to laxatives, eating a diet with adequate fiber, drinking plenty of water, and exercising can be most helpful in preventing constipation.

ATTITUDES TOWARD "PILL POPPING"

Attitudes toward taking medications are highly varied among older adults. Some may believe that all medicines are worthless and do not take their prescribed medications. Others use prescribed pills when critically ill but discard them as soon as some improvement occurs. Still others hold fixed, inaccurate ideas about what a particular pill will do for them or they discontinue them because of their side effects.

Quackery is an ever-present threat to appropriate healthcare. Certain medications, treatments, and cures appear on the market from time to time, especially on television, the internet, and social networks and via written word. They promise instant cures for a wide variety of ailments, including arthritis, diabetes, cancer, GI disorders, and other problems. Not only are these treatments expensive, but their use may cause a delay in seeking appropriate medical treatment and could even result in death.

PREVENTION OF DRUG ACCIDENTS

Lack of accurate information about medications is common in the older age group. Some believe if one pill helps, two will be sure to cure. Medications are forgotten, the treatment regimen is not understood, or the medication is not taken at the proper time and amount, especially if a number of medications is prescribed. When five or six or more pills are prescribed to be taken at different times, even a young person would be challenged. Setting the alarm clock to ring when medications are due or using pill containers with seven or more compartments may encourage compliance. Several specialized electronic pill reminders, such as E-pill, are available to assist individuals in taking their medications appropriately. These include computer-generated reminder charts and electronic medication compliance aids. There is also a medication reminder and alert watch that displays medical diagnoses, medications, allergies, emergency contacts, and vital personal information. Seeking the assistance of a relative or friend might also be helpful. Prescriptions may not be refilled, often because of cost. Sometimes pills are swapped with a friend, a common but dangerous practice. At other times they are taken along with herbs, supplements, or alcohol with which they may be incompatible.

Healthcare professionals have an obligation to ensure that older adults understand the purpose of each medication, how and when to take it, and its side or adverse reactions. A potentially dangerous situation may occur when samples are given out by the primary care provider without written information about the drug. Written information, however, is included with all prescriptions filled in most pharmacies. Reviewing the information with each person would help greatly to ensure proper understanding and increase compliance. The computer and reference books are also valuable sources of information. Those with visual impairments may benefit from large-print instructions, color coding, and improved illumination. Older adults should be encouraged to ask questions and obtain information about all prescribed and OTC medications they are taking.

MEDICARE PART D

Financial constraints are a major issue for older adults when purchasing and taking medications as prescribed. Medicare Part D, instituted January 2004, provides substantial coverage for medications. Although all drugs are not completely covered, some, usually generics, are quite affordable. There are premiums, copays, and deductibles that may need to be paid. Individuals must have Medicare coverage to sign up for Medicare Part D, and the enrollment period is October through December each year. If drug costs reach $4,130, the "donut hole" is reached and coverage ceases. When reaching the donut hole, the individual is responsible for total drug costs unless they have some other type of drug assistance. Coverage again resumes after $6,550 is spent. Gradually the "donut hole" coverage gap is expected to cease (Tabloski, 2014). While each year beneficiaries may choose a particular coverage plan, careful study and scrutiny is advised to select the best plan for each person's needs. The above protocol and numbers change from time to time; thus, it is wise to be updated regularly on specifics related to Medicare Part D.

SUMMARY

Medications are of tremendous value to older adults whose very lives depend on them. An understanding of age-related factors that influence absorption, distribution, metabolism, and excretion, as well as side effects, adverse reactions, actions, and interactions of drugs is imperative. To ensure optimal physical, psychological, and social well-being, drug dosages should be as low as possible, the number of drugs taken should be minimal, and their effectiveness and adverse reactions and interactions should be evaluated on a regular basis. With a thoughtful treatment plan, medications will prevent disease, permit healthy living, and promote a long life.

REFERENCES

Adams, M. P., & Urban, C. Q. (2013). *Pharmacology: Connections to nursing practice* (2nd ed.). Pearson.

Adams, P. M., Urban, C. Q., & Sutter, R. E. (2019). *Pharmacology: Connections to nursing practice* (4th ed.). Pearson.

American Geriatrics Society. (2019). Updated Beers Criteria for potentially inappropriate medication use in older adults. *American Geriatric Society*, *67*(4), 674–694. https://doi.org/10.1111/jgs.15767

Barclay, K., Frassetto, A., Robb, J., & Mandel, E. D. (2018). Polypharmacy in the elderly: How to reduce adverse drug events. *Clinician Reviews*, *28*(2), 38–44. https://cdn.mdedge.com/files/s3fs-public/Document/January-2018/CR02802038.PDF

Bhimji, S. S., & Pavlovich-Danis, S. J. (2012). Generic drugs: Reducing the cost of care. *Nurse.com*, *3*(12), 26–31.

Chou, K. L., Liang, K., & Mackenzie, C. S. (2011). Binge drinking and Axis 1 psychiatric disorders in community dwelling middle-aged and older adults. Results from the National Epidemiologic Survey on Alcohol and Related Conditions (NESARC). *Journal of Clinical Psychiatry*, *72*(5), 640–647. https://doi.org/10.4088/JCP.10m06207gry

Clark, J. B., Queener, S. F., & Karb, V. B. (1990). *Pharmacological basis of nursing practice* (3rd ed.). Mosby.

Cuozzo, C. A. (2007). How medications are born. *Advances for Nurses*, *8*(26), 31–32.

Ditrapano, C., & Peoples, M. (2008). Generic vs. brand name drugs. *Advance for Nurses*, *9*(10), 33–34.

Eliopoulos, C. (2018). *Gerontological nursing* (9th ed.). Wolters Kluwer.

Harding, M. M. (2017). Substance use disorders. In S. L. Lewis, B. Bucher, M. M. Heitkemper, & M. M. Harding (Eds.), *Medical-surgical nursing* (10th ed., pp. 145–158). Elsevier.

Kane, R. L., Ouslander, J. G., & Abrass, I. B. (1999). *Essentials of clinical geriatrics*. McGraw-Hill.

Karch, A. M. (2013). *Focus on nursing pharmacology* (6th ed.). Wolters Kluwer Health/Lippincott Williams & Wilkins.

LeFever Kee, J., & Hayes, E. R. (2000). *Pharmacology: A nursing process approach* (3rd ed.). Saunders.

Lehne, R. A. (2010). *Pharmacology for nursing care* (7th ed.). Saunders.

Lilley, L. L., Rainforth Collins, S., & Snyder, J. S. (2017). *Pharmacology and the nursing process* (8th ed.). Elsevier.

Locklear, M. (2017). *Side effects kill thousands but our data on them is flawed.* NewScientist Health. https://www.newscientist.com/article/2143486-side-effects-kill-thousands-but-our-data-on-them-is-flawed/

Mauk, K. (2018). *Gerontological nursing: Competencies or care* (4th ed.). Jones & Bartlett.

Metzger, L. M. (2017). Helping older adults understand medications and treatment regimens. In K. Kopera-Frye (Ed.), *Health literacy among older adults* (pp. 127–142). Springer.

Meiner, S. E. (2015). *Gerontologic nursing* (4th ed.). Mosby Elsevier.

Merrick, E. L., Horgan, C. M., Hodgkin, D., Garnick, D. W., Houghton, B. S., Garmick, D.W., & Blow, F. (2008). Unhealthy drinking patterns in older adults: Prevalence and associated characteristics. *Journal of the American Geriatrics Society*, *56*(2), 214–223. https://doi.org/10.1111/j.1532-5415.2007.01539.x

Mikulic, M. (2020). *Total OTC drug retail sales in the U.S. from 1965 to 2019.* Statista. https://www.statista.com/statistics/307237/otc-sales-in-theus/

Miller, C. (2009). *Nursing for wellness in older adults* (5th ed.). Wolters Kluwer/Lippincott Williams & Wilkins.

Rogers, B., Kirley, K., & Mounsey, A. (2013). Prescribing an antibiotic? Pair it with probiotics. *Clinician Reviews*, *23*(4), 49–50.

Rolita, L., & Freedman, M. (2008). Over-the-counter medication use in older adults. *Journal of Gerontological Nursing*, *34*(4), 8–17. https://doi.org/10.3928/00989134-20080401-08

Rosenjack Burchum, J. L. (2011). Pharmacologic management. In S. E. Meiner (Ed.), *Gerontologic nursing* (4th ed., pp. 385–399). Mosby Elsevier.

Shorr, R. I. (2007). *Drugs for the geriatric patient.* Saunders Elsevier.

Tabloski, P. A. (2014). *Gerontological nursing* (3rd ed.). Pearson.

Tache, S., Sonniehsen, A., & Ashcroft, D. (2011). Prevalence of adverse drug events in ambulatory care settings: A systemic review. *Annuals of Pharmacotherapy*, *45*(7–8), 977–989. https://doi.org/10.1345/aph.1P627

U.S. National Library of Medicine. (2020). *Drugs, supplements, and herbal information.* http://www.nlm.nih.gov/medlineplus/druginformation.html

Vallerand, A. H., & Sanoski, C. A. (2020). *Davis' drug guide for nurses* (17th ed.). F. A. Davis.

Wilson, I. B., Schoen, C., Newman, P., Strollo, M., Roger, W., Chang, H., & Safran, P. (2007). Physician-patient communication about prescription medication nonadherence: A 50-state study of America's seniors. *Journal of General Internal Medicine*, *22*(1), 6–12. https://doi.org/10.1007/s11606-006-0093-0

Yeager, J. J. (2019). Drugs and aging. In S. F. Meiner & J. J. Yeager (Eds.), *Gerontologic nursing* (6th ed., pp. 257–279). Elsevier.

21

Teaching Older Adults

INTRODUCTION

Now more than any other time in history, older adults are eager to learn. Learning is defined as the acquisition of knowledge and skills as evidenced by the individual's performance. But whether the learning is for personal growth, pleasure, or improved health functioning, certain issues influence both learning and teaching strategies used. Among these are the learner's interest in learning, readiness, motivation, self-esteem, attitudes, culture, the presence of illness, loss or grief, and the teacher's knowledge of special teaching techniques (Stanley et al., 2005). Teaching adults age 65 and older (geragogy) differs from teaching adults age 18 to 65 (andragogy) or children age 18 and younger (pedagogy). Andragogy is based on assumptions separating adults from younger persons, which include motivation to learn, self-concept, readiness to learn, and experience. Older adults are unique learners in that they are independent and goal directed. They bring life experiences to their learning and expect to participate in a learning process that is relevant and practical to them. However, certain issues may impact their ability to learn such as fear and anxiety, changes in memory, losses, such as sensory and physical, literacy level, pain, chronic illness, culture and language, among others (Ondrejka, 2018). Individuals in this age group may have psychological, cognitive, and physical limitations that could influence learning (Ondrejka, 2018). Those who support geragogy purport that older adults warrant a separate educational theory and approach because they are different in many ways from middle-agers physiologically, psychologically, and socially.

Certain myths concerning learning in the later years often influence how the ability of older adults to learn is perceived. A major myth is that older adults cannot learn. Although there is evidence of some neuronal loss in the brain with age, research now indicates that the dendrites of some neurons can continue to grow, which could possibly compensate for the neuronal loss associated with aging. The concept of neuroplasticity describes the ability the brain has to change. It refers to the resiliency in the brain allowing it to compensate for age-related changes. To potentiate plasticity and cognitive reserve it is important to regularly challenge the motor, sensory, and cognitive activities as well as maintain satisfying social interaction (Touhy, 2016). Positive neuroplasticity occurs when interneuronal connections are initiated through learning a new skill or foreign language, doing puzzles, or through other stimuli. Negative neuroplasticity occurs when the connection between the neurons is disrupted, often caused when the individual is in a non-stimulating environment (Youdin, 2016). Older adulthood should not be seen as a time when growth and cognitive development has ceased but as a time in life programmed for plasticity and the development of new and unique capacities (Touhy, 2016). Barring accidents or disease, older adults who continue to use the learning modes of a lifetime certainly have the capacity to learn and are encouraged to learn and to stimulate the brain in a variety of ways.

FACTORS INFLUENCING LEARNING IN OLDER ADULTS AND USEFUL APPROACHES FOR ENHANCING LEARNING

Intelligence

Intelligence is defined as an individual's ability to learn, reason, and understand. In healthy older people intelligence usually remains intact. Much of the early research on intelligence and aging may not be accurate due to the method of evaluation or the evaluation tool that was used. The most accurate method of describing age-related intellectual changes is found in longitudinal studies measuring age-related changes in intelligence of a specific generation that compensate for health, sensory, and educational deficits (Eliopoulos, 2018). Crystallized intelligence reflects learning and acculturation acquired by individuals over a lifetime through exposure to formal and informal learning experiences and the acquisition of new skills. It is measured by vocabulary and number skills, social judgment, general information, and word associations and is thought to remain stable or improve throughout life (Hooyman & Kiyak, 2011). Crystallized intelligence is not as dependent on the neurologic state of the individual as is fluid intelligence.

Fluid or "native" intelligence reflects neurological and physiological functioning and is not significantly affected by education, experiences, and acculturation. It is measured by the ability to do certain tasks involving symbols, figures, or words and is associated with information processing, reasoning, and abstractions. It is also linked to creative abilities such as perceiving and understanding perceptual and spatial relationships. Fluid intelligence tends to peak in young adulthood and then gradually declines. Decline can be observed as decreases in attentiveness, concentration, performance, short-term memory, and speed of learning (Hayslip & Panek, 1993; Touhy, 2012a). These decrements could be due to age-related changes in psychomotor skills, perceptual and sensory changes, slower reaction time, and cognitive processing changes (Eliopoulos, 2014).

General intelligence seems to level off in the 50s and 60s and decreases in the 70s and 80s. To compensate for changes in intelligence, instructional approaches such as pacing, use of memory aids, elaboration, and a well-planned approach to instruction are helpful when teaching older adults. Older adults, barring certain health issues, retain the ability to learn and to understand situations (Touhy, 2012a).

Reaction Time

Reaction time becomes somewhat slower in older persons, and more time is required to take in, process, and respond to information. Responding to multiple or complicated stimuli is more difficult, so materials to be learned should be presented at a slower pace, one step at a time, allowing time for the person to learn each step before going on to another.

Readiness and Motivation

Readiness and motivation are necessary if learning is to occur. Both require a positive attitude toward learning, and an awareness of the need and the value of learning. If these are absent it is important to address each possibility with the individual to help them become more receptive to learning. For example, readiness and motivation to learn to administer insulin injections may not be present when a newly diagnosed diabetic is grappling with the reality of the diagnosis and the need for daily injections. Support, clarification, and time may help the individual be more receptive to learning (Knowles et al., 2011).

Literacy Levels

Literacy is the ability to understand, use, and act on information. Healthcare status as it relates to literacy may have serious consequences for an older individual who cannot correctly understand and use health-related instructions and information (Oates & Zitnay, 2012). Indeed, the World Health Organization defines health literacy as "the cognitive and social skills which determine the motivation and ability of individuals to gain access to, understand and use information in ways which promote and maintain good health" and regards it as an important health indicator (World Health Organization, 2009). Studies have found only 41% of adults aged 65+ have intermediate or proficient levels of health literacy skills, with 30% having basic and 29% having below basic levels—the highest proportion by far compared with younger age groups (Kutner et al., 2006). Seventy percent had difficulty using print materials, 80% had difficulty with forms and charts, and 68% had difficulty with interpreting numbers and doing calculations (Kutner et al., 2006).

Older adults are especially at risk because they are more likely to cope with multiple health issues and functional limitations in addition to sensory losses and cognitive deficits. Maintaining medical regimens, diets, and medication schedules can be especially challenging under such circumstances. Such situations may have serious consequences when their healthcare condition warrants appropriate medication use, completing consent forms, eating the correct diet, and understanding written and oral instructions. The ability of the individual to communicate information effectively regarding their health status and health history may also be influenced if they have low literacy skills (Kopera-Frye, 2017). Low levels of literacy have been associated with poorer health outcomes, having poorer diabetic control, and incurring higher costs for healthcare (Davidson & Han, 2012). Furthermore, patients with low health literacy are more likely to visit an emergency room, are more likely to have more hospital stays, are less likely to follow treatment plans, and ultimately have higher mortality rates (Centers for Disease Control and Prevention, 2016).

Strategies for teaching people with low literacy skills include the following (Davidson & Han, 2012; Mauk, 2010):

1. Refrain from using technical, medical language; use simple understandable terms in the person's language.
2. Offer only limited necessary information in short words and sentences.
3. Encourage the person to repeat back the information.
4. Use pictures and diagrams.
5. Use audiovisuals in the person's preferred language; include demonstrations.
6. Written information should be presented in the person's language, using easily understood words and short sentences at or below the fifth-grade reading level.

Caregiver Teaching

Caregivers are individuals who care for their family or others on a full- or part-time basis. Approximately one in four Americans are caregivers. Many are family members, such as spouse, adult children, parents, grandparents, or brothers and sisters; others are friends or paid caregivers, with the majority being women. Services they may provide are (a) personal physical care; (b) social, emotional, financial, or spiritual care; and (c) being responsible for overseeing the individual's overall healthcare (Bucher, 2017). In many of these circumstances instruction and teaching are necessary not only for the individual being cared for but also the caregiver. The caregiver should be assessed regarding such issues as readiness to learn, literacy, primary language, and physical and mental status. The caregiver's learning needs and those of the person being cared for may differ. Some may need instruction on how best to teach.

When teaching, both the person cared for and the caregiver should be present at the same time. Offering written or audiovisual materials on how to access the internet in the home-setting is invaluable. Use of resources in the community may also greatly enhance both the person's care and the caregiver needs through the use of support groups, and social, financial, or other health support entities. Keep in mind the caregiver may be experiencing stress and depression especially as the caregiver responsibilities increase.

Miscellaneous Factors

Learning may be influenced by factors such as the meaningfulness of the material being taught, the speed at which it is presented, the manner in which it is presented, the difficulty of the material, the cautiousness of the learner, the learner's health status, and the anxiety state of the individual (Burggraf & Stanley, 1989).

Memory

Memory is defined as one's ability to store specific information and then access it as necessary (Touhy, 2016). Memory is important in learning because it enables the individual to draw on readily retrievable past experiences to use in the current learning situation. Memory includes active cognitive processes that combine new experiences with events already learned and remembered. Though the process of forgetting is similar for both young and old, more time and effort is required for older persons to memorize new information. Once the information is memorized, it takes an older adult longer to access it from memory, resulting in slower mental performance. Boredom, grieving, depression, fear of failure, and certain medications may also interfere with memory. Degenerative diseases such as Alzheimer's and Parkinson's, vascular disease, or infections of the brain, trauma, alcoholism, tumors, and toxic metal exposure may likewise impair memory (Fozard et al., 1992).

The information-processing model of memory describes the cognitive processes involved in information acquisition, information storage, and information retrieval. It explains three types of memory:

1. Immediate memory, in which the individual can retain an exact copy of the information for up to 2 seconds. Attention to this sensory information transfers it to short-term or primary memory.
2. Short-term or primary memory, in which attention and retention of information are possible for minutes to days. If attended to, or rehearsed, the information passes into long-term or remote memory.
3. Long-term or remote memory, which stores and remembers information for either an extended or limited period (Ebersole & Hess, 1998).

Research using this model indicates that age-related changes may occur in both short- and long-term, but especially short-term memory. Older adults usually show some lessened ability to move new information into long-term memory and usually have greater difficulty retrieving it. Some older persons may show losses with attentional tasks, in the ability to process information and to perceive things, and in reaction time. Most of the functions of memory, however, stay sufficient and intact. Memory-training programs using imagery, cognitive stimulation, categorization, mnemonics, reasoning, speed of process training, cognitive games, crossword puzzles, Scrabble, spaced retrieval techniques, and analysis of written materials plus repetitive practice have been shown to improve mental performance (Touhy, 2012a).

Age-associated memory impairment (AAMI) refers to the loss of memory that is regarded as normal in relation to the person's age and intellectual level. What may be observed are problems with remembering names and words, recalling new information, and the slowing of processing, storing, and recalling new information. Experiencing these may result in the older adult feeling severe anxiety and concern fearing they may be developing dementia (Touhy, 2016).

Learning Styles

Each individual has a unique learning style. They include auditory (listening), visual (pictures, reading), and physical (kinesthetic-tactile, i.e., doing things). It is not unusual that an individual will use more than one learning style to acquire new information or skills. To determine learning style question the individual about how they like to learn or have done so in the past. Some people prefer hearing new information, others seeing it. Some individuals do not read a lot and obtain most of their information from television, while others prefer using their hands to learn new things (Bucher, 2017). Visual learning is enhanced through the use of PowerPoints, graphics, pictures, charts, the computer, and books. Auditory learners respond to reading out loud, lectures, discussion video, or computer instruction. Kinesthetic-tactile learners prefer a hands-on approach, being able to touch and feel things, and the use of body movement. Assessment of the learners' preferred learning style and teaching the individual appropriately can greatly enhance their ability to understand and learn what is being taught.

Attention

Attention is the ability to concentrate despite the existence of distraction. Attention, which is closely allied to memory, is divided into (a) sustained attention, (b) selective attention, and (c) divided attention (Carroll & Linton, 2007). Sustained attention, or being able to maintain mental alertness, remains relatively intact with age when fatigue factors are eliminated. Selective and divided attention, however, show some age-related changes, especially when other stimuli are presented simultaneously. There is decline in vigilance performance, which impairs the ability to remain at attention more than 45 minutes (Eliopoulos, 2018).

Teaching sessions, therefore, should be kept short, and presentations are most effective if one topic is considered at a time and irrelevant data are eliminated. Learners should be encouraged to use their past knowledge, and every effort should be made to gain the learner's full attention (Daum, 1991).

Vision

Visual acuity, or sharpness of vision, decreases with age, as does accommodation, the ability to focus on objects at different distances. Pupil size decreases, the lens becomes less transparent and more yellow, and cataracts are common. Bright colors such as orange, red, and yellow are more easily perceived than darker colors such as blues, greens, and purples. Glare sensitivity increases and peripheral vision is reduced. (For a more detailed description of sensory changes with age, see Chapter 7.)

Useful approaches in teaching older adults who have visual impairments include the following (Meiner & Lueckenotte, 2006):

1. Identify yourself initially to get the learner's attention.
2. Face the learner when speaking.
3. Use nonverbal cues and aids along with the verbal message.

4. Make sure eyeglasses, contact lenses, or low-vision aids are clean and used properly.
5. When using printed material or audiovisuals, use large, distinct block-print typeface styles such as Impact or Times New Roman, set bold, 14 to 16 points, with contrasting colors. Black print on white or yellow paper is easily readable; avoid using green, blue, purple, or dark red paper.
6. Present one concept at a time.
7. Do not stand in front of a window or mirror, both of which produce glare.
8. The learner should not be facing a glaring light source. Preferably, lights should be located behind the individual to avoid reflection, and the light source should be glare free and adequate.
9. The learner should sit near the presenter or the source of information.
10. Use audiotapes, CDs, large-print newspapers, magazines, books, and Braille books obtained from your state's Division of Blind Services.
11. Obtain a special radio from the radio reading service (available from most public radio stations) to hear selected programs.
12. Use a multiple sensory approach in teaching, but keep in mind that too much stimulation for some learners may result in stress and less learning.

Audition

Hearing is often impaired in older persons, especially the ability to hear conversations as opposed to pure tone sounds. Sometimes speech sounds cannot be easily distinguished from other sounds, making normal conversation very difficult to follow accurately. High-pitched tones gradually become less audible, whereas low-pitched tones are usually more easily heard. High-pitched consonants such as *z*, *s*, *t*, *g*, and *f* are more difficult to distinguish because they carry less acoustic power, but these are sounds that differentiate one word from another. Background noises, especially those at the same pitch as foreground noises, are more likely to interfere with hearing.

Useful approaches in teaching older adults who have hearing impairments include the following:

1. Alert the learner to your presence, face them when speaking, and have adequate light in the teaching area.
2. Speak slowly and distinctly, and lower the pitch of your voice; do not shout or overarticulate to allow the person to read your lips.
3. Be aware if the learner's reactions indicate you are not being understood. Such reactions may include cupping the ear, a puzzled facial expression, turning the "good ear" toward the presenter, and consistent "yes" responses to questions.
4. Encourage the learner to use their hearing aid, and be sure it has functioning batteries.
5. Use appropriate gestures or facial expressions to enhance speech.
6. Stimulate multiple senses through the use of visual material, auditory messages, and touch and smell as appropriate.
7. Do not cloud the spoken message by the use of background music or noise.
8. When a question is asked by a member of the class, repeat it so all attending can understand.
9. Use a microphone as necessary. Make sure it functions well. Practice using it beforehand, because it is very frustrating for the listener to try to understand a speaker who cannot be heard clearly.
10. When using television, videotapes, or a computer, select the base tone setting, if possible, to enhance auditory perception and regulate the audio loud enough to be heard. Amplifiers or headsets are available for telephones, radio, and television. Closed-captioned television might also enhance the learning experience.

Taste, Smell, and Touch

Taste, smell, and fine discriminations of touch, temperature, and pressure change with aging, gradually becoming less efficient; however, there is substantial variation between individuals in the amount of loss experienced. These particular sensory losses usually do not have a significant effect on learning.

Nevertheless, useful approaches in teaching older adults who do have taste, smell, or touch losses include the following:

1. In dietary teaching, especially regarding restricted diets such as low salt or sugar, ensure that the individual is aware that decreased salt receptors may actually cause food to taste bland, and individuals may then use more salt or sugar than usual.
2. Encourage the use of spices, vinegar, herbs, and lemon to enhance salt-restricted diets.
3. Safety may be compromised when the individual is unaware of the presence of smoke, gas, or spoiled food.
4. When teaching manipulative skills such as checking blood sugar levels or giving oneself an injection, encourage the use of special devices easily handled to enhance finger sensations and dexterity.

Speech and Language

Even though age-related changes in language expression and understanding are minimal in most older adults, the following approaches may be helpful (Daum, 1991):

1. Use the primary language of the learner, if possible, and select words easily understood, avoiding medical jargon.
2. Use the active voice or present tense and personal pronouns.
3. Avoid using words of three or more syllables, and restrict sentences to 10 words or less, depending on the audience.
4. Encourage verbal responses to assess the learner's vocabulary and knowledge of the language.
5. Organize the content of your message and proceed from simple to complex.
6. Frequently summarize what has been presented.
7. Provide opportunities for participants to ask questions to further clarify and focus the teaching.
8. Pause from time to time to allow the learners to focus on and understand the information presented.
9. Allow learners to help set the pace of the learning session.
10. Use well-organized handout materials matching the reading level of the learners, with special attention to those with low literacy levels.
11. Use positive reinforcement and encouragement.
12. Reinforce the spoken work with written materials.
13. Do not speak down to the person using baby talk and words such as "honey" or "dearie" (Ondrejka, 2018).

Fear of failure, cautiousness, or anxiety about class participation or test taking can impair learning. Create a nonthreatening approach; avoid tests, and use reinforcement and reassurance generously.

It can be challenging to motivate the learner to become enthusiastic about learning. Teaching content therefore should be meaningful and relevant, not considered nonsense or a repetition of already known information. For health and safety teaching, it is important for the information to be taught in such a way as to integrate it into one's lifestyle over time.

Depression, quite prevalent among older adults, slows thinking and concentration, causing inattention and impaired learning ability; it also interferes with the actual performance of tasks in learning situations. Individual needs of depressed learners should be assessed, reinforcement used frequently, and learners encouraged to engage in learning skills helpful in improving their lives.

Cultural Issues

Cultural diversity is an increasing reality in the United States. Individuals age 65 and older are actually more diverse than any of the other age groups. Issues related to ethnicity, race, gender, religion, socioeconomics, lifestyle, and health status all must be considered when developing effective learning and teaching approaches. Awareness of health disparities is important, as well as an understanding of uniqueness of a particular culture and the individual, especially as related to health issues. Some individuals may not believe in healthcare given by a formal healthcare system, others may be fearful of sharing healthcare information, still others may prefer using cultural approaches to healing. Building a positive relationship through a sense of trust and respect between the caregiver and those cared for is of utmost importance.

Literacy and education levels should be initially assessed, as well as the individual's willingness to learn. Understanding the family composition, family members' ability to understand English, and their willingness and ability to assist the older person are important. Teaching materials should be developed using the individual's primary language.

Learning Environment

The physical learning environment will either facilitate or inhibit learning. Comfortable chairs arranged informally in a pleasantly decorated, well-illuminated room (without glare), with comfortable temperature control and adequate acoustics all stimulate the learning process. Extraneous visual or auditory stimuli such as kitchen noises or people moving about and talking detract from the learning process and should be eliminated as much as possible.

Other physical factors to consider are accessibility to the classroom, elimination of environmental barriers such as stairs, and easily accessible restrooms. Availability of refreshments such as water, coffee, soft drinks, and light foods also contributes to a more social environment.

Teaching Methods

Teaching methods vary depending on the needs of the individual. Family members or significant others are often present in instructional settings, especially if the class addresses an existing health problem or if the older adult has difficulty understanding the content.

One-to-One Instruction

A commonly used method of instruction involves interaction between the instructor and the learner. Sit a comfortable distance from the person, maintaining eye contact, speaking clearly, and showing sincere personal interest in and a positive attitude toward the learner. Make sure adaptive devices such as hearing aids and eyeglasses are available and in working order. Schedule the teaching when the person is not overly tired and is willing to learn. Inform them of the purpose of the session. Deliver small amounts of information at a time and request feedback. Time for questions during and after the instruction is always helpful. Repeat and reinforce the teaching as necessary. Be aware of the learners' facial responses. Adult educators obviously should possess a mastery of the subject being presented and the ability to impart it to others in a comprehensible manner.

This mode of instruction is especially helpful when teaching individuals about health matters or skills such as giving self-injections, taking blood pressure, or checking blood sugar levels. Because Medicare usually offers a limited number of visits to patients, instruction often must be given when the person is not optimally ready to learn because of the presence of pain, weakness, depression, or disability. Such situations demand creative approaches and the use of support materials such as written pamphlets, audio- or videotapes, CDs/DVDs, and computer learning. Encouraging a family member to participate in the session can also increase compliance with the treatment regimen.

Group Instruction

Group instruction usually provides a highly supportive and secure situation for most older learners. Such settings offer socialization and a chance to share thoughts and experiences with others (Mauk, 2010). However, some older adults may feel uneasy or threatened, especially if they are asked questions or expected to give a report or demonstration to the group. The latter is not recommended.

When organizing a group instruction in a healthcare facility, coordination with the nurse or education coordinator is important. Advertise the class well ahead of the date and in conspicuous places. The subject chosen to discuss must be relevant and of interest to the attendees and at a time when they are free to attend. A pleasant, friendly, comfortable, quiet environment with adequate lighting and access to a restroom is advised. Gathering a group of elders with similar needs and cognitive abilities is important, as is awareness of any sensory deficits. Using a microphone and lowering the tone of voice enhances hearing. Classes should be carefully planned focusing on one topic, encouraging participation, and offering positive reinforcement. Avoid testing or obligatory participation. Use a variety of teaching approaches, including PowerPoint, overheads, videos, demonstrations, lecture, and group work. Sharing stories and comments can be a powerful mode of learning. Save time for questions and be sure to repeat them to the group before answering them (Mauk, 2010).

Elicit an evaluation of the class either verbally or in writing if possible from the attendees and the staff. Determine their interest regarding topics for future classes. Serving water, coffee, tea, soft drinks, and snacks enhances the class and encourages attendance (Mauk, 2010).

Electronic Learning

Electronic learning, or E-learning, is an approach to teaching and learning using a variety of technologies. These include CDs, DVDs, video/webcam conferencing, online social networks, and computers, including tablets and laptops. Computer-based learning offers the individual the opportunity to learn at their own pace and to review the material as needed. Blended learning involves the use of E-learning and face-to-face instruction. Research supports computer-based education as a helpful strategy to improve both knowledge and skills in making effective healthcare decisions and producing sound health outcomes (Lewis, 1999). Older adults are becoming more computer and social network savvy and represent a rapidly increasing group of users who are very comfortable with this mode of communication and instruction (Perkins & LaMartin, 2012; Touhy, 2012b).

Generations on Line (https://generationsonline.com/) produces software and training manuals for older adults who are not knowledgeable about computers. SeniorNet (https://seniornet.org/) provides technology and computer training to thousands. Cyber Seniors, Administration on Aging, AARP, government programs, technology schools, colleges, and universities also offer computer training for older adults. The internet and other rapidly evolving technologies such as smartphones and iPads are likely to play a more pivotal role in the lives of older adults (Perkins & LaMartin, 2012). Recognizing this surge in internet usage by the older population, the National Institute on Aging (2009) issued guidelines for "senior-friendly" websites, advocating for options such as having text read aloud and multimedia formats transcribed into text form, along with basic design elements such as high-contrast color combinations, minimal scrolling, and the ability for users to enlarge text. Many possess computers; others use those of friends, the library, or senior centers. Nursing homes are becoming more involved by offering computer internet access. Administrators report that benefits to residents include mental stimulation, learning opportunities, increased family contact, and enjoyment in playing computer games. They also cite increased independence and self-worth as well as providing a sense of accomplishment for the user (Tak et al., 2007).

Many online communities are available to individuals who have a common interest or who wish to obtain certain information through interacting with one another in the community. Health information, especially concerning medications, is commonly accessed. The issues are usually considered in a question-and-answer format, and the information shared is based on the experience of the individual. A major consideration is the validity of the information shared, quack advertisements, and confidentiality issues. Those communities showing the "HON" (Health on the Net Foundation) code emblem have been checked for accuracy and legitimate facilitators. However, some caution should be used in accessing and participating in these communities, and only reputable sites are to be recommended (Moore, 2005).

Psychoeducational groups for elders can be an alternative intervention in learning behaviors and skills by combining education and psychological support. This approach is especially beneficial for individuals who are lonely, grieving, or participating in such programs as smoking cessation, exercise, weight reduction, and assertiveness training (Kraenzle et al., 2005).

SUMMARY OF TEACHING APPROACHES

Before teaching, it is wise to determine learning needs by using surveys and focus groups. Many different teaching approaches contribute to successfully teaching older adults. These include the following:

1. Older adults tend to be problem-oriented and respond better to concrete examples rather than theoretical abstractions.
2. Learning will be facilitated and more relevant if it relates to information associated with familiar topics or past experiences.
3. Overall learning goals ideally should be developed jointly by the instructor and learner, using short-term goals as achievable points along the way.
4. Genuine involvement in the learning process by both the instructor and the learner is crucial for maintaining ongoing interest and continued effort.
5. Keep in mind cultural sensitivity regarding individual customs, beliefs, and values in the course presentation and in the use of pictures and language.
6. Effective and organized use of teaching time promotes positive attitudes toward learning and makes situations more relevant.
7. Provide continual positive feedback.
8. New concepts should be presented at a comfortable pace to encourage understanding, assimilation, and application.
9. A variety of teaching aids such as videos, models, illustrations, PowerPoint, discussion, computer learning, handouts, lectures, demonstrations, reports, creative experiences, and field trips all help increase comprehension of the material.
10. Written materials should usually be at or below the fifth-grade reading level for the general population (Mauk, 2010).
11. Active participation in the class allows for clarification of newly learned ideas and concepts.
12. Because older adults learn more slowly, materials presented orally should also be available in writing and in logical, organized sequences. Small units of information are preferable because they are easier to assimilate. Offering several presentations of the same concept using

different techniques facilitates learning, but be careful not to oversimplify so as not to insult the learner.

13. Class length and meeting times should be determined by the type of group. Half-hour to 1-hour sessions are most preferable and tolerable. Be aware that some older adults rest in the afternoon, and many do not drive in the evening.
14. Sometimes learners tend to monopolize the class. Encourage a talkative learner to assist another student and clarify that it is important for each student to be heard. If the situation becomes intolerable, the instructor may need to speak to the learner in a sensitive and meaningful manner.
15. Use mnemonics training or other strategies for remembering, such as writing down information, making lists, and so on.
16. Begin with simple, easily understood information and proceed to the more complex.
17. Be aware of the literacy level of the learner.
18. Use existing abilities of the person as much as possible.
19. Focus learner attention on one single aspect of the information at a time.
20. Compensate for sensory or physical losses.
21. Be aware of the learner's readiness to learn. Be attentive to signs of boredom, tiredness, anxiety, or depression.
22. Encourage life review and reminiscence in classes concerning life and personal growth.
23. Treat older learners with respect and as mature adults capable of learning. Do not talk down to them or treat them as children.
24. Learning is more likely to take place when the older adult is not preoccupied with pressing personal issues such as illness, pain, grief, or other concerns. Attention to these issues before active involvement in the learning process is important for optimal learning.

Many learning opportunities exist for older adults. Formal classes and programs at community colleges and universities attract elders, with some states offering tuition waivers. Other learning opportunities are available at senior centers, healthcare institutions, libraries, recreation departments, and vocational schools. Distance learning is available, and the internet offers many possibilities. Road Scholar offers 5,500 educational tours in all the States and in 150 countries. Costs vary from about $400 to thousands depending on the length and location of the learning program. They offer housing, food, and the opportunity to travel and learn along with others from various parts of the United States and the world. Additional information on these and other learning topics is available on the internet.

SUMMARY

Learning in later life remains a potent force in maintaining mental alertness and physical health and for keeping in touch with life and the world. Lifelong learning is rapidly becoming routine. Increasing numbers of people have had more than one career, each requiring the acquisition of new knowledge. Individuals of all ages attend self-enhancement classes, conferences, or workshops and use computer learning. Overall, there are many more people who enter retirement with a higher level of education and who value continuing education.

REFERENCES

Bucher, L. (2017). Patient and caregiver teaching. In S. L. Lewis, L. Bucher, M. M. Heitkemper, M. M. Harding, J. Kwong, & D. Roberts (Eds.), *Medical-surgical nursing* (10th ed., pp. 46–59). Elsevier.

Burggraf, V., & Stanley, M. (1989). *Nursing the elderly: A care plan approach.* Lippincott.

Carroll, D. W., & Linton, A. D. (2007). Age-related psychological changes. In A. D. Linton & H. W. Lach (Eds.), *Concepts and practice* (3rd ed., pp. 631–684). Saunders Elsevier.

Centers for Disease Control and Prevention. (2016). *Infographic: Health literacy.* https://www.cdc.gov/cpr/infographics/healthliteracy.htm

Daum, S. G. (1991). Increasing communication effectiveness in rehabilitation programs. *Topics in Geriatric Rehabilitation*, *6*(3), 15–26. http://dx.doi.org/10.1097/00013614-199103000-00005

Davidson, H. E., & Han, L. F. (2012). Next steps for patient-centered medication information. *Aging Well*, *5*(3), 6–7. https://www.todaysgeriatricmedicine.com/archive/050712p6.shtml

Ebersole P., & Hess, P. (1998). *Toward healthy aging* (5th ed.). Mosby.

Eliopoulos, C. (2014). *Gerontological nursing* (8th ed.). Wolters Kluwer Health/Lippincott Williams & Wilkins

Eliopoulos, C. (2018). *Gerontological nursing* (9th ed.). Wolters Kluwer.

Fozard, J. L., Mullin, P. A., Giambra, L. M., Metter, E. J., & Costa, P. T. (1992). Normal and pathological age differences in memory. In J. C. Brocklehurst, R. C. Tallis, & H. M. Fillit (Eds.), *Textbook of geriatric medicine and gerontology* (4th ed., pp. 94–109). Churchill Livingstone.

Hayslip, B., Jr., & Panek, P. (1993). *Adult development and aging* (2nd ed.). Harper Collins.

Hooyman, N., & Kiyak, H. (2011). *Social gerontology: A multidisciplinary perspective.* Allyn & Bacon.

Knowles, M. S., Holton, E. F., & Swanson, R. A. (2011). *The adult learner: The definitive classic in adult education and human resource development* (7th ed.). Mosby.

Kopera-Frye, K. (2017). Health literacy 101. In K. Kopera-Frye (Ed.). *Health literacy among older adults* (pp. 1–15). Springer Publishing Company.

Kraenzle, J., Schneider, J. K., & Cook, J. H. (2005). Planning psychoeducational groups for older adults. *Journal of Gerontological Nursing, 37*(8), 33–38. https://doi.org/10.3928/0098-9134-20050801-12

Kutner, M., Greenberg, E., Jin, Y., & Paulsen, C. (2006). *The Health Literacy of America's Adults: Results from the 2003 National Assessment of Adult Literacy (NCES 2006-483).* U.S. Department of Education, National Center for Education Statistics. https://nces.ed.gov/pubs2006/2006483.pdf

Lewis, D. (1999). Computer based approaches to patient education: A review of the literature. *Journal of the American Informatics Association*, *6*(4), 277–282. https://doi.org/10.1136/jamia.1999.0060272

Mauk, K. L. (2010). Teaching older adults. In K. L. Mauk (Ed.), *Gerontological nursing: Competencies for care* (2nd ed., pp. 284–299). Jones & Bartlett.

Meiner, S. E., & Lueckenotte, A. G. (Eds.). (2006). *Gerontological nursing* (3rd ed.). Mosby Elsevier.

Moore, G. A. (2005). On-line communities: Helping "senior surfers" find health information on the web. *Journal of Gerontological Nursing, 31*(11), 42–48. https://doi.org/10.3928/0098-9134-20051101-10

National Institute on Aging. (2009). *Making your website senior friendly: Tips from the National Institute on Aging and the National Library of Medicine.* https://nnlm.gov/mar/guides/making-your-website-senior-friendly

Oates, D. J., & Zitnay, R. M. (2012). Healthy literacy's critical importance. *Aging Well, 5*(3), 34. https://www.todaysgeriatricmedicine.com/archive/050712p34.shtml

Ondrejka, D. (2018). Teaching and communication with older adults and their families. In K. L. Mauk (Ed.), *Gerontological nursing: Competencies for care* (4th ed., pp. 174–208). Jones & Bartlet.

Perkins, E. A., & LaMartin, K. M. (2012). The internet as social support for older carers of adults with intellectual disabilities. *Journal of Policy and Practice in Intellectual Disabilities, 9*, 53–62. https://doi.org/10.1111/j.1741-1130.2012.00330.x

Stanley, M., Blair, K. A., & Beare, P. G. (2005). *Gerontological nursing: Promoting successful aging with older adults* (3rd ed.). F. A. Davis.

Tak, S. H., Beck, C., & McMahon, E. (2007). Computer and internet access for long-term residents. *Journal of Gerontological Nursing, 33*(5), 32–40. https://doi.org/10.3928/00989134-20070501-06

Touhy, T. A. (2012a). Cognitive impairment. In T. A. Touhy & K. Jett (Eds.), *Ebersole & Hess' toward healthy aging: Human needs and nursing responses* (8th ed., pp. 365–390). Mosby Elsevier.

Touhy, T. A. (2012b). Communicating with older adults. In T. A. Touhy & K. Jett (Eds.), *Ebersole & Hess' toward healthy aging: Human needs and nursing responses* (8th ed., pp. 81–106). Mosby Elsevier.

Touhy, T. A. (2016). Cognition and learning. In T. A. Touhy & K. Jett (Eds.), *Toward healthy aging: Human needs and nursing response* (9th ed., pp. 54–64). Mosby Elsevier.

World Health Organization. (2009). *7th Global Conference on Health Promotion: Track Themes: Track 2: Health literacy and health behaviors.* http://www.who.int/healthpromotion/conferences/7gchp/track2/en/index.html

Youdin, R. (2016). *Psychology of aging.* Springer Publishing Company.

22

Gerontechnology

INTRODUCTION

Technology is generally defined as the practical application of knowledge, usually to make tasks or processes easier for the user and, thus, ultimately improve quality of life. Hence, *gerontechnology* can be simply described as the application of technology in the field of aging. The idea is not new; the first international conference on gerontechnology was convened in the Netherlands in 1991. The term *gerontechnology* was subsequently defined as "an interdisciplinary field of research and application involving gerontology, the scientific study of aging, and technology, the development and distribution of technologically based products, environments, and services" (Fozard et al., 2000, p. 332). Indeed, in addition to gerontology, gerontechnology utilizes biology, sociology, psychology, medicine, engineering, computer sciences, graphic design, robotics, to name just a few! The ascendance of gerontechnology is not surprising. As Charness noted, "we are in the midst of two striking trends; widespread population aging and rapid diffusion of technology" (2017, p. xxvii). To illustrate, just in terms of computer ownership and internet usage alone, in 2015 among all households, 78% had a desktop or laptop computer, 75% had a smartphone or tablet, and 77% had a broadband internet subscription (Ryan & Lewis, 2017). Though households with older adults aged 65+ lagged behind, nevertheless, 65% had a desktop or laptop, 47% had a smartphone or tablet, and 62% had broadband internet subscription (Ryan & Lewis, 2017). The baby boom generation (born between 1946 and 1964) are now reaching or have already reached 65 years. They are a generation who grew up with traditional landline telephones, black and white televisions, and no home computers. When one considers the technological advances their generation alone has witnessed, the change is nothing short of remarkable. Given the rapidity of technological advances and the rapid adoption of technology in recent years, gerontechnology is a burgeoning field.

This chapter will briefly discuss how gerontechnology is aiding research in aging, but will mainly focus on current and promising future technologies that are already making significant and practical improvements in the daily lives of older adults and their caregivers.

GERONTECHNOLOGY'S ROLE IN RESEARCH

Gerontechnology has already demonstrated the potential to transform how older adults participate in research and increase the efficiency and accuracy of how data can be collected. Research involving human subjects is dependent on the ability to recruit participants that meet eligibility requirements and obtaining appropriate sample sizes. Recruitment of research subjects is no longer restricted by advertisements in local newspapers and printed flyers, but are now greatly aided by the extensive marketing reach of social media and internet-based advertising. Longitudinal studies are critical to study the aging process; thus, the ability to maintain contact over time, and reduce the likelihood of attrition, is greatly enhanced by the use of the internet-enhanced communication not solely reliant on mail and phone methods of contact. From participating in online surveys to research studies monitoring health indicators (e.g., activity, blood pressure, fall detection), older adults may find it easier and less invasive of their time to contribute to research. Rather than attending in-person focus groups, or visits to research labs or clinics, online delivery and remote monitoring can be utilized; participation is no longer bounded by geographical location or availability of transport.

One particular novel research design embraced for gerontechnology is in the use of virtual environments (Park et al., 2017). More formally referred to as Immersive Virtual Environment Technology (IVET), it is the combination of computer software and hardware that creates artificial but very realistic environments that users perceive themselves within and can interact within that environment in real time (Persky & McBridge, 2009). As the software and hardware of such technology has advanced so has the realism of these environments. Several studies focused on older adults, including those with dementia, have implemented rehabilitation, therapy, and mobility interventions that have successfully utilized virtual environments (Bisson et al., 2007; Blackman et al., 2007; Weiss et al., 2004).

For gerontechnology to be effective and widely utilized, it is important to include older adults in the design process to carefully consider user characteristics, needs, and preferences, when products are developed (Merkel & Kucharski, 2018). Some researchers caution that the diversity of potential user groups is still not adequately reflected in research and development studies (Grates et al., 2018). As Bouma et al. highlighted, robust design is critical, and a wide range of individual differences need to be accommodated for gerontechnology to be truly adaptable (2009). This approach of a more inclusive design will result in products that can be adapted for a variety of users, rather than developing a product with a specific user in mind. For example, some cellphones are marketed to the older population for their simplicity, whereas a cellphone designed with accessibility features (e.g., that can change the contrast of the screen, sensitivity of the touchscreen, size of the text, speech to text technology, vibration alerts) can be easily modified for individual users and thus can adapt to aging-related changes in vision, hearing, and touch.

CURRENT APPLICATIONS OF GERONTECHNOLOGY

Gerontechnology can have numerous applications. To focus such broad applicability, Bouma et al. identified four goal areas in which gerontechnology can enhance the life of older individuals (2009). These are (a) *enhancement and satisfaction* to enrich daily life, (b) *prevention and engagement* (i.e., the engagement of older adults with active control over preventive measures avoiding risks, increasing exercise), (c) *compensation and assistance* for age-related restrictions with the purpose of bringing older adults' functional abilities in line with the general population, and (d) *care support and care organization* (i.e., how technology can be utilized to support formal or informal caregivers with maximum efficacy but minimum invasion of privacy). Bouma et al. further described five important life domains that are particularly relevant to older adults where these goals should be focused. These are (a) health and self-esteem, (b) housing and daily living, (c) mobility and transport, (d) communication and governance, and (e) work and leisure. Gerontechnology is fast becoming ubiquitous and indispensable in everyday life. Examples in each of these domains are discussed in the next section.

Gerontechnology's Role in Transforming Daily Life

Health and Self-Esteem

One of the most successful areas that gerontechnology has made major strides is in health and self-esteem. From apps that help people log and analyze their diets to health monitors that can track vital health data (e.g., blood pressure, temperature, pulse, blood oxygen levels, ECG data, and glucose monitors), such technology has become commonplace and extremely helpful for those living with chronic health conditions. Medication reminders are particularly helpful for older adults with multiple medications needed throughout the day, and "smart" pill dispensers can now detect when a dose has been missed and alert a caregiver. Activity trackers can monitor daily steps, movement, and the more sophisticated models are equipped with fall detection. Ambient sensors that can detect motion, pressure, object contact, and sounds are also helpful for those who live alone and wish to remain independent (Uddin et al., 2018). The increasing use of telehealth or telemedicine whereby people are able to have remote appointments with their healthcare provider has grown exponentially during the COVID-19 pandemic, and the use of remote monitoring has also expanded. Indeed, telerehabilitation has also allowed physical therapists to adapt their services for home-based sessions. Mental health and wellness providers have also increased their use of teletherapy in which virtual mental health support is provided. Both individual and group-based sessions can be easily accommodated. Such technology can help any age, of course, but older adults do derive great benefit from closer monitoring and easier sharing of clinical information with their healthcare providers, leading to better management of their health issues, and being more proactive rather than reactive in the approach to their care.

Housing and Daily Living

Though there can be concerns about privacy and autonomy, family caregivers and professional care providers are able to utilize gerontechnology to help ensure the health and safety of their aging loved ones or clients, to continue to live and age-in-place in their own homes. Indeed, so-called "smart homes" with voice-activated hubs such as Amazon's Alexa and Google's Home enable such things as remote control of thermostats, appliances, lights, and lamps that can operate on a schedule, locking and unlocking of doors, as well as important timers, alarms, or reminders. Vacuuming and mopping is a tedious chore to many and for older adults with mobility or compromised strength/grip it can be a very difficult task and a trip hazard too. Fortunately, robotic vacuum cleaners and mops are now widely available that can efficiently clean a home whenever it is convenient.

Mobility and Transport

Navigation tools can be indispensable to older adults, particularly if they are visiting a new location or are in unfamiliar territory while driving, or even walking. Electric-assisted tricycles can extend the manageable range and can assist with overcoming natural barriers (e.g., a steep hill) that may have curtailed the use of a conventional bicycle. Perhaps the most exciting prospect is the advent of autonomous self-driving vehicles, which is fast becoming a reality. Future generations who are no longer able (or simply prefer not) to drive will not have to rely on others to meet their transportation needs. Prior to the COVID-19 pandemic there was always the ability to have home deliveries for food, groceries, general household supplies in many areas. However, the pandemic has greatly increased the use of home deliveries for all kinds of products. It will be interesting to see whether people's shopping behavior will be drastically and permanently changed upon the return to postpandemic life.

Communication and Governance

It is true that we are more interconnected than ever thanks to the internet. Technology has truly transformed the way we communicate and keep connected with one another. Voice phone calls and regular mail have been augmented and, in many instances, supplanted with video-calling with the likes of FaceTime, Zoom, instant messaging, and texting. Social media platforms such as Facebook, Instagram, and TikTok allow for the sharing of messages, pictures, videos of everything from mundane daily life to life's milestone moments. Deep and meaningful relationships/friendships can be forged through this digital world. These "virtual" relationships can be as rewarding as in-person relationships and can develop into real-word relationships. As older adults' quality of life and health can be impacted by social isolation and loneliness, such use of the internet may help by fostering social support, keeping in contact, and development of social networks (Chen & Schulz, 2016; Perkins & LaMartin, 2012). Indeed, a large study of 11,000 older adults (aged 65+) in Europe found that daily or sometimes internet users were less lonely and reported less social isolation than those who never used the internet (Lelkes, 2013). Novel and intriguing forms of digital companionship are being developed whereby interactions with "socially assistive robots" and even robotic pets are providing alternatives to human relationships (Bedaf et al., 2015; Beuscher et al., 2017). Governance, in this context, refers to the ability of the government to interact with their aging citizens (Bouma et al., 2009). Channels of communication have been vastly expanded, which have benefited everyday life and in times of natural disasters or emergencies for the dissemination of pertinent information.

Work and Leisure

Technology can support a vast range of work and leisure activities. The increase of "home office" and remote work has undoubtedly enabled some older adults to continue to utilize paid employment work or have a more staggered retirement. For others, new part-time employment opportunities may have arisen. Furthermore, volunteer work opportunities can be expanded through the use of technology. Financial responsibilities such as banking, check deposits, bill payments can be automated or done online or through apps. In terms of leisure, although cable television greatly expanded choice, streaming services have further increased access to huge media libraries of film, TV, music for both leisure and educational content. Older adults can and do enjoy computer games, not just for health reasons (e.g., stroke rehabilitation/cognitive training) but for fun and recreation (Boot et al., 2020). Even those that find travel difficult can participate in virtual tours of places of interest and museums around the world.

CAVEATS OF GERONTECHNOLOGY AND THE FAMILY TECHNOLOGIST

It is important that any technology, including gerontechnologies, designed to enhance the quality of life of older adults, truly serve that purpose rather than create new unintended consequences. Privacy and autonomy concerns, a sense that "big brother" is watching and knows far too much, can make older adults fearful and distrustful of technology that may be perceived as too intrusive (such as monitors, trackers). Responsible and ethical use by corporations, providers, and family members can mitigate many of these concerns. Nevertheless, some older adults may find their lack of tech-savvyness subject to exploitation by unscrupulous actors that can take advantage of their discomfort in using technology. The cost of such innovation can result in prices of products and services being rather prohibitive. This is especially the case when new technologies are first introduced and may be beyond the financial reach to the household expenses of older adults who predominantly rely on fixed incomes. Furthermore, helpful gadgets that are purchased for older adults as well-meaning gifts often remain unopened because it was not the right technology and was offered with no support for how to use it (Huber et al., 2017). One must consider that sometimes the learning curve for adopting new technology may be far too onerous and time consuming for some older adults and their families (Huber et al., 2017).

With many families, there is usually a younger very tech-savvy relative who becomes the go-to person when an older family member is considering purchases of computers, tablets, and smartphones. These same individuals are usually their older relative's ongoing technical support for help with new software or troubleshooting issues with the internet or their computer hardware. Huber et al. (2017) describe the emergence of such individuals as being an important and distinctly new caregiving role that they have named "the family technologist." They assert that these "family technologists" can provide essential help to older adults and can help to reduce the caregiving burden felt by their family caregivers. An example given is sensor monitoring, which brings the primary caregiver much relief and freedom from worry (Huber at el., 2017). According to Huber et al., the family technologist is not usually the primary caregiver, rather they manage and maintain digital solutions bridging the gap between available technology and what can assist their family caregiving team for their older relative.

SUMMARY

Gerontechnology is a rapidly expanding field that endeavors to apply technology to improve overall quality of life for older adults. The opportunities are seemingly boundless but continuing research, thoughtful application, and ongoing support are needed for older adults to utilize and benefit from its adoption. Older adults' needs and preferences are crucial to the design process for present and future innovations. Tailored support for older adults and their caregivers may be needed if they are to fully attain the benefits that can be derived. We must also be respectful of people's right to refuse new technologies, along with being respectful of the right to privacy and autonomy. Ultimately, accessibility, user-friendliness, and affordability may be the main drivers to how quickly technology is adopted and used in all our lives. We can only ponder as to how technology will advance to support quality of life in older adults and, indeed, for all ages in the future.

REFERENCES

Bedaf, S., Gelderblom, G. J., & De Witte, L. (2015). Overview and categorization of robots supporting independent living of elderly people: What activities do they support and how far have they developed. *Assistive Technology*, *27*(2), 88–100. https://doi.org/10.1080/10400435.2014.978916

Beuscher, L. M., Fan, J., Sarkar, N., Dietrich, M. S., Newhouse, P. A., Miller, K. F., & Mion, L. C. (2017). Socially assistive robots: Measuring older adults' perceptions. *Journal of Gerontological Nursing*, *43*(12), 35–43. https://doi.org/10.3928/00989134-20170707-04

Bisson, E., Contant, B., Sveistrup, H., & Lajoie, Y. (2007). Functional balance and dual-task reaction times in older adults are improved by virtual reality and biofeedback training. *CyberPsychology & Behavior*, *10*(1), 16–23. https://doi.org/10.1089/cpb.2006.9997

Blackman, T., Van Schaik, P., & Martyr, A. (2007). Outdoor environments for people with dementia: An exploratory study using virtual reality. *Ageing & Society*, *27*(6), 811–824. https://doi.org/10.1017/S0144686X07006253

Boot, W. R., Andringa, R., Harrell, E. R., Dieciuc, M. A., & Roque, N A. (2020). Older adults and video gaming for leisure: Lessons from the Center for Research and Education on Aging and Technology Enhancement (CREATE). *Gerontechnology*, *19*(2), 138–146. https://doi.org/10.4017/gt.2020.19.2.006.00

Bouma, H., Fozard, J. L., & van Bronswijk, J. E. M. H. (2009). Gerontechnology as a field of endeavour. *Gerontechnology*, *8*(2), 68–75. https://journal.gerontechnology.org/archives/1007-1008-1-PB.pdf

Charness, N. C. (2017). Foreword. In S. Kwon (Ed.), *Gerontechnology: Research, practice, and principles of the field of technology and aging* (pp. xxvii-xxix). Springer Publishing Company.

Chen, Y. R., & Schulz, P. J. (2016). The effect of information communication technology interventions on reducing social isolation in the elderly: A systematic review. *Journal of Medical Internet Research, 18*(1), e18. https://doi.org/10.2196/jmir.4596

Fozard, J. L., Rietsema, J., Bouma, H., & Graafmans, J. A. M. (2000). Gerontechnology: Creating enabling environments for the challenges and opportunities of aging. *Educational Gerontology, 26*(4), 331–344. https://doi.org/10.1080/036012700407820

Grates, M. G., Heming, A. C., Vukoman, M., Schabsky P., & Sorgalla J. (2018). New perspectives on user participation in technology design processes: An interdisciplinary approach. *The Gerontologist, 59*(1), 45–57. https://doi.org/10.1093/geront/gny112

Huber, L. L., Watson, C., Roberto, K. A., & Walker, B. A. (2017). Aging and intra- and intergenerational contexts: The family technologist. In S. Kwon (Ed.), *Gerontechnology: Research, practice, and principles of the field of technology and aging* (pp. 57–89). Springer Publishing Company.

Lelkes, O. (2013). Happier and less isolated: Internet use in old age. *Journal of Poverty and Social Justice, 21*(1), 33–46. https://doi.org/10.1332/175982713X664047

Merkel, S., & Kucharski, A. (2018). Participatory design in gerontechnology: A systematic literature review. *The Gerontologist, 59*(1), e16-e25. https://doi.org/10.1093/geront/gny034

Park, A. J., Hwang, E., & Gutman, G. M. (2017). Methods in gerontechnology with a focus on virtual environments as a research tool in aging mobility studies. In S. Kwon (Ed.), *Gerontechnology: Research, practice, and principles in the field of technology and aging* (pp. 117–133). Springer Publishing Company.

Perkins, E. A., & LaMartin, K. M. (2012). The internet as social support for older carers of adults with intellectual disabilities. *Journal of Policy and Practice in Intellectual Disabilities, 9*(1), 53–62. https//doi.org/10.1111/j.1741-1130.2012.00330.x

Persky, S., & McBride, C. M. (2009). Immersive virtual environment technology: A promising tool for future social and behavioral genomics research and practice. *Health Communication, 24*(8), 677–682. https://doi.org/10.1080/10410230903263982

Ryan, C., & Lewis, J. M. (2017). *Computer and internet use in the United States: 2015 (American Community Survey Reports).* https://www.census.gov/content/dam/Census/library/publications/2017/acs/acs-37.pdf

Uddin, M. Z., Khaksar, W., & Torresen, J. (2018). Ambient sensors for elderly care and independent living: A survey. *Sensors, 18*(7), 2027. https://doi.org/10.3390/s18072027

Weiss, P. L., Rand, D., Katz, N., & Kizony, R. (2004). Video capture virtual reality as a flexible and effective rehabilitation tool. *Journal of Neuroengineering and Rehabilitation, 1*(1), 12. https://doi.org/10.1186/1743-0003-1-12

23

Caregiving

INTRODUCTION

A likely ramification of the increasing longevity in the general population is the greater likelihood of becoming a caregiver for an older adult. Former First Lady Rosalynn Carter once said, "There are four kinds of people in the world: Those who have been caregivers, those who currently are caregivers, those who will be caregivers, and those who will need caregivers." Caregiving is becoming a role that many of us will face. Indeed, the scope of caregiving in the United States is already considerable. Caregiving is widely acknowledged to be an unofficial health service of significant importance to the well-being of millions of older adults. Currently, more than 53 million Americans (21.3%) are estimated to be engaged in the role of informal caregiver to another adult, or child with special healthcare needs, a role for which they seldom receive any financial support or even appropriate recognition from others (AARP & the National Alliance for Caregiving [NAC], 2020).

If family caregivers were paid current market rates for the services they provide, it is estimated that the cost would be in excess of $470 billion annually (Reinhard et al., 2019). Their contribution to helping maintain the well-being of the nation's health cannot be overstated. Overall, 69% of caregivers provide their supports and undertake their caregiving role without formal assistance; just 31% of caregivers report that they receive additional support from paid caregivers (AARP & NAC, 2020). Caregivers serve millions of people while saving taxpayers billions of dollars. They are a major, and often hidden, workforce deserving of our respect and support. Indeed, in recognition of the substantial support caregivers provide, caregiver well-being is receiving far greater prominence as a major public health concern (Talley & Crews, 2007).

WHAT DO CAREGIVERS DO?

A caregiver usually performs numerous and considerably varied tasks based on the dependency of the care recipient (i.e., the person receiving care). Caregivers may initiate their involvement from a distance. For example, they may provide support or advice to their care recipient over the phone or at infrequent intervals in person. For the majority of caregivers, their role often starts with helping to maintain the home environment (e.g., cleaning, doing the laundry, shopping for groceries, helping to organize the family budget, helping with yard work, and providing transportation). In many cases, a caregiver's role expands over time to counter the increasing difficulties of the care recipients, who require more help with bathing, dressing, eating, toileting, and other basic needs. In addition to providing direct care, some caregivers also become adept at more clinical procedures, such as administering medications and injections and changing dressings.

WHO BECOMES A CAREGIVER?

The ability to become a caregiver and the tasks a caregiver is comfortable performing can vary considerably. Personality factors, in addition to family dynamics and the quality of the relationship before the onset of caregiving, are often significant factors in deciding whether to initially undertake the role of caregiver and under what circumstances it will continue. Such a decision is not one that should be made lightly. The decision to care for a loved one in this manner will result in significant sacrifices of one's time and financial resources. Some caregivers are perfectly comfortable maintaining the home environment, but not as comfortable undertaking more intimate caregiving duties for their loved one, such as bathing and toileting. Incontinence, for example, is often one of the reasons a care recipient is admitted into a nursing home.

Those who seek alternative care for their care recipient often feel guilty about making such a choice. It is important that healthcare professionals support them in their decision and recognize some people find caregiving (especially when it involves bathing and toileting of a loved one) a role they are simply unable to carry out. Similarly, one should also respect and recognize that a potential care recipient may feel very uncomfortable with the prospect of their daughter, son, or spouse becoming the caregiver. Sometimes, gender can also be an issue. A mother may feel comfortable with her daughter bathing her but not her son. Likewise, a father may be comfortable with his son bathing him but not his daughter.

Whatever the circumstances, the wishes and decisions of the caregiver and care recipient should be respected and upheld. It is not fair for either party to be placed in a position in which there is obvious discomfort. Unwanted caregiving may lead to a strained relationship, which is unfortunate, especially when there was previously no cause for difficulty.

CAREGIVING DIVERSITY

Caregivers are a diverse group of people whose duties can arise from the onset of many different illnesses and disabilities in their care recipients. With increasing age, many medical issues become more prevalent. Most caregivers are caring for older adults who may have Alzheimer's disease, chronic obstructive pulmonary disease, cardiovascular disease, diabetes, stroke, Parkinson's disease, osteoarthritis, osteoporosis, and cancer. The duration for which a caregiver may sustain their role can vary considerably depending on the care recipient's needs. Some caregivers may provide care over many years (e.g., caring for someone with Alzheimer's disease), but for others the duration is much shorter (e.g., caring for someone with a terminal illness). Former caregivers are at risk from stress as a result of long-term exhaustion from years of carrying out their roles. They have, however, had many years to psychologically prepare for the eventual loss of the care recipient. Caregivers whose help is needed for a shorter duration may undertake the role for just weeks or months, but, in that time, they must deal with the terminal prognosis of their care recipient. They must quickly adjust to their new role as caregiver and witness the rapid deterioration of their loved one. Both types of scenarios thus bring unique challenges. Another group of caregivers are those who provide care to sons or daughters and other family members with lifelong disabilities, including physical, intellectual, and sensory disabilities. Indeed, those who are already lifelong caregivers in these circumstances are finding they too are also becoming caregivers to other aging care recipients, compounding their caregiving role (Perkins & Haley, 2010).

INFLUENCE OF CAREGIVING ON PHYSICAL, PSYCHOLOGICAL, SOCIAL, AND FINANCIAL WELL-BEING

It has long been acknowledged that caregiving can be a demanding endeavor, and considerable research has been devoted to this issue. Much of the earlier research focused on the experience of caregivers to those with dementia, but no caregiver is immune from the potential adverse physical, psychological, social, and financial outcomes.

Caregiving has been linked to difficulties in physical health. Caregivers have been found to suffer from serious, clinically measurable health consequences. For example, being a caregiver can result in decreased immune system functioning (Kiecolt-Glaser et al., 1991). This can result in greater risk of infections and poorer healing of wounds (Kiecolt-Glaser et al., 1995). Caregivers have also been found to have adverse changes in blood pressure (King et al., 1994) and are at greater risk of developing cardiovascular disease (Lee et al., 2003).

Fulfilling the role of caregiver often causes considerable stress; chronic stress undoubtedly takes its toll on physical health, as detailed previously. It also impairs the psychological well-being of many caregivers. They are at substantial risk of developing clinical depression or at least suffering from numerous depressive symptoms (Schulz et al., 1995). Even if not clinically depressed, many caregivers report that they suffer from increased feelings of stress, lower levels of subjective well-being, and lower levels of self-efficacy that can ultimately affect their own daily functioning (Pinquart & Sörensen, 2003).

Caregivers are required to devote considerable time to their caregiving role. As a result, there is often a reduction in a caregiver's social interactions and personal time. Though the individual tasks of caregiving (e.g., bathing, feeding, and dressing) may be completed in minutes, rather than hours, the time demands on a caregiver can be considerable when added together. As a care recipient becomes increasingly dependent, more supervision and attention are required, which causes some caregivers to feel as if they are "on duty" all the time (Schulz et al., 2003). For example, caregivers of those with Alzheimer's disease can spend as much as 80 hours a week just performing caregiving tasks, the equivalent of two full-time jobs (Goode et al., 1998). As the illness progresses, the caregiver often has to adjust to changes in health status and the functional decline of the care recipient, which requires an increasing level of involvement for the caregiver. The caregiver's role is therefore a dynamic one that constantly evolves.

Unfortunately, caregivers' responsibilities result in their having significantly less time to spend with other family members, socialize with friends, and pursue personal hobbies; for some, their responsibilities and lack of access to respite care may prevent vacations or substantial breaks. Because they are less able to attend social activities, many caregivers run the obvious risk of becoming socially isolated, thus suffer from chronic loneliness, and may also have to cope with less outside help from others at a time when it is most needed (AARP Foundation, 2018). It is unfortunate that caregivers can, too often, become engrossed with their caregiving responsibilities and feel unworthy, or even guilty, when scheduling time away from their care recipient. Even though spending time with other family members and friends could provide welcome respite from the caregiver role, caregivers may have unwittingly severed or weakened such social ties because of repeatedly rejecting previous invitations to socialize with others.

Another issue caregivers face is the effect on their employment prospects. Some are unable to undertake paid employment because of the time they spend providing care. Up to 61% juggle their caregiving role with either full- or part-time work, and 61% also report the need to make work accommodations because of caregiving (AARP & NAC, 2020). These adjustments include disruptions in usual work schedules from needing to arrive later or leave earlier, needing time off, reducing work hours, changing jobs, or discontinuing work entirely (AARP & NAC, 2020).

They may also have to start financially supporting the care recipient, and this, coupled with the loss of potential earning, may lead to economic hardship (Langa et al., 2001). In fact, more than one in five caregivers report a high degree of financial hardship as a result of caregiving (AARP & NAC, 2020). This aspect of caregiving may not be such a problem for shorter term caregivers, but as caregiving becomes a full-time occupation in itself, it can lead to many financial difficulties. Reduction in income may affect the caregiver's ability to save, force them to reduce or cease contributions made to their pension program, and lose vital benefits such as healthcare coverage. Another scenario is that as one source of household income ceases, the spouse of a caregiver may be forced to increase their work hours to compensate and may even be forced to abandon personal retirement plans.

BENEFITS OF CAREGIVING

Although caregiving can present definite challenges, many caregivers happily continue their duties in spite of the difficulties. Caregivers will often describe their role as being personally rewarding and boosting their own self-esteem (Nijboer et al., 1999). Helping their loved one in this manner makes them feel useful and needed, can often help them to develop a more positive attitude and appreciation of life in general, and, perhaps more important, often allows for quality time that helps to strengthen the bonds of the relationship (Tarlow et al., 2004).

For the recipient, one of the greatest benefits of caregiving is that they can remain in their own home or the family caregiver's home and receive the individualized attention of the loved one (Perkins et al., 2007). The majority of caregivers become very skilled at their caregiving role, and healthcare professionals should acknowledge and recognize the expertise they develop over time. Although not diminishing the right to autonomy that a person has in determining their own treatment and care options, caregivers can also assist greatly in this endeavor and provide knowledge about current health as well as historical overview. This can provide great insight and invaluable information to clinicians and other healthcare professionals. They often have much to say, and their voices deserve to be heard, especially when planning care.

HOW HEALTHCARE PROFESSIONALS CAN SUPPORT CAREGIVERS

It is important that the whole family of the care recipient discuss caregiving responsibilities and help encourage collaboration and fair distribution of duties. Too often, a caregiver may be overwhelmed but does not make other family members aware until a crisis point is reached. Encouraging frank and open discussion on a regular basis can reduce the likelihood of this happening. Caregivers need to know it is far better to admit they need more help, rather than run the risk of exhausting themselves both physically and psychologically. Emphasizing the joy and meaning that can be derived from caregiving is important, but it is good to remind the caregiver not to forget the actual relationship between themselves and the care recipient. This can be accomplished by spending quality time together in a shared activity unrelated to a caregiving task. Reminiscence is a wonderful way to reconnect with a loved one, as is, for example, watching a favorite movie or visiting a favorite restaurant.

The use of day-care facilities for the care recipients should be encouraged because it not only gives the caregiver a much-needed break but also allows for socialization of the care recipient with other people. Furthermore, case managers and social workers, in particular, should educate caregivers and care recipients of all the services and options (e.g., respite services, Meals on Wheels, home help, and companion services) available to support the caregiver. The caregiver should be encouraged to take minibreaks or go on vacation when the opportunity arises. Make sure that the caregiver knows about available respite services or suggest that other family members be available to help. Many caregivers actually feel guilty about admitting they need a break, but it can be a beneficial change of scenery and people not only for the caregiver but also for the care recipient. It is vital that caregivers feel comfortable about taking a break because some will spend more time feeling anxious about the care recipient than truly enjoying their own vacation.

It is important that professionals who work with older adults realize the contributions that caregivers make to the care recipients' well-being. One should always be mindful that caregiving is stressful and can potentially affect the caregiver's health. Assessments of care recipients should make careful note of the caregivers' abilities and important changes in circumstances. Caregivers need to be reassured so that they understand that the stresses and strains they feel because of their role are common and understandable. Caregivers should be encouraged to use support groups where available, allowing caregivers the opportunity to vent frustrations to other caregivers who are familiar with such experiences. Two valuable resources for caregivers and professionals interested in their well-being are the websites for the Family Caregiver Alliance (www.caregiver.org) and the Rosalynn Carter Institute for Caregiving (www.rosalynncarter.org).

As the population ages, caregiving is likely to become even more widespread than it is currently. Because of their vital role, the issues facing caregivers need to be recognized and supported so that they are able to perform their duties to the best of their abilities. Ultimately, this will lead to better care for millions of care recipients.

REFERENCES

AARP Foundation. (2018). *Loneliness and social connections: A national survey of adults 45 and older.* Author. https://doi.org/10.26419/res.00246.001

AARP & the National Alliance for Caregiving. (2020). *Caregiving in the United States.* https://doi.org/10.26419/ppi.00103.003

Goode, K. T., Haley, W. E., Roth, D. L., & Ford, G. R. (1998). Predicting longitudinal changes in caregiver physical and mental health: A stress process model. *Health Psychology, 17*(2), 190–198. https://doi.org/10.1037//0278-6133.17.2.190

Kiecolt-Glaser, J. K., Dura, J. R., Speicher, C. E., Trask, O. J., & Glaser, R. (1991). Spousal caregivers of dementia victims: Longitudinal changes in immunity and health. *Psychosomatic Medicine, 53*, 345–362. https://doi.org/10.1097/00006842-199107000-00001

Kiecolt-Glaser, J. K., Marucha, P. T., Malarkey, W. B., Mercado, A. M., & Glaser, R. (1995). Slowing of wound healing by psychological stress. *The Lancet, 346*, 1194–1196. https://doi.org/10.1016/s0140-6736(95)92899-5

King, A. C., Oka, R. K., & Young, D. R. (1994). Ambulatory blood pressure and heart rate responses to the stress of work and caregiving in older women. *Journal of Gerontology: Medical Sciences, 49*, M239–M245. https://doi.org/10.1093/geronj/49.6.m239

Langa, K. M., Chernew, M. E., Kabeto, M. U., Herzog, A. R., Ofstedal, M. B., Willis, R. J., Wallace, R. B., Mucha, L. M., Straus, W. L., & Fendrick, A. M. (2001). National estimates of the quantity and cost of informal care-giving for the elderly with dementia. *Journal of General Internal Medicine, 16*, 770–778. https://doi.org/10.1111/j.1525-1497.2001.10123.x

Lee, S., Colditz, G. A., Berkman, L., & Kawachi, I. (2003). Caregiving and risk of coronary heart disease in U.S. women: A prospective study. *American Journal of Preventive Medicine, 24*, 113–119. https://doi.org/10.1016/s0749-3797(02)00582-2

Nijboer, C., Triemstra, M., Tempelaar, R., Sanderman, R., & van den Bos, G. A. (1999). Determinants of caregiving experiences and mental health of partners of cancer patients. *Cancer, 86*, 577–588. https://doi.org/10.1002/(sici)1097-0142(19990815)86:4%3C577::aid-cncr6%3E3.0.co;2-s

Perkins, E. A., & Haley, W. E. (2010). Compound caregiving: When lifelong caregivers undertake additional caregiving roles. *Rehabilitation Psychology, 55*, 409–417. https://doi.org/10.1037/a0021521

Perkins, E. A., Lynn, N., & Haley, W. E. (2007). Caregiver issues associated with wandering. In A. Nelson & D. Algase (Eds.), *Evidence-based protocols for managing wandering behaviors* (pp. 123–141). Springer Publishing Company.

Pinquart, M., & Sörensen, S. (2003). Differences between caregivers and noncaregivers in psychological health and physical health: A meta-analysis. *Psychology and Aging, 2,* 250–267. https://doi.org/10.1037/0882-7974.18.2.250

Reinhard, S. C., Friss Feinberg, L., Houser, A., Choula, R., & Evans, M. (2019). *Valuing the invaluable: 2019 update: Charting a path forward.* AARP Public Policy Institute. https://doi.org/10.26419/ppi.00082.001

Schulz, R., Mendelsohn, A. B., Haley, W. E., Mahoney, D., Allen, R. S., Zhang, S., Thompson, L., & Belle, S. H. (2003). End-of-life care and the effects of bereavement on family caregivers of persons with dementia. *New England Journal of Medicine, 349,* 1936–1942. https://doi.org/10.1056/nejmsa035373

Schulz, R., O'Brien, A. T., Bookwala, J., & Fleissner, K. (1995). Psychiatric and physical morbidity effects of dementia caregiving: Prevalence, correlates, and causes. *The Gerontologist, 35,* 771–791. https://doi.org/10.1093/geront/35.6.771

Talley, R. C., & Crews, J. E. (2007). Framing the public health of caregiving. *American Journal of Public Health, 97,* 224–228. https://doi.org/10.2105/ajph.2004.059337

Tarlow, B. J., Wisniewski, S. R., Belle, S. H., Rubert, M., Ory, M. G., & Gallagher-Thompson, D. (2004). Positive aspects of caregiving: Contributions of the REACH project to the development of new measures for Alzheimer's caregiving. *Research on Aging, 26,* 429–453. https://doi.org/10.1177%2F0164027504264493

24

Death and Grief in the Later Years

INTRODUCTION

Each day death comes to more than 7,821 Americans (Centers for Disease Control and Prevention [CDC], 2021a). It comes in many forms: suddenly while at work, slowly after an illness, during sleep or play, anywhere, any place, any time. As sure as we are born, we will die, but concern with death is often put out of mind until illness or death is upon us. Death is primarily an experience of old age, for nearly three-fourths of those who die each year are older than age 65 (CDC, 2021b). We often hear older persons say, "I'm the last one in my family, and most of my friends are gone, too."

A developmental task of the older age period is the acceptance of the inevitability of death. Losses accumulate as life progresses; children are born, reared, and leave home; parents, relatives, and friends die; retirement implies loss of a long and personally important social role; the family home is sold in favor of smaller quarters; and physical decrements of aging and disease increase in number and intensity. After reading thousands of interviews college students conducted with older persons about death and loss, it is apparent these topics are often thought about. Individuals experienced frequent losses yet few feared death, seeing it rather as a natural part of the life cycle. Some, however, expressed concern over the process of dying. Living each day to the fullest was a common theme, and most exhibited a zest for living unequaled by other age groups. Because older adults have experienced many losses, they in turn can be valuable teachers for us as we struggle with the meaning of death and loss in our lives and in the lives of others.

The discussion of death, loss, and grief is quite ignored in our society, as we are surrounded by the media's gruesome depiction of death day after day. We seem to have difficulty even saying the words *death, died,* and *dying,* cushioning their meaning by using euphemisms such as "passed away," "expired," "gone," "checked out," or "gone home." Over the years causes of death have shifted from infectious diseases and accidents to chronic illnesses. Rather than dying in minutes or days, individuals and families are challenged with months or years of disabling illness, resulting in a long, drawn-out trajectory toward death (Berry & Matzo, 2004). However, since the emergence of the coronavirus (COVID-19) pandemic this is no longer true. The pandemic has caused many people of all ages to die quickly from the virus.

Most people do not want to die in the hospital, and the number dying there is decreasing while those dying in hospice care is increasing. Hospice cares for individuals at home, hospice care centers, nursing homes, and assisted living centers. Individuals and families not receiving hospice care may not experience necessary physical and psycho-social support preceding and after death.

Children too may be deprived of experiences with death when a family member, friend, or pet dies by not being given the choice to attend a service or to be with the family when the person or pet is dying. Sometimes children are even told untruths such as "grandmother took a long trip" or "Fido ran away." Individuals may not choose to attend a wake or funeral service but rather sign a guest book via the computer. Thus, some individuals may go through life shielded from death, supported in this repudiation of reality by a death-denying society.

Dame Cicely Saunders, famous for her affiliation with St. Christopher's Hospice in England and her pioneering work with hospices in this country, believed death is not so terrible if we learn that it is an integral part of life. Death comes to us all despite our untiring efforts to evade it. The hospice movement has made dramatic differences in how Americans die and how family and significant others are supported through the dying and grieving process. Furthermore, hospices across the country are impressive role models for traditional healthcare providers as they collaborate and teach how to improve the care of the dying and grieving. Educational institutions, too, offer more courses or content on thanatology (the study of death and dying), which in turn greatly influences how people die, grieve, and perceive death.

DYING AND DEATH

Types of Death

Physical death is a biological event occurring when the heart, lungs, brain, and other vital organs cease functioning. Death is also considered a psychological and social phenomenon. *Psychological death* is associated with the death of the personality and the ability to think and is often accompanied by a gradual disengagement and withdrawal from others. *Social death*, on the other hand, relates to the social isolation commonly experienced by the dying, which greatly restricts their ability to interact meaningfully with others.

Certain conceptual frameworks describe the dying process as well as communication patterns between the dying person and others. These help to better understand the experience of dying and its effect on all involved. Several of these are discussed here.

The Dying Trajectory

The dying trajectory describes the course or path over time of the dying experience elaborated in the studies of Glaser and Strauss (1965) and Benoliel (1987). Trajectories may be rapid, moderate, slow, fluctuating, or lingering depending on the death that ensues. A lingering death can be emotionally taxing for the family as they cope with the spouse or parent dying as well as other personal and family issues. Tension may also be experienced when the person does not die when expected (Kastenbaum, 2012). Accidents or acute illnesses often result in a rapid course of dying, whereas chronic illnesses are usually accompanied by a prolonged trajectory. At the beginning of the century, deaths were usually rapid, caused by infections or communicable diseases. With the emergence of medical advances and technologies, these illnesses are far less prevalent and are being supplanted by chronic diseases. Retsinas (1988) believes older adults do not have a defined dying trajectory because their illnesses are often chronic and multiple, they adapt to the "sick role," and gradually accommodate to the losses that end in death. Kastenbaum (2012) believes this trajectory allows for the older person and family to acclimate to the dying process, to have time for life review, to do end-of-life planning, and to resolve misunderstandings.

Because more individuals are living to an older age, increasing numbers are challenged to cope with extended periods of disability and degeneration. Families and significant others, too, are increasingly called on to assume the role of family caregiver in the home setting as hospital stays become more restricted and long-term institutional care more costly.

Awareness Contexts

Glaser and Strauss (1965) describe the behavioral dynamics and communication often observed as the dying person, significant others, and healthcare personnel interact with one another. The individuals involved may move through a series of experiences they describe as awareness contexts: closed awareness context, suspected awareness context, mutual pretense context, and open awareness context. During the various awareness contexts, the patterns of interaction among patient, staff, and significant others can result in feelings of dissonance and discomfort within the individual and among individuals.

Closed Awareness Context

During this context, the terminally ill individual is not aware they are dying, although others are. Communication usually focuses on trivialities, such as the weather, or news, with those knowing of the fatal illness continually trying to keep the information from the dying person. This demands constant vigilance by family and staff lest they reveal the actual state to the individual.

Suspected Awareness Context

In suspected awareness, the dying person begins to suspect the presence of a fatal illness. Those near the person know of the condition and are careful not to relay the information to them. Efforts are made by the ill person to question others, expecting somehow to obtain more realistic information. Certain words and behaviors of the family and others will eventually help to confirm the diagnosis in the dying person's mind.

Mutual Pretense Context

During this context, both the terminally ill person and those in contact with them know the diagnosis, yet they continue to communicate as though the diagnosis does not exist and recovery will occur. Rules used to keep the secret are carefully followed, with mostly "safe" topics discussed. Healthcare personnel often assume a businesslike manner to avoid conveying the diagnosis. When the topic of death emerges in conversation, it is passed over or ignored. Some individuals may, however, gradually acknowledge the terminal diagnosis, at which time the mutual pretense breaks down.

Open Awareness Context

The family, friends, and caregivers and the dying person openly acknowledge the terminal diagnosis and openly discuss it. Other factors such as increased pain, weakness, or deterioration present a constant reminder of impending death. Despite the presence of open awareness, at times both the dying person and others continue to act and communicate as though death will not eventually occur. Openness allows for more relaxed, comfortable interactions among all those involved and creates a climate for meaningful communication throughout the dying and grieving period.

Middle Knowledge

Weisman (1972) describes middle knowledge as the dying individual vacillating between knowing a terminal condition exists and not accepting the reality. Behavior or speech does not always match knowledge, as, for instance, when a dying person speaks of trips to be taken over the next few years. It seems almost as if the stark reality of life coming to an end is too much for the human psyche to process and accept. Through this process of vacillation, middle knowledge offers periodic respite from continually facing eventual nonexistence.

The Stage Theory of Kübler-Ross

Many authors have described the dying process, and Kübler-Ross (1969) was one of the earliest. After interviewing about 200 terminally ill persons and their families, Kübler-Ross described five stages of adaptation to a terminal illness. She did not intend that these stages or defense mechanisms be rigidly applied to the dying process or that all persons pass from the first stage (denial) to the last stage (acceptance). Rather, she suggested that individuals revert back and forth in the stages of dying in their own unique fashion and time frame and that individuals may actually experience them simultaneously. Hope is interwoven throughout the dying process—for example, if the individual believes they will experience a sudden cure through prayer. Kübler-Ross believes each person copes in different ways, at different times, and in unique contexts.

Denial (shock) is the first stage that follows the disclosure of approaching death. In an attempt to come to grips with the diagnosis, other medical opinions are sought, treatment may be rejected, test results questioned, or no reference is made to the illness. Periods of denial offer some respite from the shock of impending death and allow the psyche time to adapt to the reality of the situation.

Anger, the second stage, is expressed when dying is perceived as a reality. The life and health of others then become constant reminders of one's own dying, and resentment may be shown toward persons who are not ill or dying. Anger may take the form of outbursts at family and caregivers, or discontent may be expressed with the food or the care given.

In the third stage, *bargaining,* various attempts are made to delay death. Bargaining with God through prayer is often used in an attempt to prolong life to hopefully attend a major family event, or to gain a nonpainful death. Behavior becomes admirable, and anger is usually not displayed. Hope remains, which can be a tremendous consolation.

Depression, the fourth stage, is a reaction to grieving for the impending loss of everything meaningful in life, and for life itself. As the illness progresses, the individual reviews their life, and death is seen as inevitable. Because depression is a normal reaction to loss, this is not the time to offer meaningless platitudes to the dying person, but to listen, understand, and give support.

Acceptance, the last stage, may come after the dying person has experienced the previous stages, completed unfinished business, and said their goodbyes. The dying usually prefer only a few select persons remain with them at this time, which can be more painful for close family and friends who may themselves require increased support. Death can be more difficult if we, the living, try to prevent or slow the dying process. It is not unusual for individuals to die soon after a loved one gives them verbal permission to die.

Critiques of Kübler-Ross' Stage Approach

A number of thanatologists have critiqued Kübler-Ross' stage approach. First, Kastenbaum (2012) evaluates the negative and positive aspects of the stage approach to dying. He believes stages of dying have not yet been verified and that more than five responses and moods are expressed and experienced by dying persons. Charmaz (1980) indicates that stages tend to restrict the understanding of an individual's unique experience of dying and distorts the meaning of the dying process to society. Second, Kastenbaum (2012) states there is not adequate evidence that individuals pass from stage one to stage five. Kalish (1985) shares similar views and concludes the stages represent common responses of dying but are not dependable, recognizable, discrete occurrences progressing from one stage to the next. Third, Kastenbaum (2012) believes the limitations of this approach have not yet been verified through research. Fourth, the stage approach does not distinguish between what actually happens and what is theorized should happen during the dying process. Such a framework is often naively accepted by caregivers and applied as though the dying person must pass through these stages to die in an integrated manner. Fifth, Kastenbaum (2012) is concerned lest we apply these specific stages as a determined path and neglect the uniqueness of the individual's cultural background, sex, family constellations, or personality. Corr et al. (2019) are concerned that we apply these specific stages as a determined path in which all persons die and respond to stress, neglecting the uniqueness of the social and cultural backgrounds of individuals, which include the family, cultural responses, attitudes, and encounters with others. Sixth, the stage model does not take into consideration specific characteristics of the environment such as resources available and pressures and conditions under which the person is dying (such as a hospice death versus one in intensive care).

Despite contrary research and clinical evidence, the "five stages" of coping with dying and bereavement are taught as gospel in many educational programs and courses. Thus, caregivers continue to categorize and apply the five stages without updating their information with current research and writing. Corr (2019) believes this model does not meet the standards proposed by sound theory in current thinking and if it is applied or misapplied this model may even be harmful to the person. The dying experience cannot be described in five states but must be seen in its complexity as it relates to the uniqueness and totality, physical, social, psychological, and spiritual context of each person, the culture, and ever-changing world.

Both Kalish (1985) and Kastenbaum (2012) appreciate that Kübler-Ross informed and sensitized society to the dying process but warn against accepting the stage theory blindly without addressing the many other variables influencing the uniqueness of how each person dies. Kastenbaum encourages open and honest communication with the dying person and to refrain from simplistic and oftentimes rigid use of the stage approach.

Rather than limiting human reactions and the process of dying to five stages, it is best to view the dying person in the context of all that makes them unique. Consider the various parameters of the illness causing the death, the environment, the course of the dying process, gender, psychological makeup, and cultural, developmental, social, and spiritual aspects. Past experiences with death, family dynamics, and socioeconomic level also influence the dying process. Myriad human responses may be observed in the dying person, some more commonly shared, others unique to the person. Individual responses determine how we accurately observe and therapeutically intervene.

Corr's Task-Based Model

Corr (1992) describes four primary tasks reflecting how the person responds and copes with each situation as they encounter the dying process. This model's intention is to allow individuals to participate in the dying process and to empower them as well as all who are coping with the dying (Corr et al., 2019).

The first task, *physical,* relates to body needs or basic human needs. The goal is to reduce those symptoms that cause distress to the person. Alteration in comfort such as dizziness, nausea, vomiting, or pain demands an intervention such as medication or special diet. When an individual is in pain or nauseous, they find it difficult to focus on other issues such as relating to family or tending to their own spiritual needs.

The second task, *psychological*, is coping with psychological autonomy, security, and quality of living. Autonomy relates to retaining at least some control over life. Gradually, as the dying process progresses, autonomy is challenged when others assume personal care. A sense of security is essential to the dying person and is created when around-the-clock care by competent reliable caregivers is available. Psychological richness of living is attained when one has security and autonomy. Special care of the hair, applying makeup or nail polish, a ride in the car, and a visit with a friend all contribute to quality of life and enhancement of personal dignity.

The third task, *social*, is related to maintaining and enhancing the attachments the individual has with others and with groups that have been significant to them. The person may be unable to meet with a social group but may welcome a group that offers some type of needed service. Individuals choose friends and family they wish to have near at this time.

The fourth task, *spiritual*, relates to connectedness, transcendence, and creating a sense of meaningfulness. Maintaining connections with self and others, reestablishing those that have been severed, and those with a higher being are related to this task. Meaningfulness involves searching for meaning in life, death, and suffering. Transcendence relates to that which is beyond, whether it involves a religious belief or not, while maintaining a sense of hope. In a society that is multicultural and with diverse personal belief systems, caregivers are challenged to become informed regarding a person's individual cultural and personal beliefs.

Corr et al. (2019) believe these tasks allow for autonomy and individuality and are approached uniquely by each person. Some tasks may be considered, others not, and in the individual's own time frame and sequence. The tasks are never completed by the dying person but end when the individual dies. However, the work of the tasks may continue as a guideline for those who live on coping with the person's death.

Doka's Phase Model

Doka (1993) addresses the life-threatening experience of dying as a series of phases. In the *prediagnostic phase* the individual initially becomes aware of a symptom, speaks to others, considers other options, and usually seeks medical attention. The *acute phase* is typified by crisis in trying to understand the meaning of the diagnosis and treatment and its implications for the individual. Coping skills and methods to deal with the reality of the situation and ways it affects the present and future of the person are considered. The *chronic phase* describes the tasks of struggling with the disease and its treatment. Coping with symptoms of the disease, the treatment regimen and its side effects, stress, financial concerns, personal fears, and relationships with others are among the challenges of this phase.

The *recovery phase* involves dealing with the aftereffects of the illness and treatment or disfigurement. Individuals are not the people they were before their illness; instead they carry with them concerns over a reoccurrence, reconstructing a new lifestyle and relationships with others, and returning to the work environment as a different person. The *terminal phase* relates to dealing with the reality of dying. Such issues as continuing treatment, arranging for ongoing care, forgoing life-prolonging measures, preparing a will, and planning for a funeral are considered. Individuals are coping with the finality of their lives, finding meaning in living and dying, and saying goodbye to those left behind. Each of these phases presents the challenge of many tasks to be addressed in their own unique way and time.

Hope

Hope is the ability to believe that a painful situation will improve in the future. It offers the dying person a means of making present circumstances bearable. Jett (2008) describes hope as a potent force preventing despair, as a way to survive and to die in a peaceful manner. Older adults, especially, use hope amid their many loss experiences as a means of coping with the past, present, and future. Without hope, life becomes meaningless and despair ensues.

Rondo (1984) says that initially hope is for a cure or a miracle, but later on hope focuses on other issues such as having minimal pain, a visit from a grandchild, or acceptance of one's death by loved ones. The ability to hope depends on one's perception of self-worth and effectiveness. Thus, we enhance a dying person's hope by promoting self-esteem, dignity, and control. Living each moment as fully as possible helps to maintain hope, as does faith in a religious belief or a personal philosophy.

Family and Significant Others

Large numbers of families and friends care for dying loved ones, especially because of the emergence of the hospice movement. Doka (2009) describes family responses to the stress of a terminal diagnosis of a family member. The family may be brought closer together supporting and better understanding each other, or the crisis may exacerbate old conflicts and tensions, or even cause the development of new ones. Herz-Brown (1988) describes the many variables influencing the dying person's adaptations to loss: the beliefs and values of the family, the role played by the dying person, the type of illness causing the death, whether the death is expected or sudden, the emotional functioning patterns of the family, and others. The point in the life span at which death occurs also influences adjustment to the loss. The death of a young person is often considered a greater loss than that of an older person because it is not an expected event and is not in the usual time sequence of dying. Rondo (1984) outlines other factors related to family coping such as the meaning of illness for the family, the signs and symptoms of the illness, its course and treatment, and any stigma attached to it.

Rosen (1998) describes three phases of the family's adaptation to the terminal illness of a loved one: (a) the preparatory phase, when the first symptoms appear until the initial diagnosis is made; (b) the middle phase, the time during which the dying person is cared for and the family faces the reality of the diagnosis; (c) the final phase, the period when the family accepts their loved one is dying and say their farewells to the person. Rosen believes the family response depends on the structure of the family system before the illness and the role of the dying person in that structure. The manner in which the family perceives the implications of the illness and death is also important, as is each phase of the terminal illness because it presents different challenges to the family.

HOSPICE CARE

The modern hospice movement began around the middle of the twentieth century and gained impetus through its international founder Dame Cicely Saunders and her work at St. Christopher's Hospice in England. The concept itself can be traced back to the Middle Ages, when hospices were "way stations" where people stopped for rest and treatment while on long journeys, and some remained there to die. In the early 1970s the first hospices were founded in the United States, and they have grown to more than 5,500. In the United States most hospice care is given in the home setting, but it is also available in special units in hospitals, nursing homes, care centers, assisted living facilities, and free-standing hospices. Most hospices offer a combination of these hospice services (Kuebler et al., 2007).

The patient and family are the unit of care and the dying are encouraged to live each day to the fullest. Patients admitted have a 6-month life expectancy and are dying from various terminal conditions. Honoring the wishes and desires of the dying person is central to hospice care.

Through discussion of the patient and family's needs, wishes, and values and a thorough psychosocial and physical assessment, a plan of care is developed. Using this dynamic process supported by the interdisciplinary team, physical needs of pain, symptom control, and social and psychological spiritual needs are met to allow the last days to be comfortable and as meaningful as possible for the patient and the family (Egan & Labyak, 2006). Hospices have made dramatic changes in the way people die and the manner in which their significant others adapt to the loss in a healthy manner. Older adults and their families and friends are major recipients of this humane, dignified, and supportive approach to death and loss. Hospice also provides care to terminally ill children and their families.

Hospices provide a comprehensive program for the dying person and survivors in which the dying are supported in living as fully as possible until they die. Care is available 24 hours a day, 7 days a week, and provided by an interdisciplinary team of nurses, physicians, social workers, therapists, clergy, home health aides, volunteers, and others who provide pain control, symptom control, and psychosocial and spiritual support. This team meets weekly to address the ever-changing needs of the dying person and their significant others.

Curing disease is not the goal of hospice care, but rather treating symptoms and controlling pain in an effort to make life meaningful and painless to the end. Special pain control regimens are individually prescribed using analgesics (nonnarcotic and narcotic pain relievers) alone or in combination with other drugs. Radiation therapy, pain interruption techniques, and complementary therapies such as aromatherapy, music therapy, art therapy, pet therapy, massage, guided imagery, bibliotherapy, and others may be used. Treatment modalities to extend life such as intravenous medication, resuscitation, extensive laboratory studies, and radical treatment techniques are not the major focus of care.

Supporting the social and psychological needs of the dying person and family is an important part of hospice care offered during the dying process and for a year following the death. Individual grief counseling and group counseling for widows and widowers, suicide survivors, AIDS survivors, children, and others are available. Hospice also educates health caregivers, laypersons, and the community in general, promoting positive dying and planning for the end of life.

The National Hospice and Palliative Care Organization and The Joint Commission on Accreditation of Hospitals have developed standards of care that hospices are expected to follow. National and state guidelines have also been developed to create uniformity in this new form of healthcare. In 1982, hospices became part of the federal Medicare program, which requires participating hospices to adhere to its standards. Many hospices also participate in Medicaid or other state-funded health insurance programs, as well as health maintenance organizations (HMOs) and private insurances.

Medicare offers coverage to those enrolled in Medicare Part A who elect the Medicare hospice benefit and are terminally ill with a life expectancy of 6 months or less. The Medicare hospice benefit replaces the regular standard Medicare benefit. Certain medications and rental medical equipment as well as bereavement care are provided. Depending on the level of care needed, Medicare pays per-diem rates for the care. These levels include routine home care, continuous home care, general inpatient care, and inpatient respite care. The hospice interdisciplinary team determines the level of care necessary as patient status changes (Egan & Labyak, 2006).

Family members and significant others are encouraged to play an integral part in the care of the dying person. In both institutional and home settings, visitors are welcome, as are children and pets. Every effort is made to honor the wishes and lifestyle of the dying person. After death, bereavement teams offer support for a year. Staff are present whenever possible at the time of death; they may attend the funeral and they keep in contact with grievers by telephone, letter, or visits or through individual counseling and support groups for a year. Most hospices have memorial services every few months for all who have died in the prior months, which are attended by grievers and the hospice team.

PALLIATIVE CARE

Palliative care is care focusing on individuals coping with serious illnesses. The patient and family receive care that enhances the quality of life by preventing, anticipating, and treating the patient's suffering. Care is provided by specially trained doctors, nurses, social workers, other healthcare specialists, and chaplains. Throughout the illness patient choice, autonomy, and access to information are enhanced, as well as caring for their physical, psychosocial, social, and spiritual needs (World Health Organization, 2020). The Clinical Practice Guidelines for the 4th edition encourages all healthcare providers in whatever setting to include palliative care for all individuals with serious illnesses (National Consensus Project for Quality Palliative Care, 2018).

The key recommendations include the following:

1. All individuals living with a serious illness should receive a comprehensive assessment to identify their needs and priorities for care.
2. Both families and caregivers should be assessed to determine their needs for support and education.
3. Persons with serious illnesses should have improved coordination in their transition in care from one place to another.
4. Care should be culturally inclusive.
5. Communication should be between all those who are responsible for the care of the person including the palliative care team, other health professionals, the family, and those providing community resources (Huaiquil, 2018).

Throughout the last decade, palliative care has emerged in an effort to treat individuals from the time of an initial diagnosis of a life-threatening illness until there is a life expectancy of 6 months or less, at which time hospice care may be initiated. Individuals with a broad spectrum of diagnoses are treated, such as congestive heart failure, chronic obstructive pulmonary disease, cancer, and renal disease. As the population continues to age, individuals are living longer with multiple chronic and life-threatening illnesses. Palliative care offers them hope through more effective management of symptoms and by providing psychosocial support given by an interdisciplinary palliative care team. Palliative care may be offered in hospitals, patient's homes, and community clinics. Hospice and palliative medicine are subspecialties of the American Board of Medical Specialties. National certification in hospice care is available for registered nurses and for hospice home health aides and social workers.

THERAPEUTIC APPROACHES WITH DYING PERSONS AND SIGNIFICANT OTHERS

Therapeutic approaches to take in assisting dying persons and their loved ones include the following:

1. Spend time with the dying person and their family. Our presence and concerned caring are the greatest gifts we can offer others.
2. Listen with an unbiased mind to what the dying person and family are saying and observe feelings and behaviors that accompany the words.
3. Encourage them to express denial, anger, guilt, shame, or any other feelings or emotions without judgment.
4. Respect the dying person's wishes to make their own end-of-life decisions.
5. Be open and honest in communication and behavior.
6. Be compassionate and facilitate forgiveness.

7. Offer spiritual support and call clergy as requested or needed.
8. Explain situations that may arise in simple, easily understood language.
9. Show concern through the use of touch as appropriate in each situation.
10. Prepare the family for the dying process by explaining what is happening and offering support.
11. Assist with housework, errands, caring for family members, or other tasks.
12. Offer to sit with the dying person to provide respite time for the caregiver.
13. Arrange for supportive therapy such as art, music, pet therapy, poetry, or massage as needed.
14. Refrain from using the stage approach with the dying person and survivors; rather, honor all their responses and reactions to the dying process as they are presented.
15. Explain the availability of community resources such as palliative care, hospice, and assistance from the American Cancer Society, the American Heart Association, or others.
16. Suggest using individual or group support services offered by hospice or other agencies, such as "I Can Cope" for the terminally ill person or significant others.
17. Never destroy hope as expressed by the dying person or family.
18. Assist in arranging for services requested, such as contacting a lawyer, social services, financial advisor, or funeral director.

BEREAVEMENT, GRIEF, AND MOURNING

The experience of loss involves the processes of bereavement, grief, and mourning. Even though these terms are at times used interchangeably, they are distinct in meaning. *Bereavement* describes the objective state of having experienced a loss; an individual who has lost a husband through death and is now a widow, or a child is an orphan when a parent dies. *Grief* as an intrapersonal process describes our reactions to the loss, how we think, feel, behave, and so on. It includes our thoughts, feelings, behaviors, physical, social, psychological, and spiritual responses over time. It influences all the spheres of one's life. Although it is not considered to be a defined illness, it may precipitate the development of an illness (Kastenbaum, 2012). *Mourning* usually refers to the norms, social, cultural, and ritual processes individuals use to become acclimated to the loss. Mourning behaviors or rituals are used when the flag is flown at half-staff after the president dies or when a specific service or ritual is used by a cultural group during the dying and postdeath period. It could include the wreath placed on the door of the dead person's business or a black armband worn to commemorate a death (DeSpelder & Strickland, 2020). Corr et al. (2019) include both the social and personal dimensions used in coping with loss within the definition of *mourning*.

Theoretical Approaches to Grief and Mourning

Freud (1957) describes grieving as "the work of mourning or grief work" in which the bereaved person focuses on memories of the dead person and events that led to death, which allows the griever to "decathect" or become detached from the valued person. DeSpelder and Strickland (2020) indicate, in light of more recent research, that coping with the death of a loved one is much more complicated than becoming detached from the individual and continuing on with life. It allows the griever to make the loss a part of their life over the years. The grieving process is an individualized, complex process colored by a variety of events such as the relationship with the person, cultural practices, the type of death, age of the person, coping skills of the bereaved, and so on. Bowlby (1969) believes that we as humans develop strong attachments or bonds to others probably caused by our basic need for safety and security. When these bonds are threatened or broken by death, we respond by experiencing various normal responses to the loss. Many different models of grieving exist. Parkes (1972) suggests four phases of adaptation to loss: (a) the period of numbness, (b) the phase of yearning for the loved one, (c) the phase of disorganization and despair, and lastly (d) reorganization.

Worden (2009) describes this process as the "tasks of mourning."

Task 1: To Accept the Reality of the Loss. After the death of a loved one, there is a period of searching and longing for the individual hoping they will indeed return. Some deny the reality of the death by refusing to believe it happened, others act as though the person has not died by retaining their possessions in case they return, and still others engage in selective forgetting. Arriving at accepting the reality of the loss requires time and involves emotional and intellectual acceptance.

Task 2: To Process the Pain of the Grief. The pain of grief involves the emotional, physical, and behavioral pain one goes through as grief is processed. The intensity, length, and type of pain vary from individual to individual. Some may deny their feelings and try not to feel the pain by traveling extensively or thinking only happy thoughts of the loved one. Society is not comfortable with grieving individuals, which may inhibit the process of experiencing the pain of grief. If the pain is not processed, depression may ensue and the task is not resolved. Processing the pain is necessary in attaining a healthy life.

Task 3: To Adjust to a World Without the Deceased. There are external adjustments related to the many roles the deceased played and the challenge the survivor has in assuming those roles. For the widow, the new role is that of a single person who often must assume at least some of the husband's roles, a challenge some fail to accomplish. Internal adjustments require the development of a new sense of self, a new self-identity. There are also philosophical and spiritual issues that are challenged by loss.

Task 4: To Find a Way to Remember the Deceased While Continuing the Rest of One's Journey Through Life. The essence of this task is finding a special place in one's life for the one who died where some connectedness continues but that allows for the individual to assume living their present life. The person is not counseled to withdraw emotionally from the deceased but is helped to find a place in their life for the deceased while restructuring and continuing on with life. This may be done by feeling the presence of the dead person, speaking to them, placing their pictures in the home, or memorializing them while developing other relationships and living meaningfully in the world.

Grieving has traditionally been described in psychological stages such as those of Parkes and Rondo. Bonanno (2009) believes that resilience is a dominant factor in experiencing loss and crisis. Bonanno et al. (2002) studied bereaved older people whose spouses died over time of natural causes. Forty-six percent did not show indications of anxiety, despair, or shock. They were sad but returned to their usual routines and interactions with others. The second group, called the "recovery group," represented 40% of the spouses who adapted more slowly to the loss over about a 2-year period. The last, the "chronic grief group," representing 10% to 15% of the spouses, found the loss of the spouse overwhelming. They were preoccupied with the loss and continually longed to have their loved one with them. The resilience trajectory is a newer approach to grieving that warrants more research as well as attention to its implications. The stage approach to grieving is well received and evidence of various coping mechanisms continues to be observed. How Bonanno is able to explain from the data the manner in which individuals adapt to loss in the recovery trajectory is not clear. The role of counseling is also of concern (Balk, 2013).

Holland and Neimeyer (2010) describe how individuals hope to find a sense of meaning in their life and death. Over a 2-year period they studied individuals bereaved through a natural event and those who experienced a death by a violent event. Those who lost a loved one by a natural death could more readily make sense of the loss, which helped them to better accept the loss and in a shorter time frame. Those who experienced loss through a violent death had more difficulty making sense of the loss and were more distressed. Depression and anger seemed to inhibit their ability to acknowledge the loss and to continue on with their lives. The best predictor of the manner in which individuals deal with their grief was their ability to find meaning in the loss experience. The function of time and the stage approach seemed less useful to them in the grief experience.

Common Physical and Psychosocial Reactions to Grief

Experiencing the pain of loss is a healthy and necessary process. To avoid acknowledging the various personal responses to loss merely extends the grieving period and may even result in physical and psychological illness. These responses may occur throughout the grieving period, may come and go, or may be felt more deeply at one time or another. Some are observed more during various phases of grief such as shock and numbness, which often occur soon after the death. However, if reactions such as insomnia or lack of appetite continue for months or years, it could indicate the presence of complicated grief. Corr et al. (2019) describe these as responses to loss that are distorted regarding their duration, degree, and onset. Recalling memories of the loved one is always appropriate, but excessive use of alcohol to deal with the loss is never appropriate. Monitoring by a healthcare professional is important because illnesses may develop during this time. It is important to remember that each person grieves in their own way considering their personality, ways of coping, and various cultural influences (Kastenbaum, 2012).

Physical manifestations of grief may include poor appetite, nausea, weight loss, symptoms that mimic those of the person who died, shortness of breath, arthritic pain, headache, stomach pain, fatigue, increased infections, impotence, and cold and flu symptoms. Research supports the premise that the immune system may be depressed during grieving, increasing the likelihood of acquiring an illness. Attention to physical and psychological symptoms, eating well, sleeping sufficient hours, and regular physical checkups are wise during this period of greater vulnerability.

Psychosocial responses include the decreased ability to concentrate or think, dreaming of the dead person, visual or auditory hallucinations of the dead person, guilt, yearning, anger, loneliness, depression, suicidal thoughts, low self-esteem, searching for the dead person, memory impairment, over- or underactivity, confusion, overtalkativeness, guilt feelings, restlessness, dissociation from others, impaired judgment, dependence, and changes in patterns of sexual interest and expression. Spiritual reactions include being angry with God, losing or questioning one's faith system, or gaining comfort from one's faith.

Although this list of responses to grief is not exhaustive, it may help sharpen one's ability to be more aware of the possible array of healthy human responses to loss. When these responses are experienced, individuals sometimes become frightened because they may never have known such a variety and intensity of feelings. Some even describe themselves as "going crazy," creating a sense of fear and anxiety. A grief counselor or group therapy can be of great help for many who are struggling through this difficult period of adjustment.

Continuing Bonds

DeSpelder and Strickland (2020) describe "maintaining continuing bonds" as the deceased loved one is acknowledged as part of the bereaved person's life in the relationship that continues after death. This can be observed in grieving parents who often refer to their child who died as still part of the family. Older widows, too, speak of sensing the presence of their deceased spouse, talking to them, longing to be reunited with him, and not seeking to reinvest themselves in another spousal relationship. Walter (1996) describes how the Shona of Zimbabwe keep alive the spirits of their deceased ancestors. He agrees that some individuals may wish to process grief feelings and detach themselves from the person, whereas others find it necessary to converse about the person and create a unique place for the person in their lives. Corr et al. (2019) believe that the continuing bonds premise offers an alternative to the long-held belief that grieving should end in disengaging from the deceased person.

Anniversaries, Birthdays, Holidays

The first year after the death is usually the most intense. Patterns of grief fluctuate, but certain events and occasions often precipitate more acute feelings and emotional responses. The first birthday or anniversary without the person, or other significant events such as Christmas or the Fourth of July may trigger special memories of being with the person who died. Attending social functions or hearing familiar music may also cause an upsurge of emotions. These and other significant times may cause a resurfacing of physical and psychological responses that might frighten the individual. Support from family and friends at this time can facilitate understanding of the situation and help soothe them during this stressful period.

Anticipatory Grief

Grief experienced by a dying person or a survivor before the death is called *anticipatory grief*, a term coined by Lindemann (1944), who noted that persons not only give evidence of grief after the death of a loved one but that grief is also experienced during the dying process. Rondo (2000) describes the process as *anticipatory mourning*, which is more than a reaction to the anticipated death; it involves participating in life with the person, grieving for the person they were, and planning for life without that person.

Although anticipatory grief is most often associated with the griever, the dying person, too, grieves over their eventual loss of life. As disengagement occurs, withdrawal behavior becomes evident. Individuals who undergo a slow dying trajectory are most likely to experience anticipatory grief as a "little death," such as loss of work, decreased physical functioning, or strained economic resources (Rainey, 1988).

It is believed that anticipatory grief is more commonly experienced these days because we tend to live longer with chronic life-threatening illnesses (Beder, 2002). An extended illness affords more time to think about the losses death will cause. There is more time to be anxious, time to plan, time to relate to others, and time to rethink the meaning of life and death. Anticipatory grief does not necessarily reduce the feelings of grief after death (Fulton, 2003). Rondo (1988) characterized the griever as moving closer to the dying person and giving increased attention to them while, on the other hand, moving away from them. Sometimes feelings of guilt and resentment may arise if the person's dying takes an unusually long time.

Disenfranchised Grief

Disenfranchised grief occurs when the bereaved experiences a lack of recognition and validation from others that someone close to them has died. The grief is hidden and the loss not openly substantiated, publicly supported, or mourned. Such a grief occurs when the relationship between the mourner and the dead person is not recognized, the loss is not recognized, or the griever is not recognized by others (Doka, 1989). Rondo (1993) notes that it also occurs when society tries to protect itself from anxiety, as in the case of a horrendous death, or when it seeks to punish the griever. Sometimes grief is hidden when the griever is not thought to be entitled to feelings related to the loss. This may be observed when a nurse caring for an elder becomes attached to the person and experiences grief when the person dies (Kastenbaum, 2012).

There are several instances when disenfranchised grief tends to be experienced. Parents of a child who is stillborn or parents of a child with developmental disabilities who dies may not receive the same acknowledgment of loss, even though they, too, feel the devastating loss of a child who was wanted and treasured. Likewise, many bereaved spouses of those who die of Alzheimer's disease have been a caregiver for years and gradually become disengaged from friends and family. When the death occurs, their support group is quite absent and the loss is viewed as a relief and not requiring the same level of support as another type of death. Another unacknowledged loss is the death of a pet. Animals and pets may be treasured and deeply loved, sometimes even more than a person. When the death occurs or perhaps the animal is euthanized, deep loss and grief may result, often without any or little support from others (Kastenbaum, 2012). Pet loss support groups available in many areas can be of great help to survivors. Some funeral homes even have special services for pets, including caskets, viewing hours, and cremation or burial. Pet cemeteries also provide similar options, including a gravestone. Pet loss sympathy cards are readily available wherever greeting cards are sold. If cremation is chosen, pet cremains may be placed in an urn and kept close by or worn in jewelry such as a necklace.

Complicated Grief

Complicated grief, sometimes called prolonged grief disorder, results when the individual's response to loss is unhealthy because of a variety of factors. Corr et al. (2019) attribute the response to special circumstances surrounding the death, such as multiple deaths or a traumatic loss; the psychological makeup of the individual; social issues such as lack of support; and the relationship that existed with the deceased person as well as a history of losses.

Worden (2009) describes four types of complicated grief:

1. *Chronic grief.* Individuals become aware they are not processing their grief and arriving at some kind of a reconciliation. Often they have not resolved one or more of the tasks of mourning.
2. *Delayed grief.* Individuals may have had only a minimal response to the death or it may have been suppressed or postponed. A loss experienced later on may evoke deep grief not dealt with earlier or they may experience a more intense response to the death at a later time.
3. *Exaggerated grief.* The response to the death becomes disabling and presents itself in a dysfunctional manner often related to a psychiatric disorder such as a clinical depression.
4. *Masked grief.* Physical or behavioral symptoms mask the grief, which is not attributed to the loss. These individuals may have experienced no grief after the death or it was inhibited at the time.

Kauffman (1989) discusses the "chronic grief" often seen among those living in nursing homes. For many older adults, admission to a nursing home means they will live there until they die. Kauffman describes the experience as "to live their death." He says they live with death in their helplessness and suffering, by accepting or welcoming relief from hopelessness and suffering, by resolving their fear and grief, by their strength and courage, and through their despair. Too little attention has been given to this form of grief and the manner in which it influences the individual, family, and caregivers.

DEATH AND LOSS IN THE LATER YEARS

Loss is a universal human experience that cuts across all developmental levels. Although some losses are significant, others are small and go unnoticed. Losses are predominantly an experience of older age that appear over time, such as vision and hearing loss, mobility impairments, and gradual dysfunction of body systems. Not only are there biological losses but also psychosocial losses as the social and family network support systems shrink. Death is primarily a phenomenon of older age because the likelihood of death increases sharply with aging. Superimposed on these are other losses such as those incurred by relocation, social and sexual role changes, and diminished employment income. Most older adults live productive fulfilling long lives. Over the years many have experienced multiple losses, such as loss of a spouse, children, relatives, or friends. Eventually some may have outlived nearly all their age group. They represent a unique group of "lonely oldies" who feel very alone and yearn for their loved ones who have died or are very ill. While still alive they experience a social death. Even the loving care of younger folks does not quell their sadness (Corr et al., 2019).

Losses can happen so rapidly that grieving for one loss is not completed before another occurs. Older adults may be thrust into a continual state of grieving. Kastenbaum (1969) calls this "bereavement overload." More research studies are needed to investigate the effect of bereavement overload on the adaptive mechanisms of older adults. When we consider the usual biopsychosocial human responses to a single loss, what might be the catastrophic effect of multiple losses on the aging person? Feelings of being overwhelmed, depressed, and disinterested in living and preoccupation with body functions or illnesses may result. Secondary losses, too, accompany the death of a loved one. We miss not only their physical presence but also the sound of their voice, food they cooked, or letters they wrote.

Butler (1963) first described another phenomenon common in the later years, "life review." This process may be precipitated by the person's understanding that death is near. They recall the positive and negative experiences of life and assess, evaluate, and review the past to arrive at a meaning for their life. If some life tasks are not accomplished, the person may attempt to do so at this time, eventually arriving at a sense of serenity and wisdom.

Grief reactions may be even greater among this age group because of fewer emotional involvements. In our mobile society, family and friends are dispersed around the country and world. Such isolation may be particularly felt among those older adults who migrate from a northern state to retirement communities in the sunbelt. As they grow older, friends die and fewer supports remain, especially if they have associated only with people their own age. When physical and psychological losses become overpowering, the will to live often decreases. We sometimes hear an older person say, "There is nothing to live for" or "I want to be with my dead spouse." Such statements may indicate a readiness to die or they may indicate the presence of depression. Identification and treatment of a depressed state is important because it can usually be reversed by social and psychological support, psychotherapy, and/or antidepressant medications. One must always keep in mind, however, the potential for suicide in those who are clinically depressed. Long-lasting depression occurs most commonly among older adults who lack support from family, friends, or a spiritual belief system. Loneliness is reported by Constantino (1981) and other researchers as a predominant feeling among older grievers. Having spent 40, 50, or 60 years together, a spouse can hardly forget a mate in a few weeks, years, or ever. Newspapers describe situations in which the death of one spouse is followed shortly by the death of the other.

Because more individuals are living to an older age, they experience many types of deaths. These include siblings, one or more spouses, peers, children, grandchildren, parents, and friends. Each of these involves unique relationships, many over decades, that are severed by death. For example, the death of an adult child or grandchild is not developmentally in order and may result in "survivor's guilt" (Corr et al., 2019). Such is the case in the deaths of young persons with HIV who predecease their parents and grandparents.

The older we become, the more likely we are to experience the death of a parent. Considerable numbers of older adults become caregivers for a parent over long periods. When the parent dies, the caregiver becomes an "orphan" as the parent–child bond is broken.

The bonds of friendship can be and often are as close or closer than that of a relative. We can choose our friends but not our relatives. Many families are scattered around the country and considerable numbers retire in the South. Endearing and lasting friendships of deep meaning and significance develop over the years, the loss of which can be devastating and often unacknowledged. Honoring these special and meaningful relationships by supporting the griever can validate and support the older adult experiencing such a loss.

Retirement represents many potential losses for the older person even if one looks forward to this new phase of life. Concomitant with a career is status, meaning, and interaction with colleagues and others. Retirement involves saying goodbye to a large piece of one's life, which may result in loneliness, a sense of worthlessness, and depression. Allowing one's self to grieve by sharing the losses and arriving at a new self-identity may take time and effort. Volunteering can be a very meaningful and rewarding role for retirees, offering them a reason for living and positive involvement with others.

Pets often replace family or are the "family" for many older persons. They offer unconditional love, someone to care for, and a reason to get up in the morning and to exercise. Pets assuage loneliness and provide comfort and a reason for living. Such a close human–animal bond must be honored and supported. As an older adult's health declines, caring for a pet may become difficult, paying veterinary bills overwhelming, and moving to an assisted care setting without the pet heartbreaking. Fortunately, some institutional settings allow residents to bring their pets; if not, most allow their pet to visit or welcome visits from therapy dogs. Making plans for the care of a pet after death or provisions for a pet in a will can offer some sense of peace to the individual.

Grandparents

In recent times, many grandparents have assumed the care of grandchildren. Some adult children and grandchildren live with grandparents, whereas other grandparents assume child day care regularly or from time to time. Still others who live at a distance visit often and are in touch via telephone, email, or FaceTime. When a grandchild dies, the older person experiences a unique threefold grief: for the grandchild, for their own son or daughter, and for themselves. They are often called the "forgotten grievers." Backer et al. (1994) describe them as being in a state of helplessness because often they are geographically separated from their children and are experiencing health problems themselves. Ponzetti and Johnson (1991) found that grandparents experience emotional and physical symptoms of shock, numbness, disbelief, and a need to understand why the child died. Because the child's death is "out of time," they may feel guilty and helpless by not being able to prevent it or express anger at the parents for perhaps not giving the child the best of care. Resentfulness and anger at God for allowing the death also occur (Reed, 2000). After such a loss, grandparents can offer continued support to their children in coping with the loss; they can also care for the other children, provide financial support, and assist with daily chores (Kastenbaum, 2012). Grandparents too need to share their loss with friends and family, but support is not always forthcoming because the focus is primarily on the parents and siblings. Disenfranchised grief may be experienced if their loss is not acknowledged and they lack the presence of caring, concerned persons. Individual grief counseling and grief support groups may prove beneficial in helping grandparents through such a difficult loss (Davies & Orloff, 2010).

Coronavirus

In early 2020 the COVID-19 pandemic began to invade our lives sending the world into global loss and grief. Accustomed to chronic illness causing the majority of deaths, we are faced with an acute illness affecting people of all ages but especially older adults and those with comorbidities causing many deaths. As of February 2021, more than 500,000 individuals have died in the

United States and over 2.47 million internationally (Johns Hopkins University, 2021). The virus has caused overwhelming loss experiences for all people. Not only is there illness and death to challenge each day but fear, changes in living, working, school, and relating to one another. Hope so necessary is diminished.

COVID-19 has greatly impacted dying, death, grieving, and mourning. Death occurs in a few days for some; for others it occurs after being critically ill for a longer time. Many people have died in hospital without the presence and support of family and loved ones. Long-term care settings experience multiple cases of the virus. Residents have died in isolation away from their loved ones and family. Even those residents without the virus have been unable to see their loved ones for over a year, causing continued pain, distress, and loss. Families are dealing with the death of a child, a parent, grandparents; and friends are without the usual and necessary support and are grieving intensely.

Following death due to COVID-19 the option to carry out afterdeath practices is limited. Public funerals, visitations, and gatherings are not encouraged. Connectedness with loved ones and families is greatly impaired for both the dying and bereaved. Ability to see one another's reactions of love, caring, and support can be quite absent. The influence this has on people's coping with loss, dying, and grieving can be severe. Future research will further reveal the impact of the COVID-19 virus on the physical, psychosocial, and spiritual aspects of people.

Spousal Bereavement

The death of a life partner is one of the most challenging and profound losses of a lifetime and may be felt for years to come, even forever. A spouse plays many roles, such as a lover, friend, source of financial security, confidant, partner, and so on (Dutton & Zisook, 2005). A death results in not only the loss of the spouse but also countless other secondary losses.

Becoming a widow or widower can be a very painful transition for those who are both coping with loss and attempting to redefine their self-worth. Acute loneliness may ensue as they attempt to cope with the loss, often without friends or family because most of them have already died. They face the challenge of becoming aware of their uniqueness as a worthwhile person in their own right, not solely in the role of a wife or husband. After grieving for months or longer, widows may eventually develop a support system with one another and enjoy a new life of sharing, self-fulfillment, travel, and enjoyment. Isherwood et al. (2012) found an increase in social engagement with children and in social activities during the first 6 years of widowhood. This engagement facilitated grieving as well as promoted healthy aging. However, older males in poor health and of a lower socioeconomic level and without children living nearby had a lesser opportunity for social interaction. Promoting and providing the opportunity for social contact could favorably affect grieving and living healthily. For others, the challenge of making new friends proves to be more than they can handle. Managing finances, personal business, or a household may be too great, especially if their skills are minimal. Individuals living at the poverty level may need to work at menial jobs if they lack specific skills or education. Others have little support from family, who at times expect them to be strong and to get on with life. Aldersberg and Thorne (1990) suggest widows need help in identifying who they wish to be now that they are no longer wives. Self-help programs are available in which widows can learn assertiveness skills, are introduced to dating, are taught how to develop affiliations with political or advocacy groups, and learn how to become more confident and enthusiastic about life.

Widowers seem especially vulnerable. In our culture older men often do not feel free to share their emotions but instead show a "stiff upper lip." Most are not likely to share their loss or seek help from others. Friends may not feel comfortable offering them help, and widowers are not as likely as widows to take advantage of grief counseling or grief support groups. Kalish (1985) compares widowers to widows: (a) There are fewer widowers and hence fewer male peers for support, but men are more likely to date and remarry; (b) men are usually widowed at an older age than women and are often less physically healthy; (c) men are more financially secure than women; and (d) men are not as able as women to manage a house but can earn money and perform most household maintenance.

Individual counseling and grief support groups have proven to be of great help to the widowed in assisting them through the grief process. Founded in 1986 by Phyllis Silverman, the "Widow to Widow" program offers help from peers who are bereaved for 2 years or more. Her model includes these beliefs: Grief need not have a final outcome, it can be considered to be a life transition, and people can help people. AARP sponsors "Widowed Persons Service," which helps widows by providing written and audiovisual materials and assistance in developing grief support groups. All hospices offer individual and group grief counseling as do other community agencies and churches.

OLDER ADULT SUICIDE

Suicide is defined as an action taken by an individual resulting in their death. In 2018 suicide was the 10th leading cause of death in the United States claiming over 48,000 lives, and over 1.4 million Americans attempted suicide (National Institute of Mental Health [NIMH], 2021). Older adults, who make up about 16% of the population, complete 19% of all suicides (NIMH, 2021). The incidence of suicide is increasing especially among widowed older men, and in 2018, the suicide rate for older men aged 75+ was 39.9 per 100,000 population—the highest of any age or gender category (NIMH, 2021). Generally, older adults usually do not threaten suicide but are more likely to complete the act (Touhy, 2008). Older adults are also less likely to live through a suicide attempt than a younger individual (Kastenbaum, 2012). According to CDC data (2021c), 77% of older men (aged 65+) use firearms as the primary means of suicide, whereas 37% of suicides by older women (aged 65+) are done with overdose of pills (39%) followed by firearms (37%). Older women have a much lower rate of suicide and the more children they have the lower the rate becomes (Touhy, 2012). There has been an increase in suicide rates among middle agers as well as older adults. Causes cited are major depression, negative life events, substance abuse, and affective disorders (Corr et al., 2019; DesSpelder & Strickland, 2020).

Risks for suicide include chronic illness and disability; widowhood; social isolation; drug and alcohol abuse; depression; loss of meaning in life, especially after retirement; multiple losses of family and friends; financial loss; and decline in meaningful relationships. There seem to be fewer warning signs and the use of less direct methods such as not eating or not taking medication as directed (DeSpelder & Strickland, 2020; West Ellson, 2007). Blow et al. (2004) states that in two-thirds of all completed suicides among older adults the causative factor was major depression. Both the prevalence and incidence of depression and the risk for suicide increase as one ages (Bhar & Brown, 2012). Vigilance is recommended when these risks are observed, followed by appropriate intervention by a licensed mental health professional (Crenshaw, 2013).

It is estimated that about one-half of those who complete suicide leave a suicide note. They are usually written quite near the time of the death and include a variety of messages. Fear, rejection, hostility, love, hate, and ambivalence in the relationship are but a few themes found in them. Notes illicit various responses from the receiver depending on the message. Unfortunately they are unable to respond to the message (DeSpelder & Strickland, 2020).

Older adult suicides are often underreported because the cause of death may be difficult to identify. Silent suicide is unique to this age group in which the motive to kill oneself is not as clearly identifiable when, for example, the person refuses to take prescribed medications, follow a specific diet, or therapeutic regimen (Gaynes et al., 2004). Estimates indicate the suicide rate could actually be double that reported.

A double suicide or suicide pact is more common among older adults than any other age group. Often one or both of the spouses are ill, or they may have a high intake of alcohol. Interdependence with one another without much social support is also a common finding, as is the male partner's history of attempted suicide. All suicides are not necessarily a result of a pathological condition but may represent a rational approach, carefully planned, considering current and future circumstances of life (Stillion & McDowell, 1996). Gaynes et al. (2004) report that one-half to two-thirds of those who complete suicide were seen by a healthcare professional within a month of the suicide. Careful attention to symptoms of potential suicide such as depression, alcohol ingestion, not interacting with others, being isolated, coping with chronic physical or mental illness, or being sad or restless is imperative. Use of the Geriatric Depression Scale (short form) is advised to screen for the presence of depression and to allow for therapeutic intervention (West Ellson, 2007). Being aware of behavioral, situational, or syndromatic clues is especially helpful in identifying those at risk for suicide (Holkup, 2003). Encourage older adults to speak openly about the situation, offer emotional and social support as well as plausible alternatives to ending life, and always take the person seriously. Using community resources such as a mental health center, medical care, or counseling is highly advised. A national suicide prevention hotline is available 24 hours a day, 7 days a week by calling 1-800-273-TALK (8255). All calls are confidential.

PHYSICIAN-ASSISTED SUICIDE/PHYSICIAN-ASSISTED DYING

Considerable controversy exists regarding physician-assisted suicide (PAS), a term that is also referred to as physician-assisted dying (PAD). Jack Kevorkian's actions over a period of 8 years (1990–1998) in assisting more than 100 individuals to terminate their lives was a major force in increasing the public's attention to this matter. Research on 69 autopsy reports of those deaths showed only 25% were terminally ill (Roscoe et al., 2000). PAD involves the person requesting the means (drugs) from a physician with the intention of taking the drugs to end their life. The person is responsible for taking the medicine, not the physician or family member, but the physician can offer advice or help with other issues (DeSpelder & Strickland, 2020).

Oregon voters approved the Death with Dignity Act in 1994. Similar statutes have been enacted in Washington, DC, California, Colorado, Hawaii, Maine, New Jersey, Oregon, Vermont, and Washington, and others states are considering it (Death with Dignity National Center [DWDNC], 2021). Montana does not have a law that safeguards physician-assisted death as such, but the Montana Supreme Court has ruled nothing in their state law actually prohibits a physician from honoring a terminally ill, mentally competent patient's request by prescribing medication that would hasten a patient's death (DWDNC, 2021). According to DWDNC (2021), Death with Dignity Acts stipulate that "actions taken in accordance with [the Death with Dignity Act] shall not, for any purpose, constitute suicide, assisted suicide, mercy killing, or homicide, under the law." Each state has requirements and protections to be met before the suicide can be carried out. States require that the patients receive treatment for physical, psychological, and spiritual pain before PAD is even considered. Although many people have considered PAD, many fewer have actually completed the process. This is often the case when the patient's suffering is controlled by effective treatment (Jacobs, 2013). The individual seeking death with dignity in Oregon must be 18 years of age; mentally competent; and have a fatal, irreversible illness expected to cause death within 6 months. Guidelines and safeguards must be adhered to in carrying out the process. These include three signed confidential requests for PAD, with a 48-hour waiting period between the second and third request; the patient must clearly understand the diagnosis, prognosis, and alternative treatments and be suffering with high-level pain they find intolerable. An independent physician must assess the patient, review the patient's record, and communicate with the primary care physician to determine eligibility. They must be informed they can rescind the initial request for drugs anytime. The presence of emotional distress, economic concerns, or inadequate comfort care is to be ruled out. Emotional and spiritual counseling must be made available and family approval of the request is necessary. These and other stipulations are necessary before a physician can legally prescribe a lethal dose of a drug for a patient. The law has been challenged repeatedly by the U.S. Department of Justice and is expected to be challenged in the future (Kastenbaum, 2012).

Controversy has surrounded PAD. Those in favor include the Compassion and Choices organization, which actively promotes legalizing PAD. Other arguments state that individuals should be able to make their own decisions and have the right to end their lives if pain and suffering become too great. PAD is controversial as it raises many questions regarding an individual's ethical, cultural, and religious beliefs. The American Medical Association, the American Nurses Association, and many other professional organizations continue to refer to PAD as physician-assisted suicide and oppose it because it is not consistent with their role as healers. Religious groups oppose the taking of life, whereas others believe it could become a "duty" for older adults with physical or mental disabilities to die. People with disabilities too fear they might be eliminated by PAS. Not Dead Yet (2021) is a national disability rights organization that opposes legislation of assisted suicide and euthanasia as deadly forms of discrimination of people with disabilities who may be the targets of "mercy killing." The issues surrounding PAS/PAD are not simple, even the terminology for referring to the concept is not agreed. The arguments for and against PAS/PAD abound and the issues have grave consequences for all involved, whether it be the patient, family, professionals, insurance companies, or our country as a whole. Continued study and research into the complexity of these issues is highly recommended before making a decision regarding PAD.

Out of the controversy, and as a result of this legislation, there has been a renewed effort by doctors, nurses, and healthcare professionals to better understand effective treatment of pain and symptoms as well as the psychosocial needs of the dying. Referral to hospice care has also increased because hospice professionals are experts in caring for the physical, social, psychological, and spiritual needs of the dying and their significant others.

RATIONAL SUICIDE

Rational suicide is the completion of suicide by an individual who is not mentally ill but who completes suicide for rational, lucid, and appropriate reasons (Corr et al., 2019). Rational suicide occurs more frequently these days prompting increased attention from the healthcare community. Balasubramaniam (2016, 2018) identifies reasons why rationale suicide may be the choice of some older adults. Among them are older people often die in an institutional setting away from the support of the family. Death has become more fearful as has aging. Age-related changes and the presence of healthcare issues force the individuals to redefine themselves in relation to what meaning their bodies now have to them. Many of the supports individuals had from family, friends, and involvements are diminishing, and they fear loneliness and institutionalization. Reflecting on life losses may have an impact on how they see their life and their future decisions.

A large percentage of the older population are baby boomers. As a group they value personal fulfillment, youth, and the ability to stay active and involved in older age but tend to have negative attitudes toward getting old (McCue, 2016).

Thoughts of death precipitate many deeply personal emotions. Clinicians when encountering the above scenarios in older adults should not be fearful to ask the client direct questions regarding suicide and death (Balasubramaniam, 2018). Dzeng and Pantilat (2018) recommend looking at rational suicide as a "slippery slope" involving current ethical issues that need to be addressed both ethically and clinically by all who care about and for older adults.

ADVANCE DIRECTIVES

Through the use of advance directives, competent individuals indicate orally or in writing their wishes regarding medical treatment they may or may not want in case they are unable to make such decisions or make them known to others in the future. These include the living will, durable power of attorney for healthcare, and patient care surrogate. Certain other advance directives, such as the disposition of the body, organ donation, or how the estate will be distributed, become effective upon an individual's death. The advance directives discussed here relate to an individual's medical treatment before death.

On December 1, 1991, through an Act of Congress, the Patient Self-Determination Act (PSDA) became law. It requires all institutions or agencies receiving Medicare or Medicaid funding to implement certain directives. They are as follows (Ramsey, 2004):

- To offer all patients written information about the right to make decisions concerning healthcare and to make advance directives
- To inquire of all patients being admitted for care if they have advance directives and, if they do not, to offer them written information as to how to execute such directives as the living will or durable power of attorney for healthcare
- To record the person's responses to these questions in their medical record
- To formulate written policies regarding the implementation of advance directives
- To educate the community and staff in regard to available advance directive options
- To show compliance with the requirements of the law
- To not discriminate in offering healthcare whether or not the person has advance directives

Despite the provisions in the PSDA, some individuals choose not to indicate their advance directive in writing, or even verbally, to others.

The Living Will

Developed in the 1970s, the living will was designed to allow competent individuals to record their wishes regarding specific medical treatment they do or do not want in the event they become terminally ill and unable to make healthcare decisions for themselves. Initially, the living will was not always recognized as a legal document; now, however, all 50 states and the District of Columbia have passed legislation legalizing it. Whether or not the stipulations of a living will are followed may depend on the specific diagnosis or situation at hand, the policies and procedures of the institution providing the care, and the specific laws of the state. Specific instructions regarding the withholding or withdrawing of life-prolonging procedures and the circumstances under which these are withheld plus any other wishes the person has should be indicated. The living will should state, for example, whether the person does not wish cardiopulmonary resuscitation (CPR) or to be placed on a respirator, be given antibiotics, or be given intravenous medication or tube feedings, and under what circumstances. Each state has its own requirements; thus, it is important that individuals know the laws governing advanced care planning in each state. When moving from state to state, advance directives should be checked by a lawyer to make sure they conform with the state's law. The living will should be signed and witnessed, with copies given to the lawyer, doctors, relatives, friends, or neighbors as desired. It is also recommended that its provisions be discussed with the above. Carrying a living will card with them indicating their wishes is also advised.

Medical Power of Attorney

Individuals can also designate a person to serve as a durable power of attorney for healthcare matters or as a healthcare surrogate. Each state has statutes that allow persons to designate to another person (through a written document) the power to make healthcare decisions on their behalf if they are not able to understand the situation or make the decision themselves. A considerable number of states have additional requirements that relate to decisions affecting medical treatments that sustain life such as a second medical opinion regarding the ability of the person to make or not make their own medical decisions (Dennis, 2009). The appointed person should specifically know the wishes and intent of the individual, which can vary greatly from person to person. The following are generally granted by the person to the healthcare surrogate: (a) withholding or granting the use of life-support measures or other types of medical care; (b) admitting the individual to a healthcare institution or discharging them from an institution; and (c) making other healthcare decisions not contained in the living will (Leming & Dickinson, 2007). It is recommended that individuals complete forms for an authorized living will as well as a durable power of attorney for healthcare.

The Five Wishes

The Florida Commission on Aging with Dignity, founded by Jim Towey in 1997 through a grant from the Robert Wood Johnson Foundation, authored "The Five Wishes," an alternative form for indicating advance directives. Not only does this document address living will issues, but also durable power of attorney for healthcare matters and associated psychosocial issues. "The Five Wishes" addresses (a) the person I want to make healthcare decisions for me when I can't make them myself, (b) my wish for the kind of medical treatment I want or don't want, (c) my wish for how comfortable I want to be, (d) my wish for how I want people to treat me, and (e) my wish for what I want my loved ones to know.

The latest version of "The Five Wishes" was updated in 2019 with the help of the Commission on Law and Aging. It meets legal requirements in 44 states and is used in all states along with another approved form in six states. Thus far more than 40 million people have completed the Five Wishes. The document is very user friendly and is available in 29 languages. A copy may be obtained for $5. It is also available digitally at www.fivewishes.org, Aging with Dignity (P.O. Box 1661, Tallahassee, Florida 32302-1661), or calling 888-5-WISHES (888-594-7437). A child's version "My Wishes" in English and Spanish (not a legal document) allows children to think about their choices regarding such things as pain control, healthcare, and psychosocial issues. Also available is an advanced care planning document called "Voicing My Choices" for seriously ill adolescents or young adults under the age of 21. This form is not a legal document but is helpful in considering end-of-life issues. Another useful resource is "People Planning Ahead: Communicating Healthcare and End-of-Life Wishes" (Kingsbury, 2009), designed specifically for people with intellectual and developmental disabilities.

Do-Not-Resuscitate Orders

Do-not-resuscitate (DNR) orders are another kind of advance directive. When 911 is summoned or an individual is taken to the emergency department, healthcare personnel are required to initiate CPR for a patient who has stopped breathing or whose heartbeat has ceased unless they have a valid DNR order indicating they do not wish to be resuscitated. Individuals may carry wallet cards or wear medical alert bracelets indicating they do not wish to be resuscitated, but the official paperwork must also be completed and available in these situations (DeSpelder & Strickland, 2020).

In response to the need for a physician's order for emergency medical technicians (EMTs) or other healthcare personnel to follow a patient's wishes regarding emergency care, Oregon, followed by several other states, developed the Physicians Order for Life-Sustaining Treatment (POLST). It is a pink two-sided form that indicates the patient's specific wishes regarding such issues as resuscitation and the types of treatments or medications they want or do not want. It is signed by the patient and the physician and is to be kept in a conspicuous place or carried on the person (DeSpelder & Strickland, 2020). No matter what type of advance directives one chooses, it is imperative that the family, friends, physician, clergy, and significant others know about the individual's wishes when the documents are completed.

ETHICAL WILL

An ethical will is a written method of passing down values, beliefs, feelings, traditions, blessings, and special advice to family and future generations by the dying person. Down through the centuries there is continued evidence of written ethical wills in the forms of letters or books.

There are similar themes found in ethical wills as individuals look at the past, the present, and forward into the future. The following is a suggested guide to writing an ethical will (Baines, 2006):

- Look into the past for significant stories of the family or personal stories, things that may have been learned in life as well as regrets.
- Themes of the present include sharing beliefs and values of self, your spiritual belief system, words of gratitude and caring, as well as apologies.
- Themes of the future might include leaving advice, blessings, and hopes for future generations, as well as special requests and even funeral plans.

An ethical will can offer great comfort and inspiration and become a treasure for families and loved ones.

BODY INTERNMENT

Traditionally after death most individuals choose a funeral, with the embalmed body present for a service and later buried in the ground or in an aboveground crypt. Gradually more individuals are choosing to be cremated, with the ashes kept in a receptacle or scattered in a selected place. Others choose to have a memorial service with or without the cremains present. Still other people have none of these. Considerable evidence supports the value for grievers to attend a ritual honoring the deceased person to help bring closure. Some believe it is beneficial to view the body to validate the death of the person and to support the grieving process. Although calling hours have been traditional in a funeral home, there is also the opportunity to express sympathy online at a website set up by the funeral home.

In recent years green burials are gaining credibility. Natural settings with trees, wildflowers, and grasses have been set aside across the country for green burials. Bodies not embalmed are placed in biodegradable containers and then put in the ground without a vault and covered with dirt. Traditional grave markers are not used but perhaps a stone or some simple sign to show where the person is buried is added.

THERAPEUTIC APPROACHES WITH THE BEREAVED

Therapeutic approaches to assist the bereaved include the following:

1. Become aware of the bereaved person's attitudes toward death and loss.
2. Accompany the individual on their journey through the grief experience.
3. Become comfortable using the terms *dying, dead,* and *death* rather than *passed away, expired,* or *left us.* When used in a caring manner they will help the griever focus on the reality of the loss.
4. Become aware of the normal and complicated grief manifestations and their effect on the person.
5. Assist the individual in understanding that the grieving process is a normal response to loss.
6. Refrain from categorizing grievers' responses into stages or phases. Allow the griever to experience wide range of grief manifestations as long as is necessary. No two grievers are alike.
7. Encourage the expression of feelings such as guilt, anger, and sadness.
8. Respect silence, be present to the individual, and listen actively when they wish to speak.
9. Offer no pious platitudes, such as "It was for the best," or "He's with God." Our loving, caring presence is most helpful in facilitating grief.
10. Use touch as appropriate. Touch can convey mutual sharing, love, and honest concern.
11. Listen to the person tell their story over and over again with compassion and without judgment.
12. Using alcohol as a means of assuaging grief should always be discouraged.
13. Although tranquilizers, sleeping pills, or antidepressants may be prescribed by a physician, it is wise to use them only as necessary and for a limited period. Encourage the continued use of regularly prescribed medications.
14. Encourage a medical examination after a loss and regularly thereafter. If symptoms appear, recommend seeing a healthcare provider.
15. Assess the survivors' nutritional state; assist them in obtaining and eating a healthy, well-balanced diet.
16. Promote healthy behavior such as regular exercise, stress management, and adequate sleep.
17. Become aware of the practical needs of the survivor, such as assistance with shopping, cooking, finances, caring for children, and so on.
18. Reminiscing may be helpful; through recalling the past the pain of loss may gradually be reduced, the past solidified, and memories of the loved one treasured.
19. Encourage the survivor to treasure memories but not to live only for memories.
20. Encourage the survivor to find a place in their life for the departed person, such as a revered ancestor. Create a memory wall of pictures of living and deceased family and friends that depict celebrations and memorable events.
21. Realize that grief takes time, that is, months, even years. Do not try to rush an individual through the grieving process; remember that everyone grieves differently.
22. Suggest they visit the internment site.
23. Write a personal diary during the loved one's dying and during the grieving period expressing feelings, regrets, remembrances, plans, and so on.
24. Be aware of spiritual and cultural responses and practices to death and loss. Provide persons, material, and places where meaningful ritual may take place.
25. Postpone making large decisions such as giving away valued items, moving, or even remarriage.
26. Encourage the individual to talk with others who have experienced a similar loss.

27. Evenings, holidays, anniversaries, and birthdays may be particularly difficult. Offer support especially at these times and encourage family and friends to do the same.
28. Suggest the griever maintain former social contacts and roles when they feel comfortable in doing so.
29. Encourage participation in religious rituals, group and individual prayer, and meditation if the person wishes. Religious beliefs are recognized as a strong support for persons during the dying and grieving process.
30. Maintain continued contact with the bereaved person throughout the grieving period and encourage others to do so too.
31. Recommend a professional grief counselor, psychologist, clergy, or clinical social worker if the individual shows evidence of needing such services.
32. Encourage the use of community resources such as hospice and grief support groups.

Death and Loss in the Digital Age

The use of electronic or digital media and the internet is widespread and growing rapidly among individuals of the middle and older age group. Email, social media, live streaming, FaceTime, smartphones, and other venues offer the unique ability to access others around the clock. Individuals document and share the dying and grieving experience day to day as well as their emotions with friends and others privately. Often individuals who have had a similar experience communicate with one another and receive support and help during this difficult time. Websites such as GriefNet, HealGrief, Compassionate Friends, Open to Hope, and others can be accessed for education and information (Smith & Cavuoti, 2013).

SUMMARY

Most older adults have lifelong experiences dealing with death, loss, and grief and are our best teachers regarding these issues. If they have adapted to past losses one by one in a healthy manner and have a viable support system, the likelihood is they will adapt to loss in the future. Loss is never easy no matter what age we are, but a faith system, coping strategies, and a loving support system can make a significant difference in arriving at a sense of peace and investing in others.

Being present and caring for and about persons during the dying and bereavement periods is not easy. It calls for love, commitment, and faith in the power of human and divine healing. The privilege of being a significant person to another in dying and grief touches that person and us as few other human experiences do. The rewards can be a unique appreciation for life, love, and the beauty of each person and the world around us.

REFERENCES

Aldersberg, M., & Thorne, S. (1990). Emerging from the chrysalis: Older widows in transition. *Journal of Gerontological Nursing, 16*(1), 4–8. https://doi.org/10.3928/0098-9134-19900101-03

Backer, B. A., Hannon, N. R., & Russell, N. A. (1994). *Death and dying: Understanding and care* (2nd ed.). Delmar Publishers.

Baines, B. K. (2006). *Ethical wills* (2nd ed.). Da Capo Press.

Balasubramaniam, M. (2016). A psychodynamic perspective on suicidal desire in the elderly. In R. E. McCue & M. Balasubramaniam (Eds.), *Rational suicide in the elderly: Clinical, ethical and sociocultural aspects* (pp. 149–158). Springer Publishing Company.

Balasubramaniam, M. (2018). Rational suicide: A clinical perspective. *Journal of the American Geriatrics Society, 66*(5), 999–1001. https://doi.org/10.1111/jgs.15263

Balk, D. E. (2013). Life span issues and loss, grief and mourning: Adulthood. In D. Meagher & D. Balk (Eds.), *Handbook of thanatology* (2nd ed., pp. 157–169). Routledge.

Beder, J. (2002). Grief anticipation. In R. Kastenbaum (Ed.), *Macmillan encyclopedia of death and dying* (Vol. 1, pp. 353–355). Macmillan.

Benoliel, J. Q. (1987). Health care providers and dying patients: Critical issues in terminal care. *Omega: Journal of Death and Dying, 18*(4), 341–363. https://doi.org/10.2190/4y6g-xqap-0xyn-lrw9

Berry, P. H., & Matzo, M. L. (2004). Death and aging society. In M. L. Matzo & D. W. Sherman (Eds.), *Gerontologic palliative care nursing*, pp. 31–50. Mosby.

Bhar, S. S. & Brown, G. K. (2012). Treatment of depression and suicide in older adults. *Cognitive and Behavioral Practice, 19*, 116–125. https://doi.org/10.1016/j.cbpra.2010.12.005

Blow, F. C., Brockmann, L. M., & Barry, K. L. (2004). Role of alcohol in late-life suicide. *Alcoholism, Clinical and Experimental Research, 28*(5), 48S–56S. https://doi.org/10.1097/01.alc.0000127414.15000.83?

Bonanno, G. A. (2009). *The other side of sadness: What the new side of science of bereavement tells us about after loss.* Basic Books.

Bonanno, G. A., Wortman, C. B., Lehman, D. R., Tweed, R. G., Haring, M., Sonnega, J., Carr, D., & Neese, R. M. (2002). Resilience to loss and chronic grief. A perspective study from pre-loss to 18 months post loss. *Journal of Personality and Social Psychology, 83*, 1150–1164.

Bowlby, J. (1969). *Attachment and loss: Attachment* (Vol. 1). Basic Books.

Butler, R. (1963). The life review: An interpretation of reminiscence in the aged. *Psychiatry, 26*, 67–76. https://doi.org/10.1080/00332747.1963.11023339

Centers for Disease Control and Prevention. (2021a). *Deaths and mortality.* National Center for Health Statistics. http://www.cdc.gov/nchs/fastats/deaths.htm

Centers for Disease Control and Prevention. (2021b). Older person's health. National Center for Health Statistics. https://www.cdc.gov/nchs/fastats/older-american-health.htm

Centers for Disease Control and Prevention (2021c). *Web-based Injury Statistics Query and Reporting System (WISQARS) [online].* National Center for Injury Prevention and Control. www.cdc.gov/injury/wisqars

Charmaz, K. (1980). *The social reality of death.* Addison-Wesley.

Constantino, R. E. (1981). Bereavement crisis intervention for widows in grief and mourning. *Nursing Research, 30*(6), 351–353.

Corr, C. A. (1992). A task-based approach to coping with dying. *Omega: Journal of Death and Dying, 24*(2), 81–94. https://doi.org/10.2190/CNNF-CX1P-BFXU-GGN4

Corr, C. A. (2019). The "five stages" in coping with dying and bereavement: Strengths, weaknesses, and some alternatives. *Mortality, 24*(4), 405–417. https://doi.org/10.1080/13576275.2018.1527826

Corr, C., Corr, D., & Doka, K. (2019). *Death and dying, life and living* (8th ed.). Cengage.

Crenshaw, D. A. (2013). Life span issues and assessment and intervention. In D. Meagher & D. Balk (Eds.), *Handbook of thanatology* (2nd ed., pp. 230–254). Routledge.

Davies, B. & Orloff, S. (2010). Bereavement issues and staff support. In G. Hanks, N. I. Cherny, N. A. Christakis, M. Fallon, S. Kaasa, & R. K. Rortney (Eds.), *Oxford textbook of palliative medicine* (4th ed., pp. 1361–1372). Oxford University Press.

Death with Dignity National Center. (2021). *Death with dignity acts.* https://www.deathwithdignity.org/learn/death-with-dignity-acts/

Dennis, D. (2009). *Living, dying, and grieving.* Jones & Bartlett.

DeSpelder, L. A., & Strickland, A. L. (2020). *The last dance: Encountering death and dying* (11th ed.). McGraw-Hill.

Doka, K. (1989). *Disenfranchised grief.* Lexington Books.

Doka, K. (1993). *Living with a life-threatening illness: A guide for patients, their families, and caregivers.* Lexington Books.

Doka, K. (2009). *Counseling individuals with life-threatening illness.* Springer Publishing Company.

Dutton, Y. D., & Zisook, S. (2005). Adaptation to bereavement. *Death Studies, 29*(10), 877–903. https://doi.org/10.1080/07481180500298826

Dzeng, E., & Pantilat, S. Z. (2018). Social causes of rational suicide in older adults. *Journal of the American Geriatrics Society, 66*(5), 853–855. https://doi.org/10.1111/jgs.15290

Egan, K. A., & Labyak, M. J. (2006). Hospice palliative care: A model for quality end-of-life care. In B. R. Ferrell, & N. Coyle (Eds.), *Textbook of palliative nursing* (2nd ed., pp. 13–46). Oxford University Press.

Freud, S. (1957). Mourning and melancholia. In J. Strachey (Ed. and Trans.), *The second edition of the complete works of Sigmund Freud* (Vol. 24). Hogarth Press (Original work published 1915).

Fulton, R. (2003). Anticipatory mourning: A critique of the concept. *Mortality, 8*, 342–357. https://doi.org/10.1080/13576270310001613392

Gaynes, B. N., West, S. L., Ford, C. A., Frame, P., Klein, J., Lohr, K. N., & U.S. Preventive Services Task Force. (2004). Screening for suicide risk in adults: A summary of the evidence for the U.S. Preventive Services Task Force. *Annals of Internal Medicine, 140*(10), 822–835. https://doi.org/10.7326/0003-4819-140-10-200405180-00015

Glaser, B. G., & Strauss, A. L. (1965). *Awareness of dying.* Aldine Publishing.

Herz-Brown, F. (1988). The impact of death and serious illness on the family life cycle. In B. Carter & M. McGoldrick (Eds.), *The changing family life cycle: A framework for family therapy* (pp. 457–481). Gardner Press.

Holkup, P. A. (2003). Evidence-based protocol: Elderly suicide-secondary prevention. *Journal of Gerontological Nursing, 29*(6), 6–17. https://doi.org/10.3928/0098-9134-20030601-05

Holland, J. M., & Neimeyer, R. A. (2010). An examination of stage theory of grief among individuals bereaved by natural and violent causes: A meaning-oriented contribution. *Omega Journal of Death and Dying, 61*(2), 103–120. https://doi.org/10.2190%2FOM.61.2.b

Huaiquil, A. (2018). New palliative care guidelines released. *Provider, 45*(12), 11.

Isherwood, L. M., King, D. S., & Luszcz, M. A. (2012). Longitudinal analysis of social engagement in late-life widowhood. *International Journal of Aging and Human Development, 74*(3), 211–229. https://doi.org/10.2190/AG.74.3.c

Jacobs, M. (2013). Ethical and legal issues in end-of-life decision making. In D. Meagher & D. Balk (Eds.), *Handbook of thanatology* (2nd ed., pp. 101–109). Routledge.

Jett, K. (2008). Loss, death and dying in late life. In P. Ebersole, P. Hess, T. A. Touhy, K. Jett, & A. S. Luggen (Eds.). *Toward healthy aging* (7th ed., pp. 639–665). Mosby Elsevier.

Johns Hopkins University. (2021). Covid-19 dashboard. Center for Systems Science and Engineering at Johns Hopkins University. https://www.arcgis.com/apps/opsdashboard/index.html#/bda7594740fd40299423467b48e9ecf6

Kalish, R. A. (1985). *Death, grief, and caring relationships* (2nd ed.). Brooks/Cole.

Kastenbaum, R. J. (1969). Death and bereavement in later life. In A. H. Kutscher (Ed.), *Death and bereavement* (pp. 27–54). Charles C Thomas.

Kastenbaum, R. J. (2012). *Death, society and human experience* (11th ed.). Pearson Education.

Kauffman, J. (1989). The chronic grief of the nursing home. *Forum Newsletter, 13*(1), 6–7.

Kingsbury, L. (2009). *People planning ahead: A guide to communicating healthcare and end-of-life wishes.* American Association on Intellectual and Developmental Disabilities.

Kübler-Ross, E. (1969). *On death and dying.* Macmillan.

Kuebler, K. K., Heidrich, D. E., & Esper, P. (2007). *Palliative and end-of-life care* (2nd ed.). Springer.

Leming, M. R., & Dickinson, G. E. (2007). *Understanding death, dying, and bereavement* (6th ed.). Thomson Wadsworth.

Lindemann, E. (1944). Symptomatology and management of acute grief. *American Journal of Psychiatry, 101*, 141–148. https://psycnet.apa.org/doi/10.1176/ajp.101.2.141

McCue, R. E. (2016). Baby boomers and rational suicide. In R. E. McCue & M. Balasubramaniam (Eds.), *Rational suicide in the elderly: Clinical, Ethical and sociocultural aspects.* Springer Publishing Company.

National Consensus Project for Quality Palliative Care. (2018). *Clinical practice guidelines for quality palliative care* (4th ed.). https://www.nationalcoalitionhpc.org/wp-content/uploads/2020/07/NCHPC-NCPGuidelines_4thED_web_FINAL.pdf

National Institute of Mental Health. (2021). *Suicide.* https://www.nimh.nih.gov/health/statistics/suicide.shtml

Not Dead Yet. (2021). Who we are. https://notdeadyet.org/about

Parkes, C. M. (1972). *Bereavement.* International Universities Press.

Ponzetti, J. J, & Johnson, M. A. (1991). The forgotten grievers: Grandparents reaction to the death of grandchildren. *Death Studies, 15*(2), 157–167. https://doi.org/10.1080/07481189108252420

Rainey, L. C. (1988). The experience of dying. In H. Wass, F. M. Bernardo, & R. A. Neimeyer (Eds.), *Dying: Facing the facts* (pp. 137–157). Hemisphere.

Ramsey, G. C. (2004). Legal aspects of palliative care. In M. L. Matzo & D. W. Sherman (Eds.), *Palliative care nursing: Quality care to the end of life* (2nd ed., pp. 180–218). Springer Publishing Company.

Reed, M. L. (2000). *Grandparents cry twice: Help for bereaved grandparents*. Baywood.
Retsinas, J. (1988). A theoretical reassessment of the applicability of Kübler-Ross' stages of dying. *Death Studies, 12*(3), 207–216. https://doi.org/10.1080/07481188808252237
Rondo, T. A. (1984). *Grief, dying and death: Clinical interventions for caregivers*. Research Press.
Rondo, T. A. (1988). *Grieving: How to go on living when someone you love dies*. Lexington.
Rondo, T. A. (1993). *Treatment of complicated mourning*. Research Press.
Rondo, T. A. (2000). *Clinical dimensions of anticipatory mourning*. Research Press.
Roscoe, L. A., Malphurs, J. E., Dragovic, L. J., & Cohen, D. (2000). Dr. Jack Kevorkian and cases of euthanasia in Oakland County, Michigan, 1990–1998. *The New England Journal of Medicine, 343*(23), 1735–1736. https://doi.org/10.1056/NEJM200012073432315
Rosen, E. J. (1998). *Families facing death: Family dynamics of terminal illness*. Lexington.
Smith, A. M., & Cavuoti, C. (2013). Thanatology in the digital age. In D. Meagher & D. Balk (Eds.), *Handbook of thanatology* (2nd ed., pp. 429–439). Routledge.
Stillion, J. M., & McDowell, E. E. (1996). *Suicide across the lifespan* (2nd ed.). Taylor & Francis.
Touhy, T. A. (2008). Emotional health in late life. In P. Ebersole, P. Hess, T. A. Touhy, K. Jett, & A. S. Luggen (Eds.), *Toward healthy aging* (7th ed., pp. 597–635). Mosby Elsevier.
Touhy, T. A. (2012). Mental health. In T. A. Touhy & K. Jett (Eds.), *Ebersole & Hess' toward healthy aging: Human need and nursing responses* (8th ed., pp. 338–364). Mosby Elsevier.
Walter, T. (1996). A new model of grief: Bereavement and biography. *Mortality, 1*(1), 7–25. https://doi.org/10.1080/713685822
Weisman, A. D. (1972). *On death and denying*. Behavioral Publications.
West Ellson, N. (2007). Suicide in older adults. A priority concern. *Advance for Nurse Practitioners, 15*(12), 53–68.
Worden, J. W. (2009). *Grief counseling and grief therapy: A handbook for the mental health practitioner* (4th ed.). Springer Publishing Company.
World Health Organization. (2020). *Palliative care*. https://www.who.int/news-room/fact-sheets/detail/palliative-care

Appendix A

Practical Hints for the Safety of Older Adults

VISION

Increase illumination throughout the home.

Use spot lighting for reading or work.

Good lighting is especially important in stairways (particularly at the top and bottom of stairs) and in bathrooms and kitchens.

Encourage the use of night lights, especially in bathrooms and bedrooms.

Reduce glare from windows and shiny surfaces.

Have lamps and light switches positioned so that lights can be turned on when entering a room.

Magnifying glasses are useful for threading needles, reading medication instructions, reading telephone numbers, and so on.

Large dials or marked dials should be used on appliances and telephones. The "off" position should be clearly marked on appliances.

Emergency numbers should be written in large print and kept near the telephone.

Surfaces should be painted or carpeted so that there is a clear and distinct boundary between stair steps, floor surfaces, and thresholds. Contrasting colors should be used.

Electrical cords, footstools, and other low objects should be kept out of walkways.

Furniture should not be moved to unfamiliar locations in rooms.

Robes or other loose-fitting garments should not be worn in the kitchen, where they may catch fire or get caught in appliances.

Internal and external medications should be stored separately.

Medicines should not be taken at night without turning on the light and putting on proper glasses, if needed, to see well.

Objects on the dining table should be spaced so they are not easy to knock over if vision and coordination are impaired.

Older persons should be especially cautious about moving quickly when there are small pets or small children in the home.

HEARING

Because hearing loss is embarrassing to many people, the hearing-impaired individual who only hears part of a conversation will often try to guess the rest. When a person's answers seem inappropriate for the question asked, be alert to possible hearing impairment.

Ask the question in a different way, but do not shout.

Suspicious, seemingly paranoid behavior may accompany hearing loss, so make sure the person understands what you are doing or saying and why you are doing or saying it.

Indications of hearing impairment include cupping hands to the ear, leaning forward, watching faces of speakers intently, and nodding in the affirmative whether appropriate or not.

Always speak slowly, lower the pitch of the voice, and enunciate clearly.

Face the individual so they can see your face and gestures.

Use gestures and facial expressions to enhance communication, not to distract.

Sometimes touching the person helps to get their attention before speaking, but be sure the person doesn't mind being touched.

Give basic information first; elaborate the details later.

If the message is very important, such as diet or medication information, write it down, go over it with the hearing-impaired person, and leave the memo with them for future reference.

TASTE AND SMELL

Remember that older persons may not be aware of changes in these sensory systems.

Failing to smell or taste spoiled food or failing to smell smoke or gas are dangerous. Be alert to these cues when visiting in the home.

Loss of variety in taste sensations may occur with age. Use textural and color differences in foods and separate foods on the plate so that they can be more easily differentiated.

Remember that some older adults may not be aware of body odor, such as urine or feces, or smells from use of excessive perfumes and colognes.

TOUCH AND BALANCE

Touch is very important in ambulation because information is needed from receptors in the soles of the feet to provide information necessary for secure walking or climbing.

Older adults should wear sturdy, safe shoes at all times.

Encourage the older person to walk more slowly and to use a cane or walker if needed. Some individuals may need to hold on to someone else when in a crowd or on uneven terrain.

Fasten all carpets and rugs firmly to the floor. Do not use rugs at the top of stairs or where flooring is uneven.

Mark or eliminate all uneven floor surfaces.

Install railings on all stairways, inside and out.

Use nonslip wax or carpeting on all floor surfaces.

Keep liquids, food, and other debris off floors.

Use railings on tubs and near toilets and nonslip pads in showers or tubs.

Do not use bath oils or other lubricants in the tub or shower because of the danger of falls.

Encourage older adults to use railings, furniture, and walls to help maintain balance if they are unsteady.

Teach older persons to change body position gradually—especially in rising from a lying or seated position.

Teach older persons to be cautious in looking up or turning the head quickly, because such movements may produce dizziness.

Chairs should provide firm support and have solid arms that are useful in rising from them. Pedestal tables may tip over if one presses down on them when getting up from a chair. Anchor such furniture or buy tables with four sturdy legs.

Arrange shelves so that no climbing or extreme stretching is necessary, especially to get objects frequently used. Kitchens should be arranged to conserve energy and to reduce climbing, bending, stooping, and heavy lifting, but don't encourage too little activity around the home. Some climbing, bending, stooping, and lifting are good exercises to help maintain mobility and fitness; however, overexertion and strain are to be avoided.

In general, teach the older adult to be more aware of possible dangers in the environment and to be more conscious of the need to modify or adjust behaviors depending on their limitations.

Appendix B

Resources

Academy of Nutrition and Dietetics: www.eatright.org
Administration on Aging: www.acl.gov/about-acl/administration-aging
Administration on Community Living: www.acl.gov
AIDS Information: www.aidsinfo.nih.gov
Alzheimer's Association: www.alz.org
Alzheimer's Disease Education: www.nia.nih.gov/alzheimers
American Academy of Pain Medicine: www.painmed.org
American Association for Geriatric Psychiatry: www.aagponline.org
American Association on Intellectual and Developmental Disabilities: www.aaidd.org
AARP: www.aarp.org
American Cancer Society: www.cancer.org
American Council of the Blind: www.acb.org
American Diabetes Association: www.diabetes.org
American Foundation for the Blind: www.afb.org
American Geriatrics Society: www.americangeriatrics.org
American Heart Association: www.americanheart.org
American Lung Association: www.lung.org
American Parkinson's Disease Association: www.apdaparkinson.org
American Sleep Apnea Association: www.sleepapnea.org
American Society on Aging: www.asaging.org
American Urological Association: www.auanet.org
Arthritis Foundation: www.arthritis.org
Better Hearing Institute: www.betterhearing.org
Centers for Disease Control and Prevention—Chronic Disease Prevention and Health Promotion: www.cdc.gov/nccdphp
Food and Drug Administration: www.fda.gov
Food and Nutrition Information Center: www.nal.usda.gov/fnic

Foundation for Health in Aging: www.healthinaging.org

Gerontological Society of America: www.geron.org

Glaucoma Foundation: www.glaucomafoundation.org

Hearing Loss Association of America: www.hearingloss.org

Hospice Foundation of America: www.hospicefoundation.org

National Alliance for Caregiving: www.caregiving.org

National Association for Continence: www.nafc.org

National Association of the Deaf: www.nad.org

National Cancer Institute: www.cancer.gov

National Center for Complementary and Alternative Medicine: www.nccam.nih.gov

National Center on Elder Abuse: www.ncea.acl.gov

National Center on Sleep Disorders NHLBI Information Center: www.nhlbi.nih.gov/about/ncsdr/

National Council on the Aging: www.ncoa.org

National Digestive Diseases Information Clearinghouse: www.digestive.niddk.nih.gov

National Eye Institute: www.nei.nih.gov

National Federation of the Blind: www.nfb.org

National Heart, Lung, and Blood Institute: www.nhlbi.nih.gov

National Hospice and Palliative Care Organization: www.nhpco.org

National Institute on Aging: www.nia.nih.gov

National Institute of Arthritis and Musculoskeletal and Skin Diseases: www.niams.nih.gov

National Institute on Deafness and Other Communication Disorders: www.nidcd.nih.gov

National Institutes of Health, Osteoporosis and Related Bone Diseases: www.bones.nih.gov

National Institute of Mental Health: www.nimh.nih.gov

National Institute of Neurological Disorders and Stroke: www.ninds.nih.gov

National Kidney Foundation: www.kidney.org

National Kidney and Urologic Diseases Information Clearinghouse: www.kidney.niddk.nih.gov

National Long-Term Care Ombudsman Program: www.ltcombudsman.org

National Osteoporosis Foundation: www.nof.org

National Parkinson Foundation: www.parkinson.org

Road Scholar (Elderhostel Lifelong Learning): www.roadscholar.org

Rosalynn Carter Institute for Caregiving: www.rosalynncarter.org

U.S. Department of Health and Human Services: www.hhs.gov

U.S. Government Disability Services: www.usa.gov/disability-services

Glossary

Accommodation: Adjustment of the lens of the eye that brings light rays from various distances to a focus on the retina.

Acetylcholine: A chemical transmitter released by some nerve endings.

Acid–base balance: Situation in which acidity or alkalinity (pH) of the blood is maintained between 7.35 and 7.45.

Acidosis: Excessive acidity of body fluids caused by accumulation of acids or an excessive loss of bicarbonate.

Acuity: Sharpness; distinctness; as in the various sensory systems.

Acute: Severe; having a rapid onset. Not long lasting.

Amyloid: Starchlike protein produced and deposited in tissues during certain pathological states. Found in plaque deposits in heart, blood vessels, and brain tissues.

Anaphylactic shock: A systemic allergic or hypersensitivity reaction producing life-threatening changes in circulation and breathing.

Anemia: Reduced oxygen-carrying ability of blood from too few red blood cells or abnormal hemoglobin.

Aneurysm: Blood-filled sac in an artery wall caused by weakening of the wall or dilation.

Angina: Spasmodic chest pain and a feeling of suffocation caused by a deficiency of oxygen to heart muscle.

Angioplasty: Altering the structure of a vessel, either by surgical procedure or dilating the vessel using a balloon inside the lumen.

Angiotensin-converting enzyme inhibitors (ACE inhibitors): Drugs that lower total peripheral resistance in the circulatory system.

Anorexia: Loss of appetite or desire for food.

Anoxia: Deficiency of oxygen.

Antibiotic: Natural or synthetic substances that inhibit the growth or destroy microorganisms.

Antibody: Complex glycoproteins produced by B lymphocytes in response to the presence of an antigen.

Anticholinergic: An agent that blocks parasympathetic nerve impulses.

Anticoagulant therapy: Use of drugs that interfere with blood coagulation.

Antigen: A protein marker on the surface of cells that identifies the cells as "self" or "nonself."

Antioxidant: An agent that prevents or inhibits oxidation (any process in which a substance combines with oxygen).

Aphasia: Absence or impairment of ability to communicate through speech, writing, or signs; caused by dysfunction of brain centers.

Aroma therapy: Use of aromatic oils in healing the individual and relieving some symptoms.

Arrhythmias: Irregular rhythm or loss of rhythm, especially of the heartbeat.

Arterial blood gases (ABGs): Clinically, the determination of the levels of oxygen and carbon dioxide in the blood. Important in diagnosing and treating disturbances of acid–base balance.

Arteriosclerosis: Hardening of the arteries; loss of elasticity.

Artery: A vessel that conveys blood away from the heart.
Arthritis: Inflammation in a joint or joints.
Articulation: Joint; point where two or more bones meet.
Atherosclerosis: Hardening of the arteries caused by deposits of fatty material in arterial walls.
Atrophy: A wasting or a decrease in size of an organ or tissue.
Audiogram: Record of a hearing test using an audiometer.
Autoimmune response: Production of antibodies or effector T cells that attack a person's own tissues.
Autonomic nervous system (ANS): The part of the nervous system concerned with control of involuntary body functions. Also called the visceral motor system.
Axillary: Pertaining to the axilla or armpit.
Barium swallow: x-Ray imaging of the esophagus during and after introduction of a contrast medium consisting of barium sulfate. Indicates structural abnormalities of the esophagus.
Benign: Not malignant.
Beta blocker (beta-adrenergic blocking agent): A substance that interferes with transmission of stimuli through pathways normally allowing sympathetic nervous-inhibiting stimuli to be effective.
Bile: Fluid secreted by the liver and sent to the small intestine to aid in digestion and absorption of fats.
Bilirubin: Orange or yellow pigment of bile.
Biopsy: Excision or removal of a small piece of living tissue for microscopic examination.
Blood–brain barrier: A barrier membrane between circulating blood and the brain that prevents certain substances from reaching brain tissue and cerebrospinal fluid.
Bradycardia: A heart rate less than 60 beats per minute.
BUN: Blood urea nitrogen: Nitrogen in the blood in the form of urea. An increase in BUN usually indicates decreased renal function.
Bursa: A padlike sac or cavity containing synovial fluid and found near joints, where friction is likely to occur.
Bypass surgery: Surgically installing an alternate route for blood to bypass an obstruction of a main or vital artery; a shunt.
Calcitonin: Hormone produced by the thyroid gland important in bone and calcium metabolism. Promotes a decrease in calcium levels of the blood. Sometimes called thyrocalcitonin.
Calcium antagonist (or calcium entry blockers): A group of drugs that slow the influx of calcium ions into muscle cells, resulting in decreased arterial resistance and decreased myocardial oxygen demand.
Cardiac catheterization: Passage of a tiny plastic tube into the heart through a blood vessel. Used in diagnosis of heart disorders.
Cardiac output: Volume of blood ejected in 1 minute by either of the ventricles of the heart.
Cardiotonics: Drugs that increase the tonus or steady contraction of the heart.
Caries: Decay or disintegration of soft or bony tissue or a tooth.
Cartilage: Nonvascular, tough, flexible connective tissue.
Cataract: A partial or complete clouding of the lens of the eye.
Catheter: A tube passed through the body for evacuating or injecting fluids into body cavities.
Chemotherapy: Application of chemicals that have a specific toxic effect on disease-causing microorganisms.
Cholesterol: Steroid found in animal fats as well as most body tissues. Made by the liver.
Chronic: Of long duration. A disease showing little change or slow progression. Opposite of acute.
Climacteric: The period leading to the cessation of a female's reproductive ability (menopause) and a corresponding period of lessened viable sperm production in the male.

Collagen: A fibrous, insoluble protein found in connective tissue. Represents about 30% of total body protein.

Collateral circulation: Circulation of small interconnecting blood vessels, especially when a main artery is obstructed.

Colonoscopy: Examination of the colon with an elongated tube and optical system to view the interior of the colon.

Colostomy: Surgical opening of some portion of the colon through the abdominal wall to its outside surface.

Connective tissue: A primary tissue; form and function vary extensively. Functions include support, storage, and protection.

CT scan (computed tomography): X-ray technique that produces a film representing a detailed cross-section of tissue structure. Procedure is painless and noninvasive.

Curettage: Scraping of a cavity, removal of dead tissue; also called debridement.

Cystourethroscopy: Examination of the posterior urethra and urinary bladder using a cystourethroscope.

Cytoplasm: The cellular material surrounding the cell nucleus and enclosed by the cell membrane.

Debridement: (*see* Curettage).

Defecation: Evacuation of the bowels.

Dehydration: The removal of water, as from the body or a tissue.

Dialysis: Process of diffusing blood across a semipermeable membrane to remove toxic materials and to maintain fluid, electrolyte, and acid–base balance.

Diastolic: The lower number of a blood pressure reading that reflects the relaxation phase of the heart's pumping cycle.

Dissection: Cutting of body parts for purpose of separation and studying.

Diuretic: Chemicals that increase urinary output.

Diverticula (singular: diverticulum): Sacs or pouches in the walls of a canal or organ.

Dual-photon absorptiometry (DPA): Use of two sources of radiation of different energies to measure density of a material, especially bone.

Duct: A narrow tubular vessel or channel, especially one serving to convey secretions from a gland.

Edema: Abnormal accumulation of fluid in body parts or tissues; causes swelling.

EEG (electroencephalogram): A record of the electrical activity of nerve cells in the brain.

Effectors: Nerve endings in the muscle and glands.

EKG (electrocardiogram): A record of the electrical activity of the heart.

Elastin: Connective tissue protein that is the principal component of elastic fibers.

Electrocautery: Destroying tissue by electrical current.

Electrolyte: Chemical substances (such as salts, acids, and bases) that ionize and dissociate in water and are capable of conducting an electrical current.

Embolism: Obstruction of a blood vessel by an embolus (blood clot, fatty mass, bubble of air, or other debris) floating in the blood.

Emollient: An agent that will soften and soothe when applied locally.

Endarterectomy: Surgical removal of the lining of an artery. Performed on almost any major artery that is diseased or blocked.

Endocrine: Pertains to a gland that secretes directly into the bloodstream. Ductless gland.

Endoscope: A device consisting of a tube and optical system for observing the inside of a hollow organ or a cavity. Observation done through a natural body opening or through a small incision.

Enzyme: A protein that acts as a biological catalyst to speed up chemical reactions.

Erythrocyte sedimentation rate (ESR): A laboratory test of the speed at which red blood cells settle.

Estrogen: Female sex hormone.

Etiology: Cause of a disease.

Exocrine: Gland whose secretion is carried to a particular site through a duct.

Extracorporeal shock wave lithotripsy (ESWL): A procedure to crush stones (calculi) in the bladder or urethra using vibrations of sound waves.

Fats, monounsaturated: Fatty acids with only one double or triple bond per molecule. Found primarily in almonds, pecans, cashew nuts, peanuts, and olive oil.

Fats, polyunsaturated: Fatty acids with more than one double or triple bond per molecule. Found in fish, corn, walnuts, sunflower seeds, soybeans, cottonseeds, and safflower oil.

Fats, saturated: Any number of fatty acids in which all the atoms are joined by single bonds. Chiefly of animal origin.

Fats, unsaturated: Any number of fatty acids in which some of the atoms are joined by double or triple bonds. Bonds are easily split in chemical reactions and other substances are joined to them.

Fibrillation: Quivering or spontaneous contraction of individual muscle fibers.

Fibrocystic: Consisting of fibrocysts or fibrous tumors that have undergone cystic degeneration or ones that have accumulated fluid in the tissue interspaces.

Filtration: Passage through a filter or through a material that prevents passage of certain molecules.

Folate: A salt of folic acid, a water-soluble vitamin of the B complex group.

Gallstone: A calculus or deposit formed in the gallbladder or bile duct.

Gastroscope: An endoscope for inspecting the stomach's interior.

Genogram: A format for drawing a family tree recording information about family history and the nature of relationships within a family.

Glucagon: Hormone formed in islets of Langerhans in the pancreas; raises glucose level of the blood.

Glucose: A sugar; the most important carbohydrate in body metabolism.

Glucose tolerance test: A test to determine ability to metabolize glucose.

Glycogen: Carbohydrate stored in cells for future conversion into glucose. Used in performing muscular work and liberating heat.

Goiter: Enlargement of the thyroid gland.

HDL (high-density lipoprotein): Plasma lipids, bound to albumin, consisting of lipoproteins that have more protein than low-density lipoproteins.

Heart block: Impaired transmission of impulses from atrium to ventricle resulting in arrhythmia.

Hematoma: A swelling or mass of blood (usually clotted) confined to an organ, tissue, or space and caused by a break in a blood vessel.

Hemoglobin: The red oxygen-carrying pigment found in red blood cells.

Hernia: Abnormal protrusion of an organ or body part through the containing wall of its cavity.

Homeostasis: Maintenance of a stable internal environment of the body; state of body equilibrium.

Hormone: Substance released to the blood that acts as a chemical messenger to regulate specific body functions.

Hypoxia: Oxygen deficiency.

Immunotherapy: The production or enhancement of immunity.

Incontinence: Inability to retain urine or feces through loss of sphincter control or because of cerebral or spinal lesions.

Infarct: Area of dead, deteriorating tissue resulting from a lack of blood supply.

Ion: A particle carrying an electric charge.

Ischemia: Local and temporary decrease of blood supply caused by obstruction of the circulation to a part.

Joint: A junction between bones. Usually formed of fibrous connective tissue and cartilage.
Ketosis: Accumulation of ketone bodies resulting from incomplete metabolism of fatty acids.
Lability: State of being unstable or changeable; rapidly shifting and changing emotions.
Labyrinth: System of interconnecting canals and cavities that make up the inner ear.
Laser: A device that emits intense heat and power at close range. Can be focused on a very small target. Used in surgery to divide or to cause adhesions or to destroy or fix tissue in place.
Lean body mass: The weight of the body minus the fat content.
Learning styles: Describes specific ways and individual learns, auditory, visual, and physical.
Lesion: An injury or wound; a circumscribed area of pathologically altered tissue.
Lethargy: Sluggishness; stupor.
Leukoplakia: Formation of white spots or patches on mucous membrane of tongue or cheek: May become malignant.
Ligament: Band of flexible connective tissue that binds bones together and reinforces joints.
Lipids: Any one of a group of fats or fatlike substances; organic compounds formed of carbon, hydrogen, and oxygen. Examples are fats and cholesterol.
Lipofuscin: Aging pigment; insoluble lipid pigment found in cardiac and smooth muscle cells; often seen in cells undergoing atrophy.
Lipoproteins: Proteins combined with lipid or fat components such as cholesterol and triglyceride.
Lyme disease: Inflammatory disease caused by a tick bite.
Lymph: Fluid in the lymph vessels, collected from fluid in the spaces between tissues or structures.
Malabsorption syndrome: Inadequate absorption of nutrients from the intestinal tract, especially the small intestine.
Malaise: Discomfort, uneasiness; often indicative of infection.
Malignant: Life-threatening; opposite of benign.
MAO inhibitors (monoamine oxidase inhibitors): Any of a chemically heterogeneous group of drugs used primarily to treat depression.
Melanin: Dark pigment imparting color to skin and hair. Exposure to sunlight stimulates melanin production.
Metabolism: Sum total of all chemical reactions in body cells that transform substances into energy or materials the body can use or store.
Metastasize: Movement of bacteria or body cells (especially cancer cells) from one part of the body to another.
Mindfulness: A form of meditation used to reduce stress and bring peace to the individual.
Miosis: Abnormal contraction of the pupils.
Motility: Power to move spontaneously.
Motor: Causing motion. A part or center that induces movement; for example, muscles or nerves.
MRI (magnetic resonance imaging): Medical imaging that uses a strong magnetic field to image the heart, large blood vessels, brain, and soft tissues. Does not involve exposure to radiation or injecting a contrast medium such as a dye.
Mucus: Sticky, thick, fluid secreted by mucous glands and mucous membranes that keeps the free surface of membranes moist.
Muscle: A contractile organ composed of individual muscle fibers and muscle cells.
Neuritis: Inflammation of a nerve.
Neurofibrillary tangles: Twisted neurofibrils, or tiny fibers, extending in every direction in the cytoplasm of the nerve cell body.
Neuron: Nerve cell. Consists of a cell body, an axon, and one or more dendrites (processes).
Neuropathy: Any disease of the nerves.
Neuroplasticity: Resiliency in the brain allowing it to compensate for age-related changes.
Nodule: A small aggregation or grouping of cells.

Normal pressure hydrocephalus: Enlarged ventricles of the brain with no increase in spinal fluid pressure or no demonstrable block to outflow of spinal fluid.

Occult blood: Blood in such small amounts that it is recognized only by microscopic examination or by chemical means.

Orthostatic hypotension (postural hypotension): Decrease in blood pressure on assuming an erect posture.

Pacemaker: A specialized group of heart cells that automatically generates impulses that spread to other regions of the heart. Anything that influences the rate and rhythm of occurrence of some activity or process.

Palsy: Temporary or permanent loss of sensation or loss of ability to move or to control movement.

Pathology: Having to do with disease; the study of the nature and cause of disease.

Perception: Interpretation of sensory stimuli.

Peristalsis: Progressive wavelike contractions that move food through the alimentary canal.

Peritoneum: The membrane lining the inside of the abdomen and covering the abdominal organs.

pH: The measure of the relative acidity or alkalinity of a solution. The neutral point, where a solution is neither acid nor alkaline, is a pH of 7. Increasing acidity is reflected by a number less than 7.

Phlebitis: Inflammation in a vein.

Plaque: In arteries, raised lesions associated with atherosclerosis; in the brain, lesions composed of degenerating neurons and amyloid.

Podiatrist: A health professional responsible for care and treatment of the human foot.

Polyp: A tumor with a stem that attaches a new growth.

Polypharmacy: Taking many drugs at one time, excessive use of drugs.

Presbycusis: Lessening of hearing acuity that occurs with age. Affects mainly high frequencies.

Presbyopia: The farsightedness associated with aging.

Progesterone: A hormone that helps prepare the uterus for the fertilized ovum or ova.

Proprioception: Awareness of posture, movement, and changes in equilibrium.

Psychopathology: The study of abnormal behavior and its manifestations, development, and causation.

Psychotropic drugs: Drugs affecting psychic function, behavior, or experience.

Radiation therapy: Using ionizing radiation in the treatment of malignant tumors.

Range of motion: The range of motion of a joint. Movement of joints through their available range of motion.

Rational suicide: Completion of suicide by an individual who is rational and lucid.

Reabsorption: To get back into the bloodstream a substance previously sent from the blood into spaces between body cells.

Receptor: Sensory nerve endings or a group of cells or a sense organ that, when stimulated, gives rise to a sensory impulse.

Reflex: Automatic reaction to a stimulus.

Refraction: Deflection from a straight path, as of light rays when they pass through media of different densities.

REM sleep: Rapid eye movement sleep. REM sleep periods last from a few minutes to a half hour, and alternate with non-REM periods. Dreaming occurs during REM sleep.

Remission: The period during which symptoms diminish or are reduced.

Renal: Pertaining to the kidney.

Renal clearance test: A kidney function test that evaluates the ability of the kidneys to eliminate a given substance in a standard time.

Reticular formation: Groups of cells and fibers arranged in a diffuse network throughout the brainstem.
Sclerosis: A hardening of tissue.
Sebaceous gland: Oil-secreting gland of the skin. Contains sebum.
Sebum: A fatty secretion of the sebaceous gland.
Sensory: Pertaining to sensation: Conveying impulses from sense organs to reflex or higher centers.
Serum: The watery portion of the blood after coagulation; clear watery fluid secreted by cells of a serous membrane.
Serum calcium: A test that measures the levels of calcium in the blood serum.
Serum creatinine: A test that measures the levels of creatinine in the blood. Used to diagnose kidney impairment, which causes a rise in creatinine metabolism.
Serum phosphorus: A test that measures the levels of phosphorus in the blood serum. Phosphorus is essential for bone formation.
Sigmoidoscopy: Use of a tube with an optical system (sigmoidoscope) to examine the sigmoid colon.
Single photon absorptiometry (SPA): Noninvasive test to determine bone mineral content at the wrist or vertebrae.
Sitz bath: To sit in warm, possibly medicated, water covering the hips.
Solute: A substance dissolved in a solution.
Somatic: Pertaining to the body.
Sphincter: A circular muscle surrounding an opening that acts as a valve.
Stasis: Standing; stagnation of normal flow of liquids.
Stenosis: Abnormal constriction or narrowing.
Steroid: Group of chemical substances, including certain hormones and cholesterol.
Stool softener: Substances that act as a wetting agent and thus promote soft, malleable bowel movements. Stool softeners are not laxatives.
Surgical resection: Partial excision of a bone or other structure.
Swallowing video fluoroscopy: A continuous stream of x-rays pass through the patient casting shadows of heart, lungs, and diaphragm on a fluorescent screen. Permits visualization of the swallowing process.
Synapse: Functional junction or point of close contact between neurons, or neuron and effector cells.
Syndrome: Signs and symptoms that together characterize an abnormal condition or a disease.
Synovial: Pertaining to the lubricating fluid of joints.
Systemic: Pertaining to the whole body rather than to one of its parts.
Systolic: The upper number of a blood pressure reading reflecting the contraction phase of the cardiac cycle.
Tactile: Pertaining to touch.
Tendon: Fibrous connective tissue connecting muscle to bone or muscle to muscle.
Testosterone: The principal hormone produced in the testes of males; also a hormone produced in the adrenal cortex of both human males and females.
Threshold: The smallest amount of stimulation capable of stimulating sensory receptors.
Thrombolytic therapy: Use of drugs to break up a thrombus (clot).
Thyroid function tests: Tests for evidence of increased or decreased thyroid functioning.
Topical: Pertaining to a definite area of the body; local.
Toxicology screen: Diagnostic tests measuring drug levels in the blood. Used to monitor therapeutic levels of drugs and to identify and measure toxic substances in blood.
Triglycerides: Fats and oils composed of fatty acids and glycerol; the body's most concentrated source of energy fuel; also called neutral fats.

Ulcer: Lesion or erosion of mucous membranes or lesions of the skin.

Ultrasound: Very high frequency sound that has different velocities in different tissues. It outlines the shape of various tissues and organs in the body.

Uremia: Toxic condition associated with renal insufficiency.

Urinalysis: Analysis of the urine.

Urogram, intravenous: Injected dye is excreted by the kidney and studied by x-ray during excretion.

Varicose vein: A dilated, twisted, knotted vein, usually in the leg.

Vasoconstriction: Narrowing of blood vessels.

Vasodilation: Dilation or expansion of blood vessels.

Vein: A vessel that carries blood to the heart.

Venogram: An x-ray image of the veins.

Vestibular: Concerned with equilibrium.

Viscera: Pertaining to the internal organs enclosed within a cavity; especially the abdominal organs.

Vital capacity: The volume of air that can be expelled from the lungs by forcible expiration after the deepest inspiration; total exchangeable air.

Index

B

C

D

H

I

Q

R

S

Printed in the USA
CPSIA information can be obtained
at www.ICGtesting.com
LVHW021324110124
768634LV00006B/869

9 780826 150554